Metadata Collection and Organization in Wastewater Treatment and Wastewater Resource Recovery Systems

Metadata Collection and Organization in Wastewater Treatment and Wastewater Resource Recovery Systems

Edited by

Kris Villez, Daniel Aguado, Janelcy Alferes, Queralt Plana, Maria Victoria Ruano and Oscar Samuelsson

IWA PUBLISHING

Published by **IWA Publishing**
Unit 104–105, Export Building
1 Clove Crescent
London E14 2BA, UK
Telephone: +44 (0)20 7654 5500
Fax: +44 (0)20 7654 5555
Email: publications@iwap.co.uk
Web: www.iwapublishing.com

First published 2024

British Library Cataloguing in Publication Data
A CIP catalogue record for this book is available from the British Library

ISBN: 9781789061147 (paperback)
ISBN: 9781789061154 (eBook)
ISBN: 9781789061161 (ePub)

This eBook was made Open Access in May 2024.

Doi: 10.2166/9781789061154

Contents

doi: 10.2166/9781789061154_0001

Chapter 1

Introduction

D. Aguado[1], J. Alferes[2], Q. Plana[3,4], M. V. Ruano[5], O. Samuelsson[6] and K. Villez[7]

[1]CALAGUA – Unidad Mixta UV-UPV, Institut Universitari d'Investigació d'Enginyeria de l'Aigua i Medi Ambient – IIAMA, Universitat Politècnica de València, València, Spain
[2]VITO – WaterClimateHub, Ostende, Belgium
[3]modelEAU, Université Laval, Québec, QC, Canada
[4]Eurecat – Technology Center of Catalonia, Barcelona, Spain
[5]CALAGUA – Unidad Mixta UV-UPV, Departament d'Enginyeria Química, Universitat de València, Burjassot, Spain
[6]IVL Swedish Environmental Research Institute, Stockholm, Sweden
[7]Oak Ridge National Laboratory, Oak Ridge, TN, USA

This report provides a comprehensive overview of metadata to describe sensor signals in wastewater treatment plants and methods to obtain such metadata. In this introduction, we explain the original motivation behind the MetaCO task group. This includes a description of historical challenges (data volume, data velocity) for which mature technology is now available, and newer challenges, which relate to data structure (data variety) and data quality (veracity). We conclude the chapter with an expression of gratitude to all involved.

1.1 SCOPE OF THIS REPORT: DATA AND METADATA OF INTEREST

Definitions for metadata tend to vary. In the narrowest sense, it means knowing where, when, and how data were produced. This narrow definition is often not sufficient as it may exclude a description of the motivation, implementation, operation, and condition of the data collection system. Knowing this is critical to inspire confidence in the data and any analysis derived from it. Contextual information, for example, about weather or spill events, is often also very valuable to decide whether a particular data subset is relevant for modelling, monitoring, or control. Comparing sensor signals against reference measurements or in controlled media (standard solution) often proves essential to gauge confidence in data-intensive techniques for modelling and control. Finally, data analytical techniques can often reveal patterns or trends that are considered valuable for subsequent interpretation or use of the analysed sensor signals. Based on these considerations, we consider metadata in the broadest sensor of the word and describe any of the following:

- Signal production methods, signal-producing hardware, and signal processing.
- Sensor condition and signal quality.

- Contextual information describing the conditions under which data was produced.
- Information gained about sensor signals through visual inspection or computer-aided analysis.

As this leads to a wide range of metadata to be considered, our focus is on (a) wastewater treatment application domain and (b) metadata for online sensors. This means water drainage, water production, and water supply systems, while facing similar challenges, are considered out of scope. We also limit our scope to metadata describing online sensor signals and exclude laboratory analysis and actuators (e.g., valve setpoint signals) unless warranted for didactic purposes. Finally, cyber-security and IT infrastructure integrity, while important, are not unique to wastewater treatment and are therefore excluded as well. Consequently, examples and recommendations are focused on sensor signal data in wastewater treatment applications. Definitions and recommendations may of course apply elsewhere within a wastewater utility (e.g., digital twins and models, laboratory data, and control setpoints). Inspiration may be drawn from this report in other domains of engineering as well.

1.2 INCENTIVES TO COLLECT AND ORGANIZE METADATA AND THE NEED FOR THIS REPORT

Sensor development, computational power, and data storage capacity has brought the big data era into the wastewater sector. Big data can be differentiated into four V's describing data, that is, the data volume, velocity, variety, and veracity (truthfulness). Until today, data infrastructure development in wastewater utilities has been focused on volume and velocity. Online sensors have increased the measurement sampling rates to exceed practical needs and real-time data collection is now the norm. The effect of cheap data storage capacity has further led to large databases, effectively logging the process history of water resource recovery facilities (WRRFs) in any size.

However, the variety in data types is a challenge to standardized data collection and storage routines. Data are being produced from novel and possibly wireless sensors both inside and outside the fence. The need to integrate external databases and systems with the conventional static ones becomes evident. Otherwise, data soon become outdated, which leads to data silos with limited to no plant-wide utility. Even when data sources are integrated within the fence, it remains difficult to maximize the utility of the available data. Metadata, including a quantification of the data quality, is often nonexistent so that the vast amounts of collected data quickly turn stale.

Next to new and ubiquitous sensor hardware, utilities increasingly use software sensors (soft-sensors) and models (digital twins) to produce additional signals available for plant operation and management. These may be managed in-house or through a subcontractor. Soft-sensors and digital twins critically rely on accurate and relevant data. Describing the data in a meaningful way is often the first step to make this feasible.

The veracity of data has always been important, especially when used for automatic control. So far, the control workhorse has been the proportional and integrating (PI) controller, which is simple, but sufficiently advanced to deal with the majority of WRRF control problems. In a PI-controller, the impact of a drifting sensor is predictable, and the consequences can be understood and accounted for when taking strategic control decisions. Things become more complicated when multiple signals and computations are involved. The current trend towards use of machine learning, soft-sensors, and digital twins makes it more difficult to account for the consequences of poorly managed sources of information.

The ongoing digitalization of wastewater utilities is not only driven by opportunities, but also pushed by the drive to turn wastewater treatment plants (WWTPs) into WRRFs. This transition requires more advanced technology working together with existing standard processes. Additionally, effluent limits are getting increasingly low, while greenhouse gas emissions need to be minimized as well. New sensors, data, and smart decision support systems are needed to balance these new goals with more traditional cost and energy efficiency targets. Therefore, the data-driven WRRF is here to stay.

Most water resource recovery facilities already collect large data volumes today. However, systems, routines, and knowledge to strategically handle these data are often limited. One of the contributing factors is that a systematic description of sensor signals, including sensor signal quality assessment, is often nonexistent. Such metadata are key to convert meaningless streams of data into data that are fit for purpose. They are also invaluable when crawling through volumes of historical data, trying to identify certain process conditions. Operator logbook annotations are often useful as a source of a metadata but are often not available in digital form.

For the above reasons, this technical report aims at clarifying metadata concepts and provides advice on their practical implementation in WRRF utilities to facilitate improved decision-making based on the vast number of sensor signals available on a WWTP.

1.3 GUIDE TO THIS TECHNICAL REPORT

Most of the chapters in this report can be read and understood separately. One exception is Chapter 2, which provides definitions concerning signal data that are relevant to all other chapters. Figure 1.1

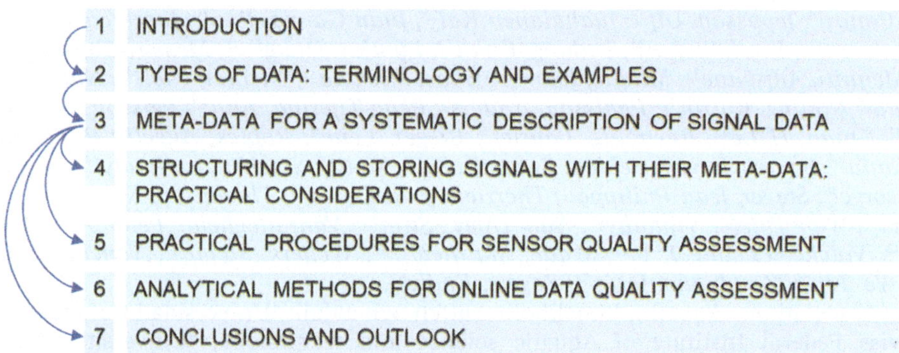

Figure 1.1 Dependencies between chapters within this report.

Table 1.1 Indication of relevance of chapters to roles within a wastewater utility.

Chapter	Chapter Title	Sensor Technician	Operational Staff	Data Management Specialist	Analyst Modeller Engineer	Plant Manager
2	Types of data: terminology and examples	×	×	×	×	×
3	Metadata for a systematic description of signal data	×	×	×	×	×
4	Structuring and storing signals with their metadata: practical considerations			×		
5	Practical procedures for sensor quality assessment	×	×		×	
6	Analytical methods for online data quality assessment		×	×	×	
7	Conclusions and outlook				×	×

shows the dependencies among the available chapters and can be used to select chapters of interest. Chapters 2–6 include similar figures to select relevant parts within the chapter.

We suspect the chapters are of interest to different roles within your organization. Table 1.1 explains the roles we identified as having the most interest in each chapter. Chapters 2 and 3 are considered relevant to most, while Chapters 4, 5, and 6 are more specialized.

1.4 THANK YOU AND CONCLUDING REMARKS

We believe this report offers a great starting point for improved data management and this could not have been possible without the contributions of many. These contributions, including many from practitioners in the field, were essential to the creation of this extensive report on metadata in wastewater treatment facilities. We are extremely grateful for this to all of them:

Åmand, Linda[1]*; Amerlink, Youri*[2]*; Andersson, Sofia*[3]*; Arteaga, Francisco*[4]*; Bach, Peter*[5]*; Baeza, Juan*[6]*; Belia, Lina*[7]*; Bellandi, Giacomo*[8]*; Blumensaat, Frank*[5]*; Bott, Charles*[9]*; Brown, Elizabeth*[10]*; Copp, John*[7]*; Corominas, Lluís*[11,12]*; Corona, Francesco*[13]*; De Gussem, Kris*[14]*; Feiyi, Li*[15]*; Ferrer, Alberto*[16]*; Filali, Ahlem*[17]*; Fletcher, Laura*[10]*; Gagnon, Ali*[9]*; Goldman, Joshua*[18]*; Grievson, Oliver*[19,20]*; Gruber, Günter*[21]*; Grundestam, Jonas1; Gubser, Nina*[22]*; Haimi, Henri*[23,24]*; Hauduc, Helene*[25]*; Heffernan, Conor*[26]*; Jeppsson, Ulf*[27]*; Juahaianen,Kai*[28]*; Juan Garcia, Pau*[19]*; Juszkiewicz, Agnieszka*[29]*; Kazemi, Pezhman*[30]*; Kroll, Stefan*[31]*; Ledung, Erika*[32]*; Lumley, Doug*[33,28]*; Massiri, Tofik*[1]*; Mazeghrane, Sofiane*[34]*; Menniti, Adrienne*[35]*; Miletic, Ivan*[36]*; Mulas, Michela*[37]*; Nicolaï, Niels*[15]*; Nilsson, Josefine*[38]*; Nilsson, Sara*[3]*; Norling, Katrin*[39]*; Pahlsson, Anders*[3]*; Pena-Tijerina, Ana*[40]*; Persson, Gabriel*[3]*; Pierce, Alexander*[41]*; Quan, H Le*[42]*; Reynolds, Laurie*[43]*; Rieger, Leiv*[44]*; Robles, Angel*[45]*; Rappmann,Tove*[28]*; Russo, Stefania*[46]*; Santos, Jorge*[47]*; Sehlén, Robert*[38]*; Seib, Matt*[48]*; Sivchenko, Nataly*[49]*; Smith, Robert*[50]*; Sprouse, George*[10]*; Steyer, Jean-Philippe*[51]*; Therrien, Jean-David*[15]*; Thompson, Ken*[44]*; Tik, Sovanna*[15]*; Torfs, Elena*[15]*; Van Daele, Timothy*[52]*; Van Hoey, Stijn*[53]*; Vanrolleghem, Peter A.*[15]*; Vargas, Jose Leornardo*[54]*; Volcke, Eveline I. P.*[42]*; Wade, Matthew*[55,56]*; Weijers, Stefan*[57]*; Yanchen, Liu*[58]*; Yang, Fenghua*[59]*; Ye, Liu*[60]*; Zambrano, Jesus*[61]*; Zegers, Emily*[62].

[1]Käppala Association; [2]Aquafin; [3]Stockholm vatten och avfall; [4]Universidad Católica de Valéncia; [5]Eawag Swiss Federal Institute of Aquatic science and technology; [6]Universitat Autònoma de Barcelona; [7]Primodal, Inc; [8]AM Team advanced modelling for process optimisation; [9]Hampton Roads Sanitation District (HRSD); [10]Metropolitan Council of the Twin Cities; [11]Catalan Institute for Water Research (ICRA-CERCA); [12]University of Girona; [13]School of Chemical Engineering, Aalto University; [14]Former Aquafin; [15]modelEAU, Université Laval; [16]Multivariate Statistical Engineering Research Group, Universitat Politècnica de València; [17]Irstea, National Research Institute of Science and Technology for Agriculture and Environment; [18]Metro Water Recovery; [19]AtkinsRéalis, Birmingham, UK; [20]Visiting Professor at the University of Exeter; [21]TU Graz, Institute of Urban Water Management and Landscape Water Engineering; [22]City of zurich, Zuerich; [23]FCG Finnish Consulting Group Ltd; [24]School of Engineering, Aalto University; [25]Dynamita; [26]Metropolitan Water Reclamation District of Greater Chicago, Chicago, IL, USA; [27]IEA, Lund University; [28]Gryaab AB; [29]Mälarenergi AB;[30]Departament d'Enginyeria Química, Universitat Rovira i Virgili; [31]R&D, Aquafin NV; [32]Esem; [33]Formerly DHI Group; [34]Suez; [35]Clean Water Services; [36]inCTRL Solutions Inc., Dundas; [37]Department of Teleinformatics Engineering, Federal University of Ceará; [38]Tekniska verken; [39]Mjölby kommun; [40]Plummer Associates, Inc; [41]Fort Worth, Texas; [42]BioCo Research Group, Department of Green Chemistry and Technology, Ghent University; [43]Aquamatix; [44]Jacobs Engineering Group Incorporation; [45]CALAGUA – Unidad Mixta UV-UPV, Departament d'Enginyeria Química, Universitat de València; [46]EcoVision Lab, Photogrammetry and Remote Sensing, ETH Zurich; [47]DMI, Nova Universtity Lisbon; [48]Madison Metropolitan Sewerage District; [49]DOSCON AS; [50]Black & Veatch; [51]INRAE, LBE, Univ Montpellier, Narbonne; [52]The Janssen Pharmaceutical Companies of Johnson & Johnson; [53]Fluves; [54]ABB; [55]UK Health Security Agency, Data Analytics & Surveillance, Canary Wharf; [56]School of Engineering, Newcastle University, Cassie Building;

[57]Waterschap De Dommel; [58]School of Environment, Tsinghua University; [59]MMSD - Milwaukee Metropolitan Sewerage District; [60]University of Queensland, School of chemical Engineering; [61]The MathWorks AB; [62]City of Toronto, Toronto.

This report is an Open Access publication thanks to the funding from the following organizations:

- *SIVL Foundation for IVL (SIVL)*
- *Svenskt Vatten Utveckling, Swedish Water & Wastewater Association*
- *Stockholm Vatten och Avfall AB*
- *Mälarenergi Vatten AB*
- *Eskilstuna Strängnäs Energi & Miljö AB*
- *Mjölby kommun*

We hope you will enjoy reading this report as much as we did when writing it. We welcome any comments and suggestions.

doi: 10.2166/9781789061154_0007

Chapter 2

Types of data: terminology and examples

Q. Plana[1,2] and K. Villez[3]

[1]modelEAU, Université Laval, Québec, QC, Canada
[2]Eurecat – Technology Center of Catalonia, Barcelona, Spain
[3]Oak Ridge National Laboratory, Oak Ridge, TN, USA

Keywords: background information, data variety, definitions, data heterogeneity, metadata, signal

Data produced at wastewater utilities are obtained through a variety of devices, including actuators, control devices, and sensors. This results in data that can be highly variable in its structure. Dealing with the resulting heterogeneity of data formats can be a challenge when storing or interpreting the data. For this reason, this chapter serves as an overview for the most important structural aspects of data typically found at a utility. The specific aims of this chapter are to:

- Introduce basic concepts for description, understanding, and management of data produced by online instruments, including sensors and actuators (e.g., valves and pumps).
- Provide definitions for the most common terms used throughout this report to describe online sensor data as well as other data sources.
- Provide practical examples to illustrate the provided definitions.

Where feasible, we relied on existing standards and references to provide applicable definitions. However, many definitions are developed specifically for this report.

2.1 HOW TO READ THIS CHAPTER

The main purpose of this chapter is to provide definitions that are essential to understand the remainder of this Scientific and Technical Report (STR). The scheme below indicates dependencies within this chapter and can be used to navigate the chapter efficiently. The elements shown in blue or white are considered essential to understand what follows in the remaining chapters. The elements shaded in grey will be particularly helpful to understand specialized needs for database technology and management of information flows.

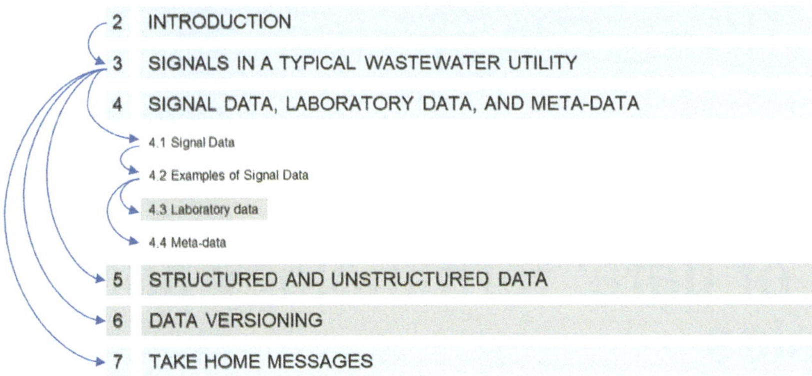

2.2 INTRODUCTION

We recommend reading this chapter from start to finish before reading any other chapter in the report. At a later stage, the chapter can be used as a reference by searching the list of definitions (Glossary) and finding the corresponding page in this chapter to find detailed explanations for each definition. Note that all examples in this chapter are shown without any metadata. As a result, the data are shown without any context and are thus stripped of their meaning. Metadata providing this context will be discussed in detail in Chapter 3.

2.3 SIGNALS IN A TYPICAL WASTEWATER UTILITY

Figure 2.1 shows a schematic representation of the hardware components in a small control system as may be found on a wastewater treatment plant. It includes two controllers, each implemented in a

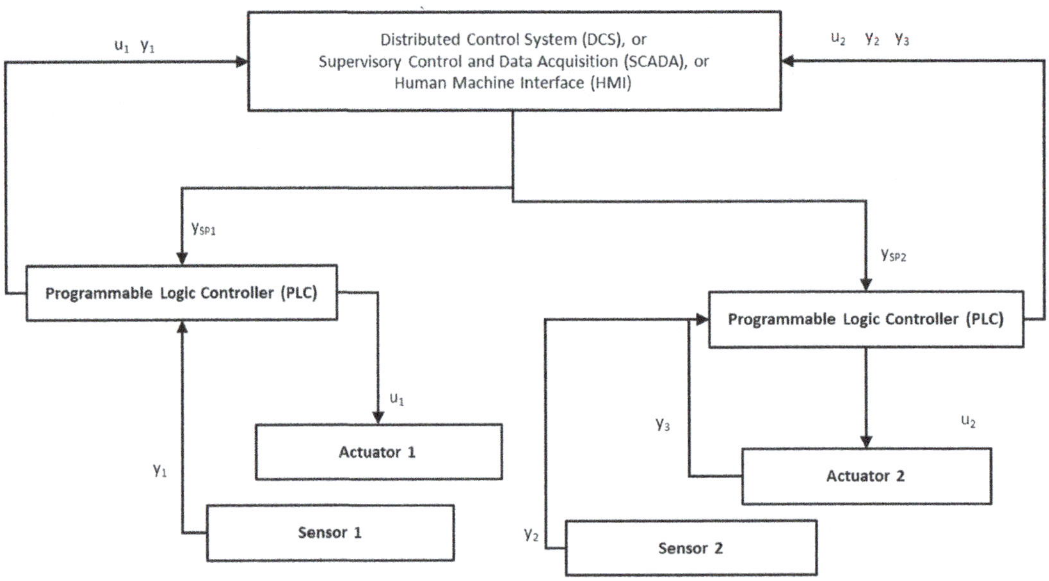

Figure 2.1 Graphical representation of a small control system. A control system of reasonable complexity will include measurements (y), actuation or control signals (u), and setpoints (y_{SP}), each of which is considered a signal.

programmable logic controller (PLC) device. The first controller takes a measurement from a sensing device, sensor 1. Common examples include sensors for flow rate, pressure, and concentrations. The measurement signal is named y_1. A setpoint for y_1 is sent from a supervisory system and named $y_{1,SP}$. The controller interprets the measurement and setpoint and produces a control action u_1. This control action is sent to an actuating device, actuator 1. Common examples include hydraulic valves and variable-frequency-drives (VFD) to manipulate blower or pump speeds. The second controller behaves similarly with a sensor signal y_2, a setpoint $y_{2,SP}$ and an actuation signal u_2. In addition, the actuator also sends a signal, y_3, to the PLC. Examples of such signals may be a position or torque measurement in a valve or a measurement of temperature, pressure, or speed in a pump. We note that actuators producing measurement signals are sometimes called smart devices (e.g., smart valve, smart pump). The PLCs report all signals to the supervisory system. Depending on the ability and complexity of the interactions with the PLCs, the supervisory system may be a distributed control system (DCS), a supervisory control and data-acquisition system (SCADA), or a human–machine interface (HMI). Consequently, every device in this system except actuator 1 can produce signals collected and interpreted by other devices within the same architecture.

For the purposes of this report, we will consider all control actions, setpoint, and measurements sent between hardware components to be signals. Note that this also includes setpoints entered manually by an operator when forwarded to other components in the control system. We will use the term *device* for any hardware component in such a system and use the term *instrument* for any hardware component that produces a signal, regardless of this signal being a measurement, setpoint, or actuation.

2.4 SIGNAL DATA, LABORATORY DATA, AND METADATA

In this section, we provide the basic terminology to describe signal data, laboratory data, and metadata. This includes definitions and general remarks in a first subsection, followed by a variety of examples showcasing how to apply the provided definitions.

2.4.1 Signal data

The next sections provide (a) definitions, (b) a graphical representation of data records, (c) notes on the evaluation of data dimensionality, and (d) additional remarks.

2.4.1.1 Data point dimensionality

The most common instruments produce data points that are of scalar form, that is, they consist of a single value (e.g., a value for conductivity). Such data points are described as **scalars** or **zero-dimensional (0D)** values. The resulting data record will consist of a timestamp and a single value (see Section 2.4.1.2 for definitions). Combining data records collected at different times leads to a signal that is considered a univariate time series.

There exist more advanced instruments that produce data points that contain more than one numeric value, for example, spectra (also: fingerprints) produced by a spectrophotometric sensor consist of multiple measurements of light absorbance, where each measurement is associated with a single wavelength. These are examples of **one-dimensional (1D) data points**. The corresponding signals are often called multivariate time series.

Multi-dimensional data points exist as well. For example, greyscale images are considered two-dimensional (2D) while coloured and hyperspectral images are considered three-dimensional (3D). Tomographic instruments exist to measure the colour intensities of an object in a 3D physical space. Even if a single tomographic image exhibits four dimensions (e.g., width, height, depth, and wavelength), it is still referred to as a single data point in this report.

2.4.1.2 Definitions

Definition D2.1: **Signal data** (also: signals) consist of values produced by an online instrument of qualities and quantities in a physical system of interest. Signal data consists of series of data records.

Definition D2.2: A **data record** is a pair or triplet of constituting elements, which are (i) a timestamp (required), (ii) a data point (required), and (iii) a qualifier (optional).

Definition D2.3: A **data point** consists of one or more values that satisfy the following three requirements:

(1) All values within the data point are produced by the same instrument.
(2) All values are associated with the same time of measurement or, where relevant, the same time of water sample collection.
(3) All values describe the same type of variable and are produced by the same measurement principle.

Definition D2.4: A **qualifier** is a single number or text produced by an instrument indicating a level of trust in the timestamp and/or data point.

A qualifier may be produced by an instrument to indicate to the data user that data are absent, missing, anomalous, faulty, or mistrusted for any reason. We recommend adhering to existing standards to store data qualifiers. For example, the OPC Foundation (2003) specifies an 8-bit description to indicate data quality as part of the OPC Data Access standard, which is broadly supported by commercial DCS, SCADA, and HMI systems.

The above definitions are illustrated in Figure 2.2. A signal is composed of data records, each of which is composed of a timestamp, a data point, and a qualifier. Where required in this report, we will apply the following symbolic representation of the same concept for signals with 0D values:

$$R_j = (t_j, y_j, q_j)$$

where R_j is the jth data record in a signal, t_j is the timestamp in the jth data record, and y_j is the value for the measured variable in the jth data record. When the data points are 1D, 2D, or high-dimensional, we underline the text and/or use a bold font type according to the following conventions:

- 1D (scalar): $R_j = (t_j, \mathbf{y}_j, q_j)$
- 2D (matrix): $R_j = (t_j, \mathbf{Y}_j, q_j)$
- 3D and higher (tensor): $R_j = (t_j, \underline{\mathbf{Y}}_j, q_j)$

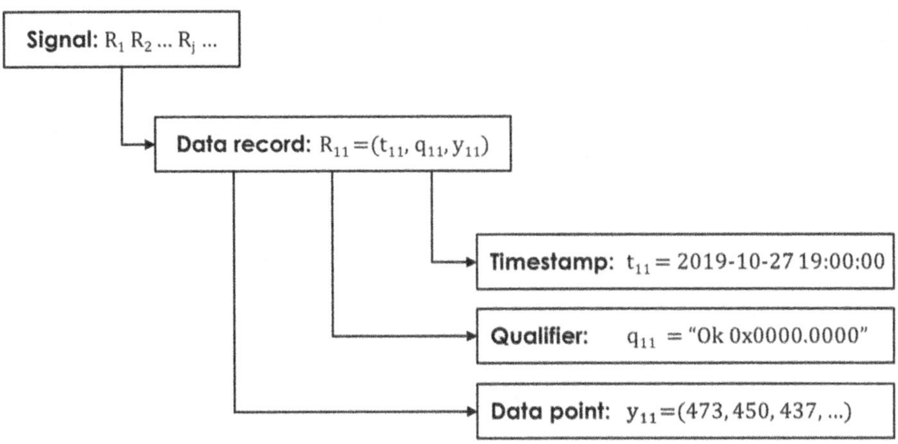

Figure 2.2 Graphical representation of Definitions D2.1, D2.2, D2.3, and D2.4. The provided values correspond to the 11th data record in Table 2.2.

2.4.1.3 Absent and missing data

A lack of access to valuable data points is a commonly observed challenge when dealing with data sets collected in practice. Causes for this are profuse. Frequent causes include (a) failure to initiate or complete data collection steps, (b) exceeding the measurement range of an instrument, (c) communication loss, or (d) inappropriate frequency of data collection. In this report, we distinguish between absent data and missing data with the following definitions.

Definition D2.5: A data record, timestamp, or data point is considered **absent** if having access to this information is considered valuable at the time of data analysis but no procedure to collect this information was ever initiated.

Definition D2.6: A data record, timestamp, or data point is considered **missing** if a procedure to collect it was initiated but failed to produce the intended numerical result.

The major difference between absent data and missing data is the notion of intent. If collection of information was intended but failed, then the information is considered missing. If no attempt to collect the information was made and the utility of this information only becomes apparent later, then it is considered absent.

There are several good reasons to store or archive information about missing data, that is, (a) to not omit information about the intent to collect data and (b) to not replace missing information with imputed values for long-term storage purposes. We also encourage the storing of information about absent and missing data in a robust yet accessible manner. Examples 2.1, 2.2, and 2.3 below are used to illustrate absent and missing data. Knowledge about the presence of missing data points may provide valuable information. For example, this can occur when the value of the measured variable is out of measurement range. In this case, knowing that the measured value was below or above a certain limit can be useful to estimate and impute it later. This is an example of a missing-not-at-random (MNAR) pattern (e.g., Dong & Peng, 2013; Kang, 2013; Osman *et al.*, 2018). In contrast, the presence of missing data caused by disruption of power supply to an instrument will generally not correlate with the value of the measured variable. This is then described as a missing-at-random (MAR) or missing-completely-at-random (MCAR) pattern (see, e.g., Bhaskaran & Smeeth, 2014), for which different imputation methods are considered best.

To enable optimal diagnostics and data imputation later, we recommend that missing data indicators are used systematically for long-term storage. This however requires that all elements of the digital infrastructure affecting data registration, communication, and storage must be designed to accept values in the form of text or a specialized code. Many modern instruments, including sensors and actuators, offer the automatic generation of metadata, often reflecting the root causes of missing data. Collecting and storing such metadata can be used to facilitate both data management and sensor management.

2.4.1.4 How to store information describing missing data

Given that absent data correspond to conditions where there was no intent to collect the data, there is no particular need to store information about absent data. Assuming information about missing data is stored in a comprehensive manner, one can categorize data that is unavailable as absent if no corresponding information about missing data is available. Missing data on the other hand should be documented to indicate failure to record data when this was the intention.

The wrong way to describe missing data: In many cases, a specialized numeric code is used to indicate a piece of missing information. For example, a negative number for a positively ranged number, like a concentration value. This can be useful when a single analog connection is used for communication. However, it should be avoided when feasible for two reasons: (1) misinterpretation of a numeric value as a meaningful value is very likely, especially if the data analyst is foreign to the field of application, and (2) the consequences of using conventional data processing steps, thus ignoring missing values, are likely to become visible only in the late stages of projects relying on data, in turn leading to delays, errors, and incurrence of unforeseen costs. In any case, the potential presence of specialized values in data set indicating missing data should be discussed early-on in a project to avoid egregious errors or overconfidence in data-based results.

A good way to describe missing data: Where feasible, a missing data point should be represented with a specific non-numeric code, such as *not-a-number, NA, NaN, below-detection-limit,* and *out-of-range.* One can add a qualifier (see above) or an additional value within the data point to do so (e.g., 0 for an accepted value and 1 for a rejected value). This is the recommended approach as it is much harder to mistake this kind of code for an accepted numeric value since calculated results produce the same textual description and does not propagate through calculations producing useful information. This is commercially available in most DCS and SCADA systems and is part of the open platform communications (OPC) standard.

Another good way to describe missing data: Another way that facilitates an unambiguous description of missing data in a signal (signal A) is to store a separate signal (signal B). For example, one can store separate pairs consisting of a timestamp and a data point, where the data point is a numeric indicator describing missingness. For example, a zero-valued data point in signal B can indicate that the corresponding data point in signal A is missing while a value equal to 1 could be used to signify that the corresponding data point in signal A was collected as intended. This concept is already included in OPC Data Access standards (OPC Foundation, 2003) and supported by most commercially available control systems, including DCS and SCADA systems.

A note on storing absent data: Since absent data are data that were never intended to be collected, there is no need to store such absences explicitly. Moreover, it may be impractical to define a unique timestamp for data records that were never intended to be collected. This however assumes that missing data are properly recorded so that the distinction between absent and missing data is easy to make later.

A note on storing imputed data: Many methods used in data analytics cannot handle absent or missing data by default. This is commonly addressed by estimation of the absent or missing data values (see Chapter 6, Section 6.6.2). The process of adding information to account for absent and missing data is known as data imputation. An original data set without any additions or modifications should always be available. The main reason is that imputation procedures are subject to change. Moreover, an imputation procedure is often used for a particular project or task and may not be valid for another. To do this effectively, data sets where absent or missing data have been imputed should include metadata to clearly identify the imputed values and the procedures used for imputation. This is discussed in more detail below in the section on data versioning (Section 6).

2.4.1.5 Additional remarks

Use of the term *sample*: In the fields of computer science, data analytics, machine learning, and statistics, it is very common to refer to data records and data points as samples. However, a sample may also refer to a representative amount of a physical substance collected for the purpose of measurement. In statistics, it may refer to a set of values obtained through sampling of a distribution. For clarity, this report uses the term sample only to refer to a representative volume of aqueous media. In addition, we remove ambiguity by specifying the nature of the medium wherever needed, for example, by writing *wastewater sample* instead of just *sample.*

Nature and origin of the numeric value(s): Values contained in a data point may be quantitative (e.g., a concentration, a valve position), qualitative (e.g., the on/off state of a pump), or indicative of missing information (e.g., *not-a-number, out-of-range, below-detection-limit, limit-of-quantification*). Signal data may be produced by any instrument, including actuators (e.g., a valve torque measurement), sensors (e.g., dissolved oxygen sensor), or control devices (e.g., a PLC).

Multi-variable instruments or multi-sensors: We note that many instruments can produce measurements for multiple variables. For example, consider a dissolved oxygen sensor that measures temperature and oxygen concentration. In this case, the temperature signal and the dissolved oxygen are treated as separate signals, each being univariate time series and containing scalar (0D) data points. The reason for this is that the metadata describing these two signals, although produced by the same instrument, are sufficiently different to warrant separate treatment during data manipulation and storage. Note that this is the main motivation for the third condition in Definition D2.3.

2.4.2 Examples of signal data

To demonstrate how the definitions above can be applied, we provide four examples. The first and second examples consist of a setpoint regulated by a controller and a corresponding sensor signal. The third and fourth examples consist of sensor signals that take a more complex form.

2.4.2.1 Example E2.1: air flow rate setpoint

The first example of a simple signal is that of an air flow rate setpoint. A setpoint signal is shown in the top panel of Figure 2.3 as red points. The setpoint signal consists of data records including one timestamp and one setpoint value. Some of these points are shown in the two columns on the left-hand side of Table 2.1. This type of signal is a univariate time series and is extremely common.

2.4.2.2 Example E2.2: air flow rate measurement

The second example of a simple signal is that of an air flow rate measurement. An exemplary signal is shown in the top panel of Figure 2.3 as black points. In this case the sensor signal follows the setpoint signal closely. The sensor signal, like the setpoint signal in this figure, consists of data records including one timestamp and one measured value. Some of these points are shown in the two columns on the

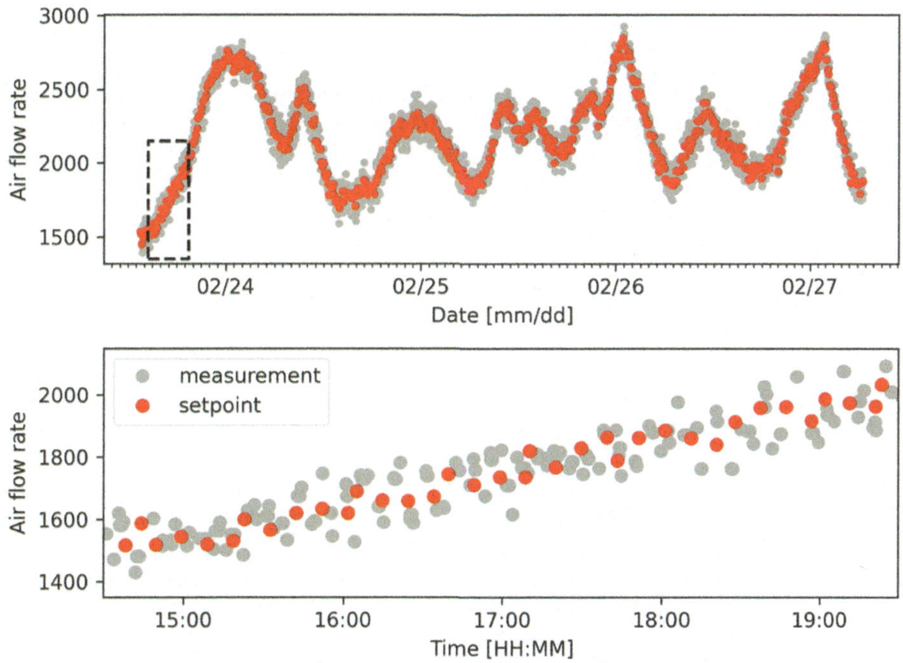

Figure 2.3 Examples E2.1 and E2.2: two univariate signals. A control setpoint (red) for an air flow rate and a corresponding measurement (grey). Top panel: Four-day time window. Bottom panel: A focused view on the data records matching the dash-lined box in the top panel. Notice that (a) the data records are recorded in an asynchronous manner and that (b) the rate of data record generation is different for the two signals. Attempts to synchronize this type of data should be documented and implanted in a reversible manner.

Table 2.1 Examples E2.1 and E2.2: two univariate signals.

Univariate Signal A (E2.1)		Univariate Signal B (E2.2)	
Setpoint Time	**Air Flow Setpoint**	**Measurement Time**	**Air Flow Measurement**
t	*y*	*t*	*y*
n/a	n/a	2020-02-23 14:31:01	1555.17
n/a	n/a	2020-02-23 14:33:52	1473.6
n/a	n/a	2020-02-23 14:35:44	1621.91
n/a	n/a	2020-02-23 14:36:09	1583.18
n/a	n/a	2020-02-23 14:36:48	1583.18
n/a	n/a	2020-02-23 14:37:20	1593.9
2020-02-23 14:38:05	1517.25	n/a	n/a
n/a	n/a	2020-02-23 14:42:00	1430.47
n/a	n/a	2020-02-23 14:42:32	1482.39
n/a	n/a	2020-02-23 14:43:19	1482.39
2020-02-23 14:44:05	1589.78	n/a	n/a
n/a	n/a	2020-02-23 14:48:41	1604.06
n/a	n/a	2020-02-23 14:48:48	1604.06
2020-02-23 14:49:33	1519.64	n/a	n/a
n/a	n/a	2020-02-23 14:52:17	1516.99
n/a	n/a	2020-02-23 14:52:23	1534.57
n/a	n/a	2020-02-23 14:55:30	1534.57
n/a	n/a	2020-02-23 14:55:43	1521.66
n/a	n/a	2020-02-23 14:56:56	1585.66
2020-02-23 14:59:14	1546.29	n/a	n/a

The first signal is a control signal and contains data records consisting of a timestamp generated by the controller and a setpoint for an air flow rate. The second signal is a sensor signal and contains data records consisting of a timestamp generated by the sensor and the measured value of the air flow rate. The two signals are (a) recorded asynchronously, (b) are recorded at different frequencies, and (c) exhibit time intervals that vary over time. Note that the data set is commonly observed in practice. There are no missing data here, i.e., the measurement availability is as intended for this application. Due to the lack of synchronization, there are however absent data as there are no records in the two signals with matching timestamps. These are indicated as *n/a* (not applicable).

right-hand side of Table 2.1. Once again, this type of signal is a univariate time series. The bottom panel of Figure 2.3 illustrates why the setpoint and measurement signal are treated as distinct signals. The timestamps of the two signals do not match because the data records are generated in an asynchronous manner. Note also how the rate of data record generation is different for the two signals and changes over time. The data records in Table 2.1 are organized so that each row corresponds to unique timestamp, for either the setpoint signal or the sensor signal. No special symbol is used in the table cells without numeric values. This is because the information in these cells is considered absent. Indeed, there are no missing data here as the data collection system worked as intended (Definition D.4).

2.4.2.3 Example E2.3: ultraviolet-visible light absorbance spectrophotometry

Spectrophotometric sensors are typically used for monitoring one or more water quality variables. For example, they include total suspended solids (TSS), chemical oxygen demand (COD), nitrite, and nitrate. Importantly, most spectrophotometric sensors are equipped with software sensors, which translate the light absorbance measurements into values for the target variables (e.g., see Lepot *et al.*, 2016; Mašić *et al.*, 2015; Rieger *et al.*, 2006). The produced data records consist of a timestamp and a data point, where the data point corresponds to a spectrum. Each spectrum, sometimes referred to as a fingerprint, consists of multiple absorbance values, with each absorbance value corresponding to a wavelength in the ultraviolet–visible range. Figure 2.4 shows 24 data records produced by a spectrophotometer. Each line corresponds to a single data record. The number of absorbance values in a single data point is 221. Table 2.2 tabulates the 24 records shown in Figure 2.4. This table shows the timestamp (t), the qualifier (q), and the multivariate 1D data points (y). This table contains values equal to *nan*, indicating missing data (following Definition D2.6).

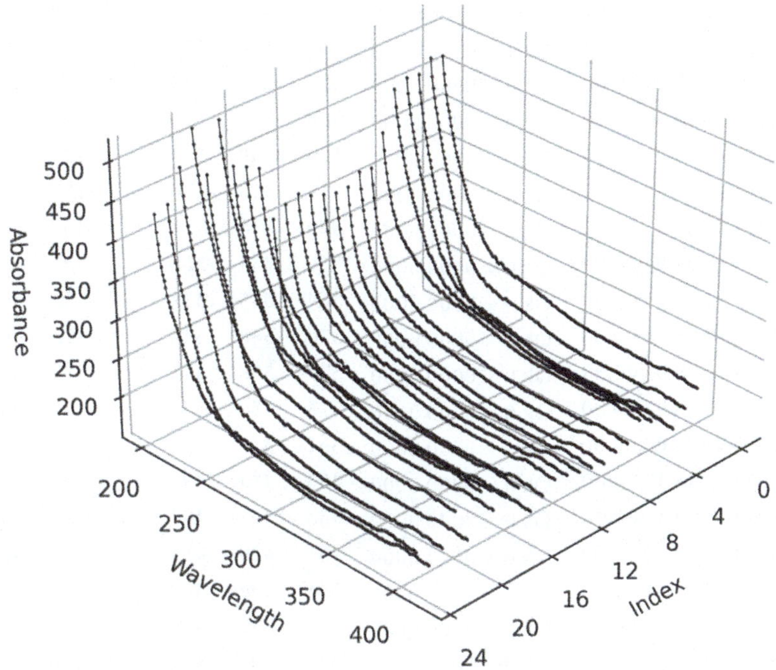

Figure 2.4 Example E2.3: spectrophotometric sensors deliver data that is 1D. Each data record consists of a timestamp (t) and a corresponding data point, which is a set of absorbance measurements (y) taken for a fixed set of light wavelengths (λ). Each data point is 1D (length: 221) and is shown as a dotted line.

2.4.2.4 Example E2.4: flow cytometry

Online flow cytometry represents a relatively recent development for online water quality monitoring purposes (see, e.g., Besmer *et al.*, 2014; Wang *et al.*, 2010). In a typical flow cytometric instrument, a water sample is analysed by measuring the intensity of light obtained through light scattering and fluorescence effects generated by individual particles. This is executed in two steps. First, the particles are dyed with a fluorescent stain. Second, laser light of a specific wavelength (e.g., blue: 488 nm) is sent to a particle followed immediately by measurement of the intensity of the light returned by the particle through scattering or through emission at another specific wavelength (e.g., green: 533 nm, red: 670 nm). It is possible to register light intensity at a multitude of wavelengths and at different angles. As a result, measurement of a single water sample results in a multi-dimensional table, with rows corresponding to individual particles and columns corresponding to distinct light detectors. This means each data record consists of a timestamp, reflecting the time of water sample collection, and a 2D array of light intensity measurements associated with that timestamp. Figure 2.5 shows the values for two light intensities in two data records, with each panel in the figure corresponding to a single

Table 2.2 Example E2.3: spectrophotometric sensors deliver structured data as illustrated with this table.

Record Index	Time	Qualifier	Data Point							
j	t	q	y							
			0	1	2	...	201	202	203	204
0	2019-10-27 19:00:00	Ok 0 × 0000.0000	473	450	437		197	nan	nan	nan
1	2019-10-27 20:00:00	Ok 0 × 0000.0000	477	453	440		179	nan	nan	nan
2	2019-10-27 21:00:00	Ok 0 × 0000.0000	462	440	426		160	nan	nan	nan
3	2019-10-27 22:00:00	Ok 0 × 0000.0000	465	446	432		179	nan	nan	nan
4	2019-10-27 23:00:00	Ok 0 × 0000.0000	457	435	423		187	nan	nan	nan
5	2019-10-28 00:00:00	Ok 0 × 0000.0000	408	385	370		196	nan	nan	nan
6	2019-10-28 01:00:00	Ok 0 × 0000.0000	369	348	331		174	nan	nan	nan
7	2019-10-28 02:00:00	Ok 0 × 0000.0000	371	350	336		172	nan	nan	nan
8	2019-10-28 03:00:00	Ok 0 × 0000.0000	359	336	324		165	nan	nan	nan
9	2019-10-28 04:00:00	Ok 0 × 0000.0000	359	339	327		169	nan	nan	nan
10	2019-10-28 05:00:00	Ok 0 × 0000.0000	364	345	332		172	nan	nan	nan
11	2019-10-28 06:00:00	Ok 0 × 0000.0000	370	348	334		176	nan	nan	nan
12	2019-10-28 07:00:00	Ok 0 × 0000.0000	379	358	343		179	nan	nan	nan
13	2019-10-28 08:00:00	Ok 0 × 0000.0000	373	350	337		167	nan	nan	nan
14	2019-10-28 09:00:00	Ok 0 × 0000.0000	361	337	323		155	nan	nan	nan
15	2019-10-28 10:00:00	Ok 0 × 0000.0000	433	410	397		195	nan	nan	nan
16	2019-10-28 11:00:00	Ok 0 × 0000.0000	443	422	407		191	nan	nan	nan
17	2019-10-28 12:00:00	Ok 0 × 0000.0000	451	429	415		183	nan	nan	nan
18	2019-10-28 13:00:00	Ok 0 × 0000.0000	516	496	482		217	nan	nan	nan
19	2019-1028 14:00:00	Ok 0 × 0000.0000	455	431	418		163	nan	nan	nan
20	2019-10-28 15:00:00	Ok 0 × 0000.0000	520	501	490		209	nan	nan	nan
21	2019-10-28 16:00:00	Ok 0 × 0000.0000	479	459	445		171	nan	nan	nan
22	2019-10-28 17:00:00	Ok 0 × 0000.0000	440	419	406		157	nan	nan	nan
23	2019-10-28 18:00:00	Ok 0 × 0000.0000	435	414	399		183	nan	nan	nan

Each data point is 1D. Each data record consists of a time (*t*), a qualifier (*q*), and a data point, which is a series of absorbance measurements (*y*) taken for a fixed set of light wavelengths. The column position of each absorbance value in the table corresponds to a single wavelength while the row position corresponds to a unique time of measurement. Note that only a subset of the absorbance values is shown and that missing values (nan) are present.

water sample. The two light intensities shown here correspond to the forward scatter count (FSC) and the side scatter count (SSC). However, the same data records include additional measurements such as fluorescent light emitted by a green stain (fluorescein isothiocyanate, FITC) and a red stain (peridinin-chlorophyll-protein, PerCP), which are both excited by means of blue laser light. For more details on the applied measurement principles and practical implementation, we refer the reader to specialized literature (Besmer *et al.*, 2014; Hammes & Egli, 2010).

2.4.3 Laboratory data

While this report is primarily focused on sensor data, we provide a definition of laboratory data here to distinguish such data from signal data.

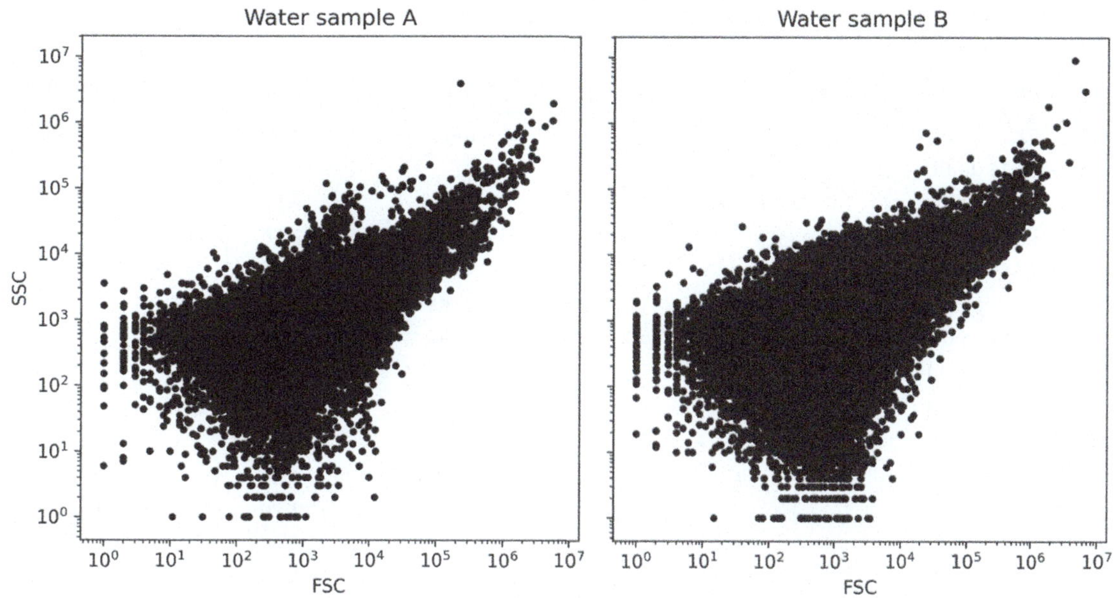

Figure 2.5 Example E2.4: two data records from a flow cytometry signal. The timestamp is displayed on top of the figure panels while the measurements for two light emission intensities are shown as a function of each other. The measured values are FSC and SSC. Left panel: One data record with measurement values for 56.358 particles obtained on 9 May 2019. Right panel: One data record with measurement values for 164,809 particles obtained on 23 May 2019.

2.4.3.1 Definition

Definition D2.7: **Laboratory data** (also: laboratory measurements) consist of values that are (a) recorded through a sequence of steps that include at least one manual water sample preparation step and (b) describe observed values of qualities and quantities in a physical system of interest.

Like signal data, laboratory data consist of series of data records, with definitions the same as in Section 2.4.1.2 (Definitions D2.2 and D2.3). Production of laboratory data may involve an analytical instrument or analytical test.

2.4.3.2 Example E2.5: laboratory measurements

Table 2.3 lists some laboratory measurements as an illustration. All measurements have been obtained in a laboratory with manual water sample preparation, thus satisfying Definition D2.7. Although the values in each row are associated with a single water sample, the concentration measurements of each variable (NH_4^+-N, NO_3^--N, PO_4^+-P, TOC) are treated as individual data points, following Definition D2.3. The table contains 61 data points, including 8 values that are missing. The missing value indicators are caused by the measured variable being below the limit of detection (indicated by '<LOD'). Following Definition D2.6, we consider the measured values missing. The indication '<LOD' offers diagnostic information about the root cause. This LOD is the minimum value for the concentration in a water sample that is expected to produce a positive value (i.e., higher than zero) for the measurement. This LOD is typically unique to a measurement principle and applied analytical protocol (e.g., applied dilution). For a detailed review of this topic and precise definitions, we refer to Currie (1995, 1999) and ISO (1997).

The LOD is useful to understand what the maximal value for the corresponding concentration might have been, even though a precise measurement may not be available. We note that the LOD

Table 2.3 Example E2.5: a table containing a set of laboratory measurements including absent data, missing data, and textual information.

Date/Time	NH_4^+-N	NO_3^--N	PO_4^+-P	TOC	Comment
2018-11-02 11:00:00		9.3	0.60	2.90	
2018-11-06 14:00:00				3.40	
2018-11-13 11:30:00	<LOD	5.9	1.30	3.12	
2018-11-15 09:30:00	<LOD	5.3	0.40	2.36	
2018-11-19 12:20:00	0.041	11.0	1.20	5.87	
2018-11-21 15:00:00	3.050	11.5	1.20	8.40	
2018-11-26 14:15:00	<LOD	5.2	0.56	3.40	
2018-11-26 16:00:00				4.70	
2018-11-28 14:30:00	<LOD	6.6	1.00	3.00	
2018-12-03 14:00:00	<LOD	5.8	1.10	6.80	
2018-12-06 15:30:00	0.038	8.2	0.74	4.50	
2018-12-06 12:00:00				2.20	Batch 1.2.a.1
2018-12-06 12:00:00				2.56	Batch 1.2.a.2
2018-12-06 12:00:00				3.91	Batch 1.2.b.1
2018-12-06 12:00:00				3.13	Batch 1.2.b.2
2018-12-06 12:00:00				2.08	Batch 1.2.c.1
2018-12-06 12:00:00				2.51	Batch 1.2.c.2
2018-12-06 12:00:00				2.59	Batch 6.1
2018-12-06 12:00:00				2.96	Batch 6.2
2018-12-08 10:00:00	<LOD	5.4	0.36	2.75	
2018-12-10 09:50:00	<LOD	5.0	0.44	3.44	
2018-12-12 11:30:00	<LOD	8.0	1.00	2.90	
2018-12-18 14:00:00	2.170	5.8	2.40	10.40	

itself may vary from one water sample to another (e.g., due to changes in applied dilution). In an ideal world, it is therefore recorded individually for each measurement where and when relevant. Alternatively, one can record the applied dilution factor and the analytical method that was used so that the LOD for the undiluted water sample can be computed based on a known LOD for the diluted specimen. Naturally, the metadata describing laboratory data may be varied and typically includes references to water sample collection, water sample preparation, and analytical protocols.

There are also 31 empty cells in Table 2.5. Note how there is no indication in these cells regarding missingness. As a result, these (unmeasured) values are considered absent. When paired with the date, these data points form 61 data records in total, including data records where the data point is missing, however not counting any absent data.

2.4.4 Metadata

In this chapter, we briefly discuss metadata, mainly to provide a definition. As there is virtually no limit to the variety and scale of metadata that could be considered, Chapter 3 of this report is entirely devoted to the description and use of metadata.

Definition D2.8: **Metadata** are defined as the pieces of information that are used to describe signal data or laboratory data.

Metadata can be used to describe multiple aspects relevant to the use or interpretation of signal data or laboratory data, including but not limited to:

(1) How, where, when, and why the data was generated and/or modified.
(2) How the data can or should be interpreted.
(3) The quality of the data.
(4) Any contextual information regarding the origins, purpose, or use of the data.

More specific examples of metadata are given in Chapter 3.

2.5 STRUCTURED AND UNSTRUCTURED DATA

Most signal data contain data points that have a constant size, that is, they have a constant number of dimensions and a constant number of values along every dimension. Data points that are predictable in size are convenient to handle through computer-aided means. Their predictable form (number of dimensions) and size (lengths along every dimension) facilitates many data-related tasks including communication, long-term storage, database querying and navigation, data analytics, data visualization, and modelling.

Recent developments in online instrumentation have also enabled the production of signal data that does not follow a constant-size format. Examples shown below include data in the form of text or measurements of particle properties. Removing constant-size constraints makes it easier to navigate and interpret by visual inspection but require specialized treatment to enable its effective use in data analysis, modelling, or automation. For this reason, we provide definitions to distinguish between structured data, which exhibit a constant-size format, and unstructured data, whose format is more flexible but requires specialized handling for long-term storage and use.

Most signal data are structured with few exemptions. In contrast, the metadata available on a typical plant comes in both structured and unstructured forms. We define structured and unstructured data below and illustrate the definitions by means of a range of examples.

2.5.1 Structured data

In this subsection, we provide a definition for structured data and discuss the consequences of using such data, followed by a series of examples.

2.5.1.1 Definition

Definition D2.9: A signal is said to contain **structured data** when every data point in every data record satisfies the following requirements:

(1) The number of dimensions is constant.
(2) The number of values along all dimensions is constant.
(3) The position of a value along every dimension is meaningful and its meaning does not change over time.
(4) All data point values are restricted to belong to a single number type, such as boolean, integer, or continuous.

The third requirement implies that changing the position of a value within a data point is likely to change the validity of the data point. In other words, the position of a value matters. This idea is further explored below.

2.5.1.2 Additional remarks

Array and tabular formats. The major advantage of structured data is that they can easily be stored in an array or table with fixed lengths along every dimension. The position along each dimension (e.g., row position, column position) has a unique meaning so that comparing values appearing in the same row or column is a meaningful activity. Furthermore, data can easily be visualized in a chart or on a real-time dashboard. The ease of processing also ensures that one can proceed quickly to further analysis without many pre-processing steps. This ease of use is a hallmark property of structured data.

Non-numerical data can be structured. Structured data does not require the values to be numerical. For example, qualifiers (see Section 2.4.1.2 for definitions) are often a text of a predefined and constant length. Similarly, structured signal data may contain specialized codes for absent or missing data as is discussed next.

Missing data. Absent or missing values in a data point have no consequences for the application of the definition above. However, structured data requires that a specialized symbol or text is used to indicate an absent or missing value (e.g., *NaN*) to ensure that number of values along all dimensions remains constant. In addition, the use of an accepted symbol or text facilitates the identification and handling of absent and missing values during data analysis.

2.5.1.3 Examples E2.1 and E2.2 (ctd.): univariate signals

Univariate signals or time series, that is, data containing scalar (i.e., 0D) data points, are basic examples of structured data as shown in Examples E2.1 and E2.2.

2.5.1.4 Example E2.3 (ctd.): ultraviolet-visible light absorbance spectrophotometry

The data produced by spectrophotometric sensors is classified as structured because (a) each data point, that is, spectrum, is always 1D, (b) the number of absorbance measurements in a single data point is constant for a given sensor, and (c) the individual absorbance measurements in the same position correspond to the same light wavelength. The structured nature of the data permits an easy arrangement of all data records in a tabular format, as shown in Table 2.2. The structured nature of the spectrophotometric measurements also allows converting the values into a heatmap with rectangular tiles. Figure 2.6 shows 24 spectra in this form. These are the same spectra used to produce Figure 2.3 and Table 2.2.

2.5.1.5 Example E2.5 (ctd.): concentration measurements

The concentration data from Example 2.5 (Table 2.3) and the associated timestamps are classified as structured data. It is easy to see that every data point is scalar (0D) and that all conditions in Definition D2.9 are applicable. Note that the presence of absent values (empty cells) and missing values (<LOD) has no influence on this classification.

2.5.2 Unstructured data

In this subsection, we provide a definition for unstructured data and discuss the consequences of using unstructured data, followed by a series of examples for the purpose of illustration.

2.5.2.1 Definition

Definition D2.10: Data are said to be **unstructured** as soon as one of these conditions is satisfied:

(1) The number of dimensions of the data points can vary.
(2) The number of values along one or more dimensions can vary.
(3) The position of a value along one or more dimensions can be changed without consequences for further analysis.
(4) Data point values may belong to multiple number types, such as boolean, integer, or float.

2.5.2.2 Managing unstructured data

Collecting unstructured data has two important consequences:

- Storing unstructured data requires a specialized approach to data management, given that the data cannot be fitted meaningfully in a simple tabular structure. Some types of data storage technologies that can handle unstructured data are discussed in Chapter 4.

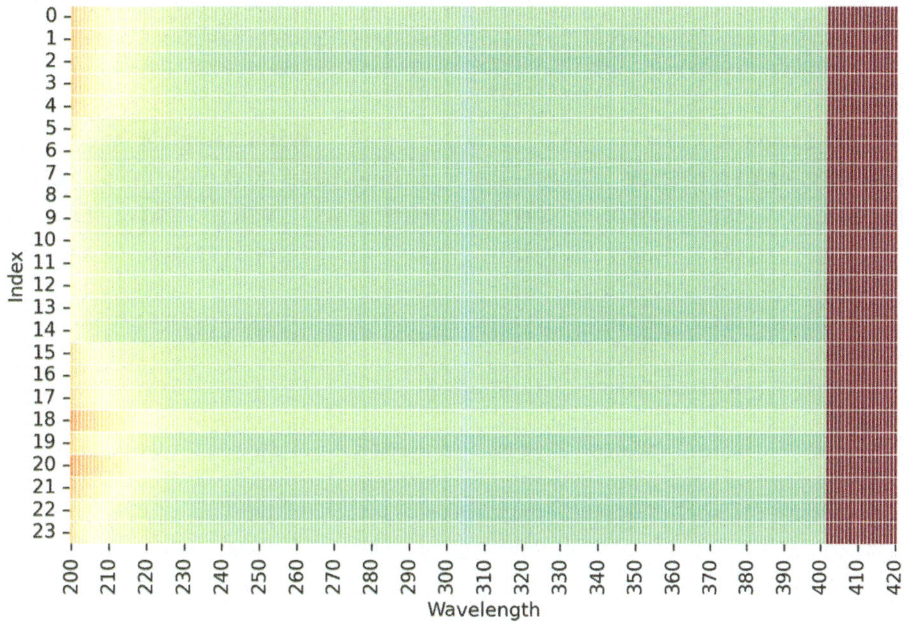

Figure 2.6 Example E2.3: spectrophotometric sensors deliver data that is 1D with a fixed length (here: 221). As a result, one can easily visualize the measured values in a heatmap with a regular grid. Cells are marked with white lines. Colour indicates the measured absorbance value as a function of time (t) and wavelength (λ). Missing data are indicated in red.

- Using commonly available models and tools for data analysis require that the unstructured data are transformed into a structured format first. This kind of data pre-processing is discussed in detail in Chapter 6 (Section 6.6.7).

2.5.2.3 Additional remarks

Tabular format. Even if unstructured data may appear in tabular format, it is unlikely that comparing values appearing in the same row or column is a meaningful activity.

Typical cases. Most unstructured data sources satisfy conditions 2 and/or 3 and systematic approaches to handle this exist. Unstructured data for which condition 1 applies are rare. As a hypothetical example, one could consider that a camera producing images with settings that change over time (some grey-scale images and some coloured images) will lead to image data that satisfies the first condition. There are no methods known to the authors to handle this case in a systematic manner.

Absent and missing data. Presence of absent or missing data does not influence whether data are considered structured or not. In general, the use of unstructured data format enables storing data by simply removing entries corresponding to absent or missing data, thus avoiding the need for a special value or code indicating absence or missingness. This may save some storage space. However, this means that the number of values can change along one or more dimensions. Since most data-analytical tools and models require structured data with fixed dimensions and number of values along each dimension, we do not recommend the use of unstructured data with the purpose of minimizing storage space or to avoid use of specialized symbols to indicate absence or missingness of data. Furthermore, removing information about missing data can be detrimental, as mentioned previously.

Table 2.4 Example E2.4: flow cytometry data. Measurement values for the first and last five particles for the data record visualized in the left and right panel of Figure 2.5, respectively, for water samples A and B. These data are classified as unstructured data given that the number of rows in the data records changes with every water sample.

Water Sample A					Water Sample B				
Timestamp: 2019-05-09T11:35:32Z					Timestamp: 2019-05-23T14:34:24Z				
Particle No.	FSC	SSC	FITC	PerCP	Particle No.	FSC	SSC	FITC	PerCP
0	968	384	4802	270	0	509	299	7542	837
1	3428	770	32,189	2459	1	0	283	1368	162
2	0	181	28,152	2421	2	6895	3691	86,812	9478
3	259	0	1353	373	3	0	161	1273	183
4	7984	5331	89,288	12,484	4	3990	3487	101,834	6708
...	...	...	...	...	...	...	...	...	...
56353	3203	1440	32,005	2495	164,805	302	1083	26,686	3764
56354	275	18	656	253	164,806	651	0	32,965	3659
56355	55,324	4066	1009	2743	164,807	953	359	32,600	2995
56356	129	0	4703	766	164,808	9741	1557	1583	192
56357	507	2103	40,058	5055	164,809	7353	4327	41,676	5346

2.5.2.4 *Example E2.4 (ctd.): flow cytometry*

Table 2.4 shows the values for four light intensities measured in two water samples. Each water sample corresponds to a single data record. Each data record consists of one timestamp and recorded measurements for particles in the water sample. Values are shown for the first five and the last five particles in the corresponding data records. The most important observation is that the number of particles in each sample is not the same (56,358 and 164,810). Figure 2.7 shows the number of particles as a function of time for a signal containing 137 data records in total. Clearly, the length along the first dimension of the data records (particle number) varies from one data record to the next. This means condition 2 in Definition D2.10 applies, and that the data are classified as unstructured data. Note that the length along the second dimension (emission wavelength) is constant. The order by which the particles are subjected to excitation and emission is of no influence during data interpretation. Indeed, one can shuffle the row position without consequences for further analysis of these data. This means condition 3 in Definition D2.10 applies as well. In Chapter 6 (Section 6.6.7), this same data set is used to illustrate three data pre-processing methods aimed at converting this type of unstructured data into a structured form.

2.5.2.5 *Example E2.5 (ctd.): text*

Text is an extremely efficient and versatile tool for communication and storage of information. A lot of descriptive information is expressed in the form of text, especially when a human-interpretable form is needed. Examples on typical wastewater treatment plants include self-contained documents, such as manuals, protocols, and reports, as well as annotations added to describe and isolate subsets of larger data sets (e.g., text describing an outlier, or applied signal corrections). In the field of natural language processing, a subfield of machine learning, text is typically considered 1D, that is, as a list of words separated by spaces and other delimiters. In this form, textual information has no fixed number of characters or words, such that condition 2 in Definition D2.10 applies.

The comments in the last column of Example 2.5 (Table 2.3) are classified as unstructured data. The free-text comments vary in length, ranging from 0 characters (no comment) to 13 characters. This variety in length is the reason these comments are considered unstructured data. Note that

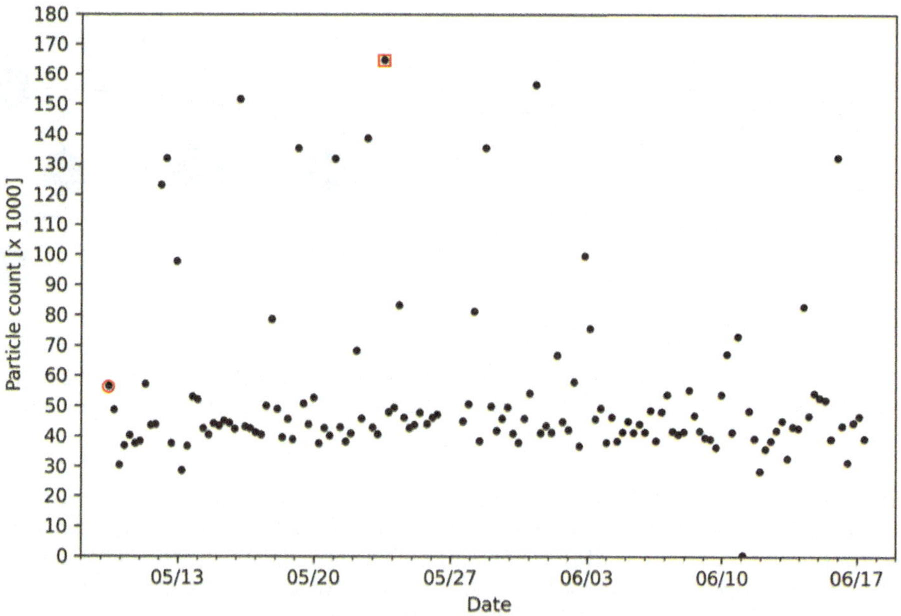

Figure 2.7 Example E2.4: flow cytometry data. Number of particles for each data record in a flow cytometry signal as a function of time. The circle and square correspond to water samples A and B in Figure 2.5.

textual data are not automatically considered unstructured. When the values have a fixed length as in Example 2.3 (see Table 2.2), they are considered structured. Similarly, labels used to annotate data may take on a fixed-length description and are then considered structured.

2.5.2.6 Example E2.6: incidence matrix to represent plant layout and balance equations

It is common for wastewater treatment plants to experience changes in their configuration and operation on a regular basis. This includes retrofitting a plant by additional treatment steps or modifications of the control system. In general, information about such changes that is useful for data interpretation resides in a diverse set of documents, such as a piping and instrumentation diagrams (P&ID), a construction and/or product specification, manuals, standard operating protocol (SOP), and other records. Most plants also include several devices that are operated manually (e.g., valves, switches). In addition, most control systems permit operators to activate or de-activate subsystems (e.g., for seasonal use of a treatment lane or maintenance) and switch between different control configurations (e.g., to adjust to different operational goals or allow for sensor maintenance). These manual and semi-automatic changes in the system are often recorded in an ad hoc way or not at all.

The metadata describing changes in plant construction and configuration, control system design and configuration, and manual and semi-automatic changes in operation are typically unstructured and do not facilitate a systematic use for analysis by default. At the same time, it often proves essential to have easy and complete access to these metadata for reliability in data-based decision making. One approach that may help is to provide an abstract, mathematical description of the plant configuration or the control system and keep track of this description as time progresses. This is illustrated next with a graph designed to represent flow balances.

A graph is a mathematical object consisting of nodes and edges and can be used to summarize the networked structure of a system comprising distinct units. Figure 2.8 shows a scheme of a small wastewater treatment plant on the left and the corresponding graph on the right. It is greatly simplified

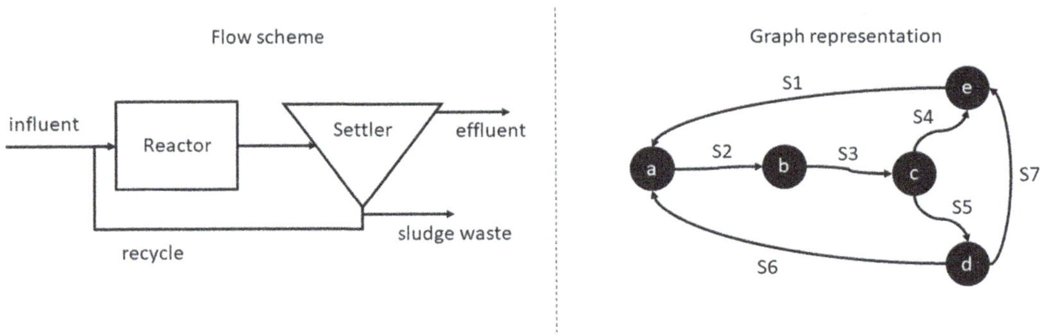

Figure 2.8 Example 2.6: using graphs to describe flow patterns and mass balances in a mathematically precise form. Left: a simple plant layout – right: graph of the plant layout, providing all information needed to construct mass balances.

relative to real-world cases for illustrative purposes. The graph is obtained by first listing its nodes. These correspond to the locations in the flowsheet of the plant where two or more streams meet. In the example, these are (a) combiner of influent and recycle stream, (b) biological reactor, (c) settler, and (d) splitter between recycle and sludge waste stream. An environmental node (e) is added to connect all streams crossing the system boundary, in this case the plant fence. The nodes are shown as circles in the right panel of Figure 2.8. Then, edges are added to represent every stream. These are shown as arrows in the right panel of Figure 2.8. The direction of these arrows corresponds to positive values for the corresponding flow rate. For example, positive values for the influent flow rate signify a net positive flow from the outside environment (e) to the combiner (a). There are seven streams in this simple case (S1–S7). To use this graph efficiently in computing systems, it can be stored in the form of an incidence matrix, **M**. This is illustrated in Table 2.5. The matrix contains a row for every node and a column for every stream. For a given node (row in the matrix), a value of +1 is placed for edges directed towards the node and a value of −1 is placed for edges directed away from the node. Streams not connected to the considered node are given a value of zero in the corresponding row. When executed correctly, the sum of all elements in any column will sum to zero.

A practical advantage of the graph-based approach is that flow balances can be defined easily. For example, the hydraulic balance error over the settler can be written as follows:

$$e_3 = \sum_{j=1}^{7} \mathbf{M}_{3,j} \cdot q_j$$

Table 2.5 Example E2.6: The incidence matrix is a compact representation of the plant flowsheet and the equations representing flow and mass balances.

		Edge / Stream						
Physical Element	**Node**	**S1**	**S2**	**S3**	**S4**	**S5**	**S6**	**S7**
Combiner	a	+1	−1	0	0	0	+1	0
Reactor	b	0	+1	−1	0	0	0	0
Settler	c	0	0	+1	−1	−1	0	0
Splitter	d	0	0	0	0	+1	−1	−1
Environment	e	−1	0	0	+1	0	0	+1

where q_j is the hydraulic flow rate in stream j. Thus, the incidence matrix **M** provides a compact yet complete representation of all flow balances. This can be extended straightforwardly to specify and use mass and energy balances as well (see, e.g., Spindler, 2014). Graphs are extremely useful to evaluate which variables (flow rates and concentrations) can be estimated from available measurements. In turn, this can be used to optimally place flow rate and concentration sensors in view of data quality assessment and control in mind (Villez *et al.*, 2016, 2020). In their matrix form (**M**), graphs are convenient for data reconciliation and fault detection purposes (see, e.g., Le *et al.*, 2018,2022; Puig *et al.*, 2008).

As the flowsheet of a typical wastewater plant may be subject to frequent changes, it follows that the graph and the incidence matrix will change over time. For example, the number of nodes may change as process units are taken in or out of operation. The number of edges may change as recycle streams may be taken in or out of operation. These changes in dimensions means that graphs should typically be considered unstructured data (condition 2 in Definition D2.10).

Systematic definition and use of graphs can also offer great advantages also for data-analytic purposes. A historical record of such graphs stored in an adequate database will enable highly automated identification of meaningful operational periods during which the plant and control configuration remained fixed and allows for data reconciliation despite changes in plant layout or operations. In turn, this could enable an effective cost–benefit analysis between different modes of operation or control configurations. In the long term, we also expect metadata taking the form of graphs to be useful to maintain plant models online with limited human supervision (i.e., digital twin).

Are graphs metadata? Graphs and incidence matrices are often discussed and stored separately from signal data. As a separate source of information, graphs are rarely mentioned in the context of metadata. As the above example illustrates, graphs can be used to document how multiple variables on a plant, and by extension their measurements, are related to each other. For this reason, we suggest including plant and control layouts information as potential metadata to describe actuator, sensor, and setpoint data.

In the future, we expect on-the-fly reconfiguration of plant layout and their control system to become more common. As the history of modifications of both the plant and control systems may not be obvious after the fact and as plant models may require historical data spanning multiple configurations, collecting such information in an automatic and systematic manner is considered valuable.

2.6 DATA VERSIONING

It is typical to modify signal data after collection to improve quality (e.g., remove noise) and its fit for a particular purpose (e.g., change units, aggregation, or averaging). Under ideal conditions, all steps used to modify signal data are traceable in a way that allows scrutiny and, where feasible, reversibility. This can be achieved with **data versioning**, which is the activity of saving and tracking any modification to the signal data collected. This is like the *Track Changes* feature in popular text editing software. Most typically, data versioning is based on the same principles as software version control. When executed, this enables navigation forward and backward in the temporal history of the data set.

2.6.1 Dealing with dynamic data sets

A data set will evolve over time and/or over the course of a project when more data or information become available. Thus, data versioning should be applied to data sets that are modified through any of the following steps:

(1) Data are added routinely (e.g., through sensor signal registration).
(2) Data are removed routinely (e.g., through outlier removal).
(3) Data are modified routinely (e.g., through noise filtering).

To this end, it is best to identify a central repository for all data subject to any of these steps. One way of doing this is explained below.

2.6.2 Definitions

In this subsection we provide the definitions of the different data versions to be tracked together with an example.

2.6.2.1 Original or raw data

Definition D2.11: **Original data** are the data as produced by a signal source, for example, sensors.

Most often, signals produced by sensors do not adhere to a format suitable for long-term storage since it includes signal processing in the instruments or they may not be accessible. For this reason, signal data are rarely stored in its raw, original form.

We note that many instruments will apply some modification of the raw data, for example, denoising, before it becomes available for logging, thus preventing collection of data in unprocessed, raw form. In this case, we recommend documenting the applied procedures as best as possible in the form of metadata. This is explained in Chapter 3 (Section 3.2.4).

2.6.2.2 Primary data

Definition D2.12: **Primary data** are the data consisting of the most accurate, comprehensive, and up-to-date version of a given set of data. This data set is used as a *single-source-of-truth*, that is, a data set that every data user accepts as the best source of information.

The existence of a clearly identified source of primary data (e.g., a central database) ensures that all users of the data are accessing the same information. Proper management of primary data aims to achieve the following objectives:

(1) Primary data capture a complete history of every data record in the database, including (a) all modifications executed to convert the original, raw data into the stored data and (b) all modifications executed after registration in this repository. Typical modifications applied before storage include but are not limited to formatting, change of unit, imputation, and annotation. See Chapter 3 for a more complete overview and Chapter 6 for an overview of commonly used methods to modify data after collection.
(2) Minimize the amount of valuable information that is lost. To this end, the authors believe that certain modifications of the data should be avoided in the primary data. These include any step that achieves (a) noise reduction, (b) data removal, or (c) aggregation. If applied anyway, we recommend that the primary data contain a copy of the same data for which these steps were not applied. There are two reasons for this. First, many modification procedures are subject to change but may not be reversible once applied. Second, different purposes of the data are likely to require different modifications.
(3) Continuous updating of the data through addition of new data records and careful updating in view of improved data quality.

All above examples (Example 2.1 until 2.6) are considered for storage as primary data. For certain projects, this is not the most convenient form of data. For example, the oxygen concentration data in Examples 2.1 and 2.2 have asynchronous timestamps, which complicate further analysis. Similarly, the flow cytometry data (Example 2.4) cannot be handled directly through conventional data-analytical means due to the changing number of measured values per water sample. This kind of inconvenience of the primary data should be addressed through the generation of secondary data, not through irreversible modification of the primary data. Reasons for this are two-fold. First, pre-processing procedures to derive structured variants of unstructured data are subject to change. One should thus be able to apply a different or new procedure later. Second, pre-processing steps aimed at providing structure (e.g., synchronization, binning) will lead to a loss of information.

2.6.2.3 Secondary data

Definition D2.13: **Secondary data** are the data that have been generated by selection and modification of primary data for a specific context or purpose.

Secondary data are typically generated for a specific project (e.g., plant upgrade) or task (e.g., controller tuning). To make the primary data useful for practical purposes, it may be necessary to apply several data processing steps that alter the information content. For example, one may want to check oxygen controller performance by comparing the signals from Examples 2.1 and 2.2. Since the timestamps do not match in these two signals, one may obtain synchronized timestamps by interpolation or averaging. The applied steps to produce secondary data are often unique and exclusive to a specific task, study, or project. They may include aggregation, filtering, imputation, smoothing, data point removal, and more. Chapter 6 of this report contains a comprehensive list of data processing steps that may be used during generation of secondary data.

In contrast to primary data, secondary data are typically not updated continuously. Instead, a secondary data set is likely static within the timeframe of a single task or project. This also means that secondary data can quickly become outdated and out of sync with existing primary data. For this reason, we recommend the following practices:

(1) Never add data to a secondary data set if it is not made part of the primary data first.
(2) Assign an expiry date to every secondary data set.
(3) Document the process by which secondary data are derived from primary data in such a way that the production of secondary data can be repeated. See Chapter 6 for typical data processing steps that should be available in the form of well-managed computer code or documented in detail otherwise.

2.6.2.4 Data traceability: documenting the history from original to secondary data

Definition D2.14: Data **traceability** consists of the application of systems that enable monitoring of the entire process of data collection, storage, and processing for a proper and an effective use.

Documenting and tracking any modification of the original data all the way to the production of secondary data can help with the assurance of the quality and validity of any decision or recommendation based on data. In addition, this allows to reproduce data sets without uncertainty about the process by which they were generated.

Figure 2.9 gives an example of how data traceability can be achieved using the definitions given above. Data produced in the laboratory, sensor signals, control signals produced by a supervisory control and data acquisition (SCADA or DCS) system, and weather-related data are ingested into a single data repository containing a comprehensive data set called the primary data, which act as a single, commonly accepted source of information. This primary data are used to derive other data sets for specific tasks or projects. For example, one data set is built specifically for the purpose of control tuning. This data set will likely exhibit a high temporal resolution and synchronized timestamps. A second data set is generated to support the sludge management on the plant. This leads to a data set of low temporal resolution focused on a few variables only. It may include some variables derived from other measurements, such as an estimate of the sludge residence time (SRT) of the plant. Finally, a third data set consists of aggregated data for the purpose of yearly reporting. This includes data at an extremely low temporal resolution.

Ideally, the procedures to generate derived data sets are available as clearly documented code or procedures, thus allowing the production of the derived data sets from the primary data using information in the code or procedure only. A well-documented code or procedure will thus include and completely specify every step in the generation of these secondary data sets. This means that the controller tuning can be inspected later by simple modifications of the corresponding script. Similarly, an update in the procedure to compute SRT can be implemented by execution of an updated script. Finally, the script to produce the data for this year's annual report can be reused easily for next year's report.

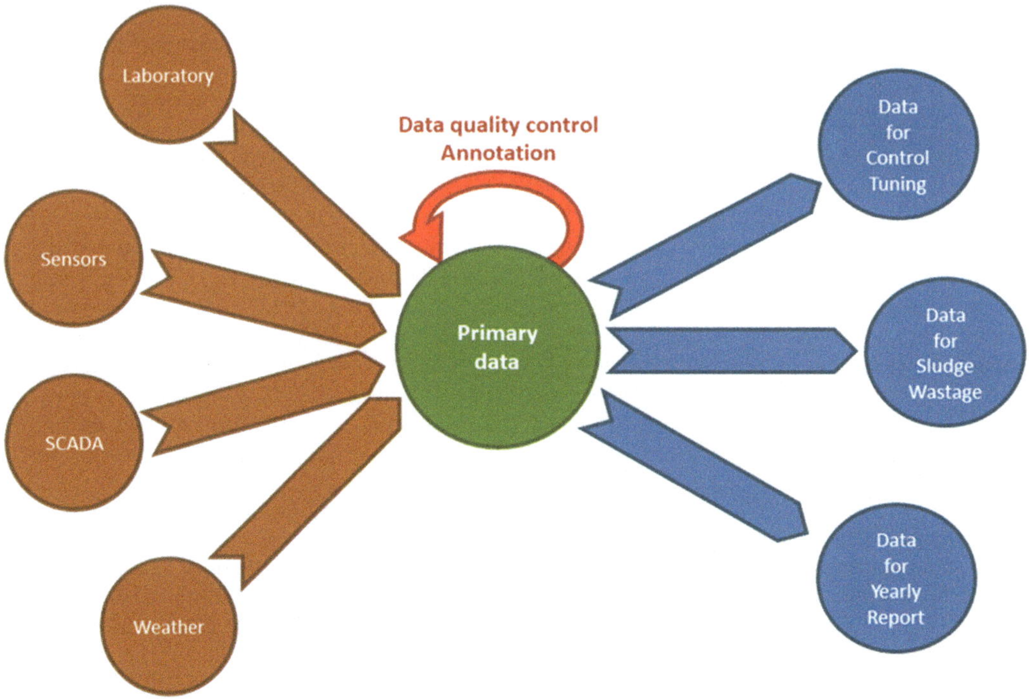

Figure 2.9 Data flow diagram including the data versions (in red) and the data modifications (in blue).

2.6.3 Data versioning vs software versioning: similarities and differences

Data versioning and software versioning allow tracking of any changes in any data set or any software code. Both types of version control are used to ensure the reliability and reproducibility of computer-aided decision-making processes. Today, software versioning is a widespread practice while data versioning is not. Practically, software versioning can be executed effectively with relatively small data storage requirements. In contrast, the storage requirements for data versioning today are large. In time, more advanced solutions for data versioning are expected to reduce these requirements while improving the ease of use.

2.7 TAKE-HOME MESSAGE

- We distinguish between signal data, laboratory data, and metadata in this book.
- Most sensor signals produce structured data, and this will likely remain so.
- Many types of metadata and specialized sensor signals are unstructured and therefore more difficult to manage and analyse by conventional means.
- Responsible data management practices include:
    - Keeping original data in unmodified form (i.e., primary data)
    - Defining a *single-source-of-truth*, that is, a unique and commonly referenced location for data of accepted quality.
    - Establishment of documented and repeatable procedures for modification of data for analytical, modelling, or automation purposes.

ACKNOWLEDGEMENTS

Examples 2.1, 2.2, and 2.3 were provided by Ali Gagnon (HRSD). Example 2.4 was provided by Michael Besmer (onCyt Microbiology AG, Dübendorf, Switzerland). Example 2.5 was provided by Angelika Hess and Eberhard Morgenroth (Eawag, Dübendorf, Switzerland). Example 2.6 was inspired by discussion with Adrienne Menniti (Clean Water Services, Hillsboro, OR, USA). This chapter was edited by Ali Gagnon (HRSD) and Matthew Wade (Newcastle University).

REFERENCES

Besmer M. D., Weissbrodt D. G., Kratochvil B. E., Sigrist J. A., Weyland M. S. and Hammes F. (2014). The feasibility of automated online flow cytometry for in-situ monitoring of microbial dynamics in aquatic ecosystems. *Frontiers in Microbiology*, **5**, 265. https://doi.org/10.3389/fmicb.2014.00265

Bhaskaran K. and Smeeth L. (2014). What is the difference between missing completely at random and missing at random? *International Journal of Epidemiology*, **43**(4), 1336–1339. https://doi.org/10.1093/ije/dyu080

Currie L. A. (1995). Nomenclature in evaluation of analytical methods including detection and quantification capabilities (IUPAC recommendations 1995). *Pure and Applied Chemistry*, **67**(10), 1699–1723. https://doi.org/10.1351/pac199567101699

Currie L. A. (1999). Detection and quantification limits: origins and historical overview. *Analytica Chimica Acta*, **391**(2), 127–134. https://doi.org/10.1016/S0003-2670(99)00105-1

Dong Y. and Peng C. Y. J. (2013). Principled missing data methods for researchers. *SpringerPlus*, **2**(1), 222. https://doi.org/10.1186/2193-1801-2-222

Hammes F. and Egli T. (2010). Cytometric methods for measuring bacteria in water: advantages, pitfalls and applications. *Analytical and Bioanalytical Chemistry*, **397**, 1083–1095. https://doi.org/10.1007/s00216-010-3646-3

ISO (1997). ISO 11843-1:1997, Capability of detection – Part 1: Terms and definitions. 10pp.

Kang N (2013). The prevention and handling of the missing data. *Korean Journal of Anesthesiology*, **64**(5), 402–406. 10.4097/kjae.2013.64.5.402

Le Q. H., Verheijen P. J., van Loosdrecht M. C. and Volcke E. I. (2018). Experimental design for evaluating WWTP data by linear mass balances. *Water Research*, **142**, 415–425. https://doi.org/10.1016/j.watres.2018.05.026

Le Q. H., Verheijen P. J., van Loosdrecht M. C. and Volcke E. I. (2022). Application of data reconciliation to a dynamically operated wastewater treatment process with off-gas measurements. *Environmental Science: Water Research & Technology*, **8**, 2114–2125. https://doi.org/10.1039/D2EW00006G

Lepot M., Torres A., Hofer T., Caradot N., Gruber G., Aubin J. B. and Bertrand-Krajewski J. L. (2016). Calibration of UV/Vis spectrophotometers: a review and comparison of different methods to estimate TSS and total and dissolved COD concentrations in sewers, WWTPs and rivers. *Water Research*, **101**, 519–534. https://doi.org/10.1016/j.watres.2016.05.070

Mašić A., Santos A. T., Etter B., Udert K. M. and Villez K. (2015). Estimation of nitrite in source-separated nitrified urine with UV spectrophotometry. *Water Research*, **85**, 244–254. https://doi.org/10.1016/j.watres.2015.08.031

OPC Foundation (2003). OPC Data Access Custom Interface Specification. Version 3.0. 190pp.

Osman M. S., Abu-Mahfouz A. M. and Page P. R. (2018). A survey on data imputation techniques: water distribution system as a use case. *IEEE Access*, **6**, 63279–63291. https://doi.org/10.1109/ACCESS.2018.2877269

Puig S., Van Loosdrecht M. C. M., Colprim J. and Meijer S. C. F. (2008). Data evaluation of full-scale wastewater treatment plants by mass balance. *Water Research*, **42**(18), 4645–4655. https://doi.org/10.1016/j.watres.2008.08.009

Rieger L., Langergraber G. and Siegrist H. (2006). Uncertainties of spectral in situ measurements in wastewater using different calibration approaches. *Water Science and Technology*, **53**(12), 187–197. https://doi.org/10.2166/wst.2006.421

Spindler A. (2014). Structural redundancy of data from wastewater treatment systems. Determination of individual balance equations. *Water Research*, **57**, 193–201. https://doi.org/10.1016/j.watres.2014.03.042

Villez K., Vanrolleghem P. A. and Corominas L. (2016). Optimal flow sensor placement on wastewater treatment plants. *Water Research*, **101**, 75–83. https://doi.org/10.1016/j.watres.2016.05.068

Villez K., Vanrolleghem P. A. and Corominas L. (2020). A general-purpose method for Pareto optimal placement of flow rate and concentration sensors in networked systems – With application to wastewater treatment plants. *Computers & Chemical Engineering*, **139**, 106880. https://doi.org/10.1016/j.compchemeng.2020.106880

Wang Y., Hammes F., De Roy K., Verstraete W. and Boon N. (2010). Past, present and future applications of flow cytometry in aquatic microbiology. *Trends in Biotechnology*, **28**(8), 416–424. https://doi.org/10.1016/j.tibtech.2010.04.006

doi: 10.2166/9781789061154_0031

Chapter 3

Metadata for a systematic description of signal data

J. Alferes[1], J. A. Baeza[2], N. Nicolai[3], J-D. Therrien[3], Q. Plana[3,4] and K. Villez[5]

[1]*VITO – WaterClimateHub, Ostende, Belgium*
[2]*Universitat Autònoma de Barcelona, Cerdanyola del Vallès (Barcelona), Spain*
[3]*modelEAU, Université Laval, Québec, QC, Canada*
[4]*Eurecat – Technology Center of Catalonia, Barcelona, Spain*
[5]*Oak Ridge National Laboratory, Oak Ridge, TN, USA*

Keywords: annotation, contextual information, data quality, metadata, sensor history

This chapter aims to provide a comprehensive overview of metadata types that may be useful during system design, optimization, and automation. Metadata are grouped into three main categories: (a) metadata describing signal generation, (b) metadata describing signal quality, and (c) contextual information in the form of annotations. Each of these categories is introduced and explained in three separate sections. Importantly, this chapter mainly answers *what* is considered metadata. To a lesser degree, recommendations are made regarding the selection of metadata for long-term storage. Chapter 4 will explain *where* and *how to store* metadata. Chapters 5 and 6 explain *how to collect* certain metadata through dedicated sensor validation tests (Chapter 5) or algorithmic analysis (Chapter 6).

3.1 HOW TO READ THIS CHAPTER

The main purpose of this chapter is to provide the overview and definitions on the types of metadata and their purpose within the data pipeline. Three metadata categories are recognized:

- Metadata describing signal generation, including acquisition and registration.
- Metadata for quality assessment of sensor signals.
- Metadata providing contextual information relevant for later use.

The scheme below indicates the structure of the chapter and the dependencies between sections. Sections marked in blue or white are considered essential to understand the remainder of STR, specifically concerning (a) data organization and storage (Chapter 4) and (b) production of metadata through careful testing (Chapter 5) or signal analytics (Chapter 6). The sections marked in grey

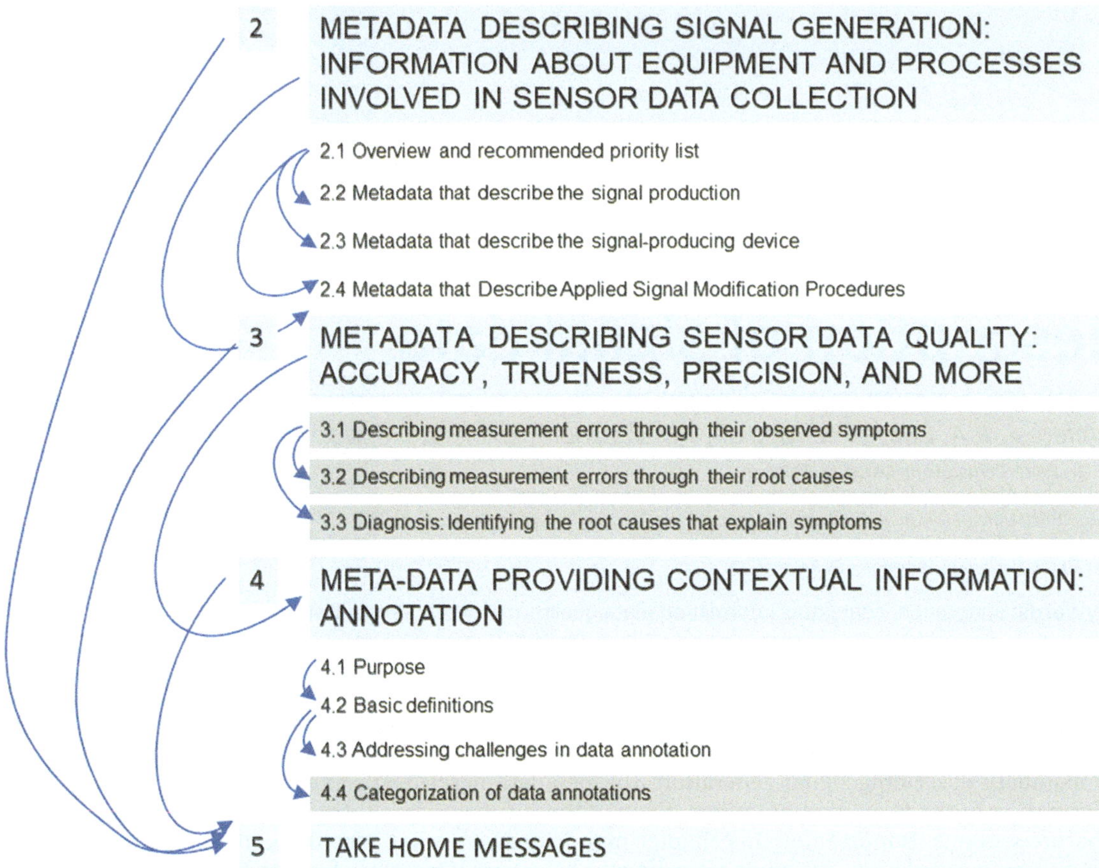

provide details relevant to understand the specifics of standard operational procedures (SOPs) for sensor validation (Chapter 5) and analytical methods for data quality assessment (Chapter 6).

3.2 METADATA DESCRIBING SIGNAL GENERATION: INFORMATION ABOUT EQUIPMENT AND PROCESSES FOR SENSOR DATA COLLECTION

In what follows, we describe types of metadata that document the routines used for data production and the results produced through these routines. Such routines are often named data acquisition, data collection, data pipelines, or data-generating processes. This includes documentation of the routines in use (e.g., data pipelines) as well as the metadata produced through such routines (e.g., calibration curve). In principle, the metadata considered in this section can be described before any signal data is collected. This is the main difference between the metadata described here and other metadata described in Sections 3.2 and 3.3. In practice, information may change during sensor purchasing, installation, and deployment. Regular verification of these metadata may therefore help to ensure they are up-to-date. While often limited, this type of metadata is considered straightforward to identify, collect, and store.

The following section starts with an overview of all the metadata belonging to the category of signal generation, using a dissolved oxygen sensor and a pH sensor as examples. Through the same examples, we assign a priority level to each type of metadata. After this section, each of the metadata types is explained in detail.

3.2.1 Overview and recommended priority list

Table 3.1 gives an overview of typical metadata that may be collected to describe the signal generation process for a dissolved oxygen sensor and a pH sensor. The table includes an indication of how frequently this type of metadata may be reviewed and updated, as well as a priority level according to the following descriptions:

- **Level 1 – Basic:** A signal cannot be understood without this information. Collecting this type of metadata after the data collection is considered hard. Therefore, this level of metadata should carry the highest priority.
- **Level 2 – Advanced:** These metadata are extremely useful for sensor validation, especially when using automated tools for data quality assurance and control, as described in Chapter 6.
- **Level 3 – Future proof:** These metadata are particularly useful for cases where the future use of the collected data may differ substantially from the current purposes.

We note that the current level assignments are based on a consensus between the authors of this chapter. Depending on the purpose of data collection, different priority levels may be assigned in your organization. We recommend discussing priority levels within your organization and with potential stakeholders before identifying methods to collect and manage the metadata.

3.2.2 Metadata that describes the signal production

The metadata described herein provides the necessary context to understand and assign meaning to values produced by a sensor. In what follows, we provide a comprehensive list of types of metadata belonging to this category. We provide a definition and motivation for each so that you can consider its utility for your purposes. Where relevant, we add additional remarks and examples.

3.2.2.1 Timestamp origin

Definition D3.1: The **timestamp origin** is the device that produces the timestamp in a data record (see Chapter 2, Section 2.4.1.2).

Motivation: Knowing the timestamp origin can be of great help when diagnosing situations where the clocks of the devices do not align. For instance, some devices enable the use of daylight saving time while others do not. While timestamp standards exist to prevent misinterpretation (ISO 2019*a*, *b*), adoption is often limited in practice, thereby reducing trust in the produced data or increasing the cost of their use.

Additional remarks: In sensing devices equipped with digital communication, the timestamp is often produced by the sensing device using its internal clock. In devices connected through an analog connection, the timestamp is often created by the system configured for the data logging (e.g., SCADA or DCS).

3.2.2.2 Temporal resolution

Definition D3.2: The **temporal resolution** is the time interval between subsequent data records produced by a sensor.

Motivation: The temporal resolution is often critical to determining whether the sensor will capture certain phenomena. If a phenomenon of interest occurs in a shorter timeframe than the temporal resolution of the sensor, it may not be captured or may be captured poorly.

Additional remarks: The temporal resolution is typically constant for sensors powered with a permanent power supply. The use of sensors powered with batteries or intermittent power supplies (e.g., solar) may lead to changing temporal resolutions. In this case, the data user may benefit from knowing the lowest temporal resolution (i.e. the shortest time interval) acceptable in view of battery or sensor management.

Table 3.1 Typical metadata for signal documentation.

Section	Metadata Type	Examples		Review Frequency	Priority Level
3.2.2	**Metadata that describes the signal production**				
		Dissolved oxygen sensor	pH sensor		
3.2.2.1	*Timestamp origin*	Analog-to-digital converter integrated into the PLC device	SCADA system	Years	Basic
3.2.2.2	*Temporal resolution*	15s	1s	Years	Advanced
3.2.2.3	*Numeric resolution*	0.3×10^{-3} (16-bit)	Not available	Years	Future-proof
3.2.2.4	*Sensor location*	Aeration basin 3 in lane 2	Primary clarifier outlet	Months – Year	Basic
3.2.2.5	*Measurement unit*	g/m^3	Not applicable (dimensionless)	Years	Basic
3.2.2.6	*Measurement range*	0–20	2–12	Years	Basic
3.2.2.7	*Variable range*	2–6	5–7	Months – Year	Advanced
3.2.2.8	*Transmission mode (analog/digital/manual)*	Digital: Modbus	Analog: 4–20 mA	Years	Advanced
3.2.2.9	*Measurement principle*	Luminescence-based sensor	Ion-selective electrode	Years	Future-proof
3.2.2.10	*Purpose of measurement*	Automatic feedback control of dissolved oxygen concentration.	Warning for industrial/toxic influent	Months – Year	Future-proof
3.2.3	**Metadata that Describe the Signal-Producing Device**				
3.2.3.1	*Product specifications and guides*	*Reference/link to stored manual, datasheet and/or installation guide*		Years	Advanced
3.2.3.2	*Manufacturer, model name, and serial*	Hach, 9020000, 9FT3K45G	E+H, CPS11D, R2EKGEQT	Years	Basic
3.2.3.3	*Unique identifier*	DO20210001	PH18065	Years	Basic
3.2.3.4	*Device history*	*Reference/link to purchase order, record of installation, historical sensor calibration curves and validation results*		Weeks – Month	Advanced
3.2.4	**Metadata that Describe Signal Modification**				
3.2.4.1	*Change of units*	None applied	None applied	Years	Advanced
3.2.4.2	*Lossy modifications (aggregation, averaging, filtering, compression)*	None applied	Median filter applied to five consecutive measurements to avoid false alarms *(add a reference to code repository or software manual)*	Months – Year	Future-proof
3.2.4.3	*Data augmentation (imputation, interpolation, synchronization)*	Missing data imputed with the last known value to ensure continuous operation of the controller	None applied	Months – Year	Future-proof
3.2.4.4	*Data fusion*	None applied	Internal: Nernst equation applied to compute pH from raw potential and temperature measurements	Months – Year	Future-proof

Note on multivariate sensors: Multivariable instruments or multi-sensors may hold several sensing elements within a single enclosure, each recording a different physical quantity. Depending on the device, certain metadata will be the same for all signals produced by the sensor (e.g., temporal resolution, location), while others may be different (e.g., measurement unit, measurement range). Also check the related notes in Section 2.4.1.5 of Chapter 2.

3.2.2.3 Numeric resolution

Definition D3.3: The **numeric resolution** is the smallest increment in the monitored variable that can be recorded after digitization (i.e., analog-to-digital conversion).

Motivation: The effects of digitization are often barely noticeable. However, knowing the numeric resolution can help to assess data quality as well as to define adequate performance metrics for model-based or algorithmic predictions. Generally, one cannot expect a prediction to be more precise than the measurement resolution.

Additional remarks: Due to the binary nature of modern computing and data storage systems, data points are stored in digital form. Digitization involves using a discrete (i.e., not continuous) representation for both the timestamp (digitized) and the data point (quantized). The numeric resolution is the parameter describing this process. The effect of rounding is also known as quantization error (see Section 3.3.1.2).

3.2.2.4 Sensor location

Definition D3.4: The **sensor location** is the physical position of the sensing element of the sensor.

Motivation: Knowing the exact position of a sensor is key to signal interpretation. In addition, it has multiple advantages during data quality assessment and control.

- It can help to detect, isolate, and diagnose anomalous data points that may be produced through maintenance actions, such as sensor cleaning or calibration.
- It can provide context to the recorded data points. For example, the data points produced by an ammonia sensor located in an aeration basin will typically cover a different range than one located in the effluent of the wastewater treatment system.
- It can help monitor sensing devices, particularly by describing and quantifying the exposure to harsh measurement environments.

Additional remarks: The position of a sensor can be described in at least three different forms:

(1) **Position relative to a physical reference point.** As text describing the relative position to the measurement environment. For example, *in basin 3 of aeration lane 2* or *at 10 m on the waste recycle stream pipe*. Such descriptions may include relative position indications, such as *middle of the basin*, *depth = 2 m*, or *at exit gate*. This form is easy to enter, discuss, and interpret by a human operator. However, this type of information may require additional metadata (e.g., a flow sheet or other process diagrams) to facilitate efficient analysis.

(2) **Position relative to design schemes.** As part of a P&ID or similar drawings or schemes. Modern, computer-aided design software for P&IDs allows for augmenting or linking elements in the scheme with metadata describing the depicted devices to support the sensor location information visually. These file formats facilitate automatic interpretation for computer-based data analytics (Udugama *et al.*, 2021).

(3) **Geospatial coordinates.** As a set of geospatial coordinates. This format may be particularly useful for mobile sensors, for example, for sensors on drones (submerged, floating, or flying) or robotic arms. These coordinates may be produced through a globally referenced system (e.g., GPS) or a locally referenced system, such as distance measurements based on high-definition cameras, laser light, or radio signals (e.g., Bluetooth, RFID). Naturally, a locally referenced location measurement system will require its own metadata to ensure accuracy of and trust of the geospatial coordinates produced by it.

While most sensors today are positioned in a fixed position over long periods, maintenance actions often require a change in the exact position of the sensor. A position measurement system may enable automatic logging of periods of time where the data were not representative of the intended measurement location. Moreover, recent research has explored the use of mobile sensors, whether they are submerged, floating, flying, or attached to robotic systems. We suspect that automatic tracking of sensing device position will offer distinct advantages over systems requiring human-in-the-loop tracking of sensor location and placement.

Some newer sensing systems, like the drones mentioned above, exhibit frequent or continuous changes in measuring position. For this reason, one should consider storing the metadata as time series. A similar approach to signal data (see Chapter 2) may be taken to store this type of metadata.

As a final note, the placement of redundant instruments requires special care. The first form of position metadata described above (position relative to plant components) is vulnerable to reporting the position of multiple sensors as the same. For certain tasks or projects, proper differentiation between the exact location of redundant sensors may be crucial.

3.2.2.5 Measurement unit

Definition D3.5: The **measurement unit** is the chosen standard measurement by which the magnitude of the sensor reading is expressed.

Motivation: The measurement unit gives physical meaning to the numerical value generated by the sensor. Storing the unit as metadata ensures that the data can be interpreted correctly.

3.2.2.6 Measurement range

Definition D3.6: The **measurement range** is the interval defined by the minimum and maximum values that can be produced by a sensor that is configured and operated correctly.

Motivation: Knowing the measurement range helps detect and isolate malfunctioning sensors. Among others, an out-of-range value may be related to damaged parts (e.g., sensor membrane), wrongful programming of the relationship between an analog signal and measured variable range in the DCS or SCADA, or wrongful calibration.

Additional remarks: Note that the measurement range can be different from the output range of the analog signal (e.g., a pH sensor can be configured to output a 4–20 mA signal for a linearly corresponding pH range between 4 and 10, even though the measurement range is 0–15). The measurement range only depends on the sensor and not on its location or the time of measurement. Most often, the measurement range exceeds the range of values expected for a particular location on a plant, time of day, or season (variable range, see below).

3.2.2.7 Variable range

Definition D3.7: The **variable range** is the interval defined by the minimum and maximum value that is acceptable or typically observed during normal operation, either in general or for a particular context (e.g., location and/or time).

Motivation: Knowing the measurement range helps detect, isolate, and diagnose malfunctions. An out-of-variable-range data point may relate to a fault in equipment (e.g., blower), process (e.g., toxic load), or sensors (e.g., fouling).

Additional remarks: Note that the variable range describes the monitored process rather than the sensor itself and can be defined for a specific process type (e.g., anaerobic digestion) or specified for a particular plant, process unit, time of day, and/or season.

3.2.2.8 Transmission mode (analog/digital/manual)

Definition D3.8: The **transmission mode** is the way a signal produced by a sensor is carried to the physical storage device (e.g., PLC, SCADA).

Motivation: Understanding available communication protocols and their implications for data quality and management is valuable in view of the generation of interpretable and future-proof data sets. Knowing the transmission mode also allows technicians to focus efforts when troubleshooting sensor communications.

Additional remarks: Automated, high-frequency signal communications can be roughly characterized into two broad categories:

(1) *Analog transmission:* The measured value is carried through a continuous electrical current (often varying between 4 and 20 mA) which can be interpreted by systems downstream of the sensor (e.g., PLC, SCADA). This kind of transmission is simple, which facilitates maintenance. However, classical analog signals do not permit the transmission of metadata, such as the internal status of the sensing device. Special precautions must be taken to avoid interference with other signals (e.g., radio) for transmission over long distances. Protocols, such as HART (highway addressable remote transducer), enable the addition of complementary digital information through frequency modulation (HART Communication Foundation, 2013).
(2) *Digital transmission:* The measured value is first digitized (analog-to-digital conversion) and then transmitted over a digital cable or network using a defined communication protocol. Converting to digital transmission often requires the training of local staff. However, this enables transmission and automatic logging of additional information, such as the sensor's internal status (e.g., calibration mode, battery state) or parameters (e.g., calibration data, calibration curve). Modbus, Profibus, and Profinet communications all belong to this category. A challenge unique to digital transmission is that timestamps on the sensor devices may not be synchronized with PLC timestamps (see Section 3.2.2.1).

In addition to the above automated transmission modes, it is also possible to enter data through manual entry:

(3) *Manual transmission:* The measured value is copied from a device (e.g., handheld device) or written record (e.g., calibration sheet) and typed into a long-term storage device or system, usually through a dedicated graphical user interface (GUI). At least one step in the transmission requires human action.

When multiple modes of transmission are used, we recommend to either:

(1) Describe the complete data pipeline as a series of modes.
(2) Describe the least performing mode in the data pipeline, using this order (low to high): manual, analog, digital.

3.2.2.9 Measurement principle

Definition D3.9: The **measurement principle** of a sensor is a description of the fundamental physicochemical relationship between the variable of interest and the raw signal produced by the sensor.

Motivation: Knowledge of the applied measurement principle and its implications for signal quality and maintenance cost is key to adequate data quality assurance and quality control.

Additional remarks: Not every measurement principle is equally accurate, precise, or timely. In addition, differences in measurement principle often coincide with differences in installation requirements (e.g., depth, angle), sensor maintenance requirements (e.g., frequency of part replacement), and measurement ranges. For example, today's ammonia sensors based on ion-selective electrodes do not measure accurately below 1 mg NH_3-N/L. In contrast, an ammonia analyser using the colorimetric principle is considered accurate below 1 mg NH_3-N/L.

3.2.2.10 Purpose of measurement

Definition D3.10: The **purpose of measurement** describes the way(s) the signal is intended to be used.

Motivation: A description of a sensor's original and later purposes can assist in the prioritization of maintenance actions and helps a data analyst in deciding whether data records produced by a sensor are useful for a particular task.

Additional remarks: Signal data are often collected for one purpose (e.g., operator supervision) and later used for another (e.g., modelling, automation). Importantly, different purposes often warrant distinct quality assessment and quality control practices, leading to significant differences in the resulting data quality. The purpose of measurement may also indicate whether a particular measurement is meant to be temporary, for example, as part of a measurement campaign, or intended to be available indefinitely.

3.2.3 Metadata that describes the signal-producing device

The metadata related to the signal-producing device includes information about the device's identification, measurement principle, technical characteristics, and history.

3.2.3.1 Product specifications and guides

Definition D3.11: **Product specifications and guides** consist of documentation detailing the installation, use, and maintenance of an instrument.

Motivation: Product specifications and guides help to ensure correct data quality assessment and quality controls are in place and that the sensor is installed and used as intended.

Additional remarks: Information describing proper sensor installation, use, and maintenance is typically provided by the manufacturer. We strongly recommend collecting this information at the time of purchase to avoid inconsistencies due to later changes reflecting product updates. This information may be contained in one or more documents or electronic files, often including:

- Specification sheet
- Installation guide
- Operator's manual
- Datasheet
- Quick-start guide
- 3D design file (e.g., CAD)

Critical pieces of information to look out for include:

- Required power supply
- Ingress protection (IP) rating of the device or requirements for dust-proof or water-proof installation
- Available communication protocols (analog, digital)
- Number of ingoing and outgoing channels
- Inspection and maintenance requirements, including replacement of parts (cap, filter, joint, membrane) and consumables (e.g., chemical reagent)

3.2.3.2 Manufacturer, model name, and serial

Definition D3.12: **Manufacturer information** consists of the instrument brand or make, model number, and serial number.

Motivation: Having access to the make, model, and serial number produced by the manufacturer allows for looking up manuals and specifications with the manufacturer. Moreover, having access to the reference used by the manufacturer facilitates quick ordering of new or spare parts with the same or compatible properties.

3.2.3.3 Unique identifier

Definition D3.13: A **unique identifier** is a name or number unique to a particular instrument or component thereof, allowing a detailed description of the signal source.

Motivation: Keeping track of the sensor inventory is made easier by keeping a record of the information that uniquely identifies a sensor and any part that has been modified or replaced. It allows tying multiple pieces of information together, including manufacturer information, sensor life cycle tracking, and the data produced by the instrument. Finally, it enables a searchable overview of all instruments available within a plant.

Additional remarks: A unique identifier can often be created by combining the make (manufacturer), model number, and serial number (see above). Individual parts may also be identified in the same way.

3.2.3.4 Device history

Definition D3.14: The **device history** consists of all documentation describing important events in the history of the signal-producing device.

Motivation: A well-documented sensor history is helpful to assure adequate yet cost-effective sensor management practices, including selecting sensors for purchasing and optimized sensor calibration, validation, and verification practices. It may also be extremely helpful in gauging the quality of the sensor signals as one may obtain information about difficult-to-detect issues such as calibration errors, sensor wear-and-tear, and symptoms of drift.

Additional remarks: The device history includes the documentation of important elements of a sensor's life cycle, including but not limited to:

- The time of purchase, delivery, commissioning, and decommissioning.
- Periods of time corresponding to an out-of-order or out-of-use status.
- Any maintenance action by the manufacturer or plant's staff, such as cleaning, on-site or factory calibration, validation, testing and sensor or part replacement.
- A history of calibration curves and/or the measurements used to produce such curves.

Chapter 5 explores the generation of metadata falling into this category, specifically the production of validation data through reference measurements.

Note on calibration curves: In the most common scenario, sensor calibration results in a modification of the calibration curve stored on an electronic device, such as a measurement amplifier, transformer, or transmitter. In many cases, only the current calibration curve is available on the device. However, many modern instruments now feature automatic logging of the calibration history. In addition, it is often feasible to export the current and historical calibration curves from the device to another system, such as a historian or a database.

3.2.4 Metadata that describes applied signal modification procedures

The data produced by the sensor and transmitted downstream to a PLC, SCADA, historian, or any other database may be modified significantly in the process. The reasons the original signal may be modified prior to storage may include:

- Changing the original unit of measurement to an S.I. unit or an accepted standard.
- Removing unwanted features in the signal, such as outliers, spikes, or interpolation, to mitigate the presence of absent and/or missing data, either through manual or automated procedures (see Sections 3.2.4.2 and 3.2.4.3).
- Synchronization of the timestamp with a reference clock. We recommend that time synchronization after data generation is avoided as much as possible. In an ideal world, all clocks in a data-acquisition system point to the same reference clock.

To make both original and modified data records as future-proof as possible, we recommend that irreversible changes (e.g., denoising, outlier removal) are handled by storing the original signal in addition to the modified signal. This way, a data analyst can use unmodified data if needed. Reversible changes (e.g., change of unit, timestamp modification) do not require such duplication as long as the modification itself is recorded adequately. By using the types of metadata described below, one can make sure that a future user can understand the impact of the processing on the final data stream and, where feasible and useful, revert the applied modifications. We refer the reader to Section 2.6 in Chapter 2 for available strategies and tools to store distinct versions of the same data.

Remarks on time of availability: Note that this section concerns the *procedures* used to produce a modified version of the original data records. In principle, these can be documented before generating the modified data. In addition, we consider the metadata describing such procedures to be descriptive of the modified signals and not of the original signal. In principle, this means that the metadata describing these procedures can be collected before the data they describe is generated. This is true regardless of (a) whether the procedure is automated or manual and (b) whether it is executed in real-time, online, or offline. Consequently, we group these types of metadata with other metadata describing the signal-generating process.

3.2.4.1 Change of units

Definition D3.15: A **change of units** is the deliberate change of units from the unit of measurement provided by the instrument to a unit considered appropriate for long-term storage and use.

Motivation: Documentation of any unit change between sensor and data storage avoids issues with inconsistent units and can save considerable time in data-intensive projects.

Additional remarks: A common transformation is the conversion of a measurement to a different unit than the one used by the sensor. This transformation can take the form of the application of a simple conversion factor (e.g., converting grams per litre to milligrams per litre). However, the conversion may sometimes imply additional assumptions about the measurement context. An example would be the conversion of a TSS measurement taken by the sensor in nephelometric turbidity units (NTU) into the deemed equivalent solids concentration expressed in milligrams per litre. In such a case, the physical meaning of the expressed value changes with the conversion, which means that the conversion requires the use of a correlative model. In such cases, one should at least store the applied model and historical versions of it. Where feasible, one should store the unmodified signal as well.

3.2.4.2 Lossy modifications (aggregation, averaging, filtering, compression)

Definition D3.16: A **lossy modification** of a signal is any procedure that transforms the original signal into a derivative signal such that the original signal cannot be reconstructed perfectly from the derivative signal, that is, some information is lost in the process.

Motivation: Documentation of lossy modifications of a signal can help a data analyst decide whether a particular signal is fit-for-purpose.

Additional remarks: To understand lossy modifications, consider first a change of units as explained above. Such a change is reversible since one can always change back to the original unit, provided the procedure is documented well. We name it a lossless modification as a result. In contrast, lossy modifications lead to changes that cannot be undone. For example, computing an hourly average from a signal measured every minute is irreversible. Indeed, one cannot compute the original signal given the hourly averages. For this reason, this is considered an irreversible or lossy modification. Unless the original signal and modified signal are stored separately, information in the original signal will be lost without means for recovery. Lossy modifications may also introduce unwanted artefacts in the modified signal. For example, nonlinear compression or filtering methods may introduce oscillations at frequencies not originally present in a signal.

Types of irreversible data processing steps include filtering (most often for denoising), downsampling (e.g., keeping every 10th data record in an original signal), or aggregation (e.g., computing daily or weekly averages). Lossy modifications can occur at many places within a data pipeline, for example, within the instrument, during ingestion into a historian or database, during querying from a historian or database, or during data analysis. For lossless and lossy modifications frequently deployed in historians and databases, please also check Sections 4.4.4 and 4.4.5 in Chapter 4. Lossy modifications often exhibit a number of tuning parameters (e.g. averaging window). A complete description of lossy modifications will include the history of the applied tuning parameter values.

Note on smart sensors: Many sensors come with their own computing capabilities and can interpret the raw signal in a variety of ways. On-device computing may be used for calibration, validation, denoising, and outlier removal. Modern sensors may also include the production of derivative information based on a model (soft-sensing). For example, UV–VIS spectrometers may be equipped with a model describing the relationship between the recorded signal (absorbance spectrum) and a variable of interest (e.g., COD, nitrate). Unfortunately, information about how the on-device data processing functions is often not available to the data user. For example, it may be proprietary or shielded from view. It may be of interest to inspect the manual carefully or to contact the manufacturer to document such modifications. In any case, the need to access on-device information is ideally discussed with the manufacturer prior to purchasing the sensor.

3.2.4.3 Data augmentation (imputation, interpolation, synchronization)

Data imputation consists of including new, computer-generated data records into a signal to account for absent or missing data. This includes techniques for upsampling (e.g., interpolation) and imputation (e.g., of absent or missing data). Most imputation techniques use an interpolated value to estimate absent or missing data (see Chapter 6). More advanced methods consist of model-based imputation, such as techniques based on the Kalman smoother or specialized time series models (see, e.g., Alavi *et al.*, 2006; Ba-Alawi *et al.*, 2022; De Mulder *et al.*, 2018; Osman *et al.*, 2018). Similarly, two signals with asynchronous timestamps may be synchronized through simple interpolation or model-based methods.

Note that imputation and interpolation steps may be irreversible when poorly documented. Indeed, it may be difficult to infer which data records are considered original and which were added later, thus preventing recovery of the original signal. It is generally preferred that the augmented data signal is stored separately while the original signal is kept untouched. Ideally, the metadata of the augmented signal includes the tuning parameters, if any, that were applied in the data augmentation procedure.

Motivation: Given that the introduction of artificial data records is likely to modify the statistical properties of the dataset, the data end-user must be aware of the data records that were obtained through imputation, mainly to assign an adequate level of trust in these data points. If not, this may lead to over-confidence in the data-based results or decisions.

3.2.4.4 Data fusion and soft-sensors

Many signals for interest are obtained by the combination of multiple signals. Consider a pH measurement as an example. Most typically, the pH is not measured directly. Instead, the sensor records a temperature signal and an electrode potential. Based on the Nernst equation, a simple model relating the pH, potential, and temperature to each other, one can compute the pH. In the case of pH measurements, this data fusion happens inside the sensing equipment. The parameters of the Nernst equation are typically stored on the instrument itself and usually require a dedicated effort to become available for data analysis. In other cases, a model may be established by the end user. When so, the method to compute the new signal is often named a soft-sensor. A simple case is the computation of a mass flow rate (solids mass flow) based on a hydraulic flow rate measurement and a concentration measurement (e.g., TSS). Tracking the history of such soft-sensors should be considered, especially if the deployed model is updated on a regular basis (see also Section 3.2.3.4).

Motivation: Understanding how multiple signals are combined to produce a signal of interest can help diagnose a malfunctioning soft-sensor. The documentation often highlights which of the relevant sensors need to be prioritized for visual inspection or maintenance.

3.3 METADATA DESCRIBING SENSOR DATA QUALITY: ACCURACY, TRUENESS, PRECISION AND MORE

To gauge the usefulness of the data for a particular purpose, it is important that the signal quality is assessed and that measurement errors are identified and, where appropriate, quantified and corrected. Measurement errors have multiple causes, including hardware malfunctions, human error, and changes in the measured environment.

A clear example of the problems caused by an inadequate quality signal is when this measurement is used to control a process variable. For instance, in pH control, a pH sensor that responds slowly to a change in the bulk pH value will imply slow changes in the manipulated variable leading to a delayed response to changes in the pH. This may lead to reduced process performance and longer times to reach new setpoints.

Measurement errors may have several causes. For example, a calibration error may cause a sensitivity (see Chapter 5) that is lower than the true sensor's sensitivity so that the measured variations in the pH are smaller than the ones found in the real system. It may also be caused by damage to the sensor. Importantly, a calibration error can affect trueness and precision but will rarely affect response time. In contrast, fouling and scaling typically affect precision and response time but are unlikely to change the sensor's trueness. It is thus important to describe measurement errors in some detail to isolate the root cause.

The number of techniques available for data quality assessment, including detection and identification of faults, diagnosis, and data reconciliation, is vast (see, e.g., Venkatasubramanian, 2003*a*, *b*, *c*). One unfortunate aspect of the body of research is that root causes and symptoms are often conflated. To benefit from the available techniques, it is best to clearly distinguish between identified causes for an anomaly or fault and the symptoms that led to its detection. This section aims to help by providing a list of concepts that can be used to this end, tracing a crisp delineation between (a) a description of observed measurement errors and (b) potential root causes. Specifically, we aim to:

- Describe typical measures to quantify the symptoms of measurement error (Section 3.3.1).
- Describe and classify known root causes of measurement error for sensors used in the wastewater sector (Section 3.3.2).
- Illustrate the important challenge of using quantified symptoms of measurement error to identify the root cause of measurement error (Section 3.3.3).

3.3.1 Describing measurement errors through their observed symptoms

This section introduces some terminology used for quantifying symptoms of improved or reduced data quality. These are frequently used to evaluate analytical measurement methods and are regularly applied to sensor signals.

In what follows, we differentiate descriptions of data quality that assume static conditions and those that specifically account for sensor dynamics or variable data quality. The first subsection defines the term error explicitly. The next two subsections describe basic and advanced descriptions of data quality under static conditions. The third subsection focuses on dynamic symptoms.

3.3.1.1 Errors as deviations from reference values

We take the following definition of error from Isermann and Ballé (1997):

Definition D3.17: An **error** is a deviation between a measured or computed value (of an output variable) and the true, specified, or theoretically correct value.

With y a measured value and y_{ref} the true value, the measurement error is quantified as

$$y - y_{\text{ref}} \tag{3.1}$$

The reference value used to evaluate the error is an accurate value for the measured variable obtained through a separate mechanism, founded in scientific theory or principles. One common mechanism is to place the sensor into a medium for which the true value of the measured variable is known through experimental controls. For instance, the trueness of a pH probe can be evaluated by placing the pH probe in a solution with a known pH value (e.g., buffered solution or calibration medium) at standard pressure and temperature. Another mechanism is to obtain a redundant, high-quality measurement of the measured variable through a trusted measurement method. For example, the trueness of an ammonia ion-selective electrode is typically evaluated through analytical measurement of the concentration. Most standards do not account for the reference measurements being subject to error themselves. In general, applicable standards imply that the accuracy of the reference measurements is orders of magnitude better than the sensor signal being assessed. Chapter 5, Section 5.5.2 describes methods to obtain high-quality reference measurements that may satisfy this requirement.

3.3.1.2 Static description of data quality – basics: accuracy, trueness, and precision

Basic descriptions of sensor data quality include accuracy, bias, and precision. We describe next what these are and illustrate some ways they can be quantified. Assessing accuracy, bias, and precision requires static conditions. Therefore, the following definitions and methods assume that several error values are available within a timespan that is short enough for the quality of sensor measurements to be unchanged, that is, one posits that the sensor remains in the same state.

Accuracy, trueness, and precision are typically evaluated by comparing multiple measurements obtained under the same conditions (y) against a single reference value (y_{ref}). With N measurements (y_i) one obtains N measurement errors:

$$e_i = y_i - y_{\text{ref}}, \quad i = 1,2,\ldots N \tag{3.2}$$

Creating the conditions under which a single reference value can be assumed offers the best chances to obtain good quantification of accuracy, trueness, and precision. It often means that a sensor needs to be isolated from the effect of confounding factors. Chapter 5 offers several standard operating procedures to do so.

The provision of a single reference value may not always be feasible or practical. As a result, it may be difficult to obtain multiple measurements under the exact same circumstances. To account for this difficulty, one could opt to quantify accuracy by pairing each measurement produced by the sensor of interest (y_i) with a corresponding reference value ($y_{\text{ref},i}$). Then, the measurement errors are computed as the pair-wise deviations:

$$e_i = y_i - y_{\text{ref},i}, \quad i = 1,2,\ldots N \tag{3.3}$$

We now turn to the definition and discussion of accuracy, trueness, and precision.

Accuracy

Definition D3.18: **Accuracy** is the degree of closeness to a test result with the accepted value of reference (ISO, 2006).

Motivation: Assessing accuracy is usually the first step to making evidence-based decisions on sensor maintenance frequency and replacement. This is a valid approach, given that increases in random, systematic, and gross errors (definitions below) all contribute to a decrease in accuracy. It also supports deciding whether the data quality is sufficient for a particular purpose. However, accuracy is considered inadequate to formulate corrective actions or improved sensor management. To make this possible, accuracy is typically decomposed into trueness and precision (definitions below).

Quantification: Accuracy is typically evaluated by comparing multiple measurements obtained under the same conditions (y) against a reference value (y_{ref}). With N measurements (y_i) one obtains N measurement errors and then quantifies accuracy as the root mean square error (RMSE):

$$\text{RMSE} = \sqrt{\frac{1}{N}\sum_{i=1}^{N} e_i^2} \tag{3.4}$$

A high value for RMSE will indicate poor accuracy. Figure 3.1 illustrates the evaluation of accuracy. A pH electrode is placed in calibration solutions with known pH values of 4, 7, and 4. The accuracy is then computed by selecting periods with a stable signal and computing accuracy accordingly. The accuracy is very poor in this particular case, with RMSE values of 0.84 at pH 4 and 0.77 at pH 7.

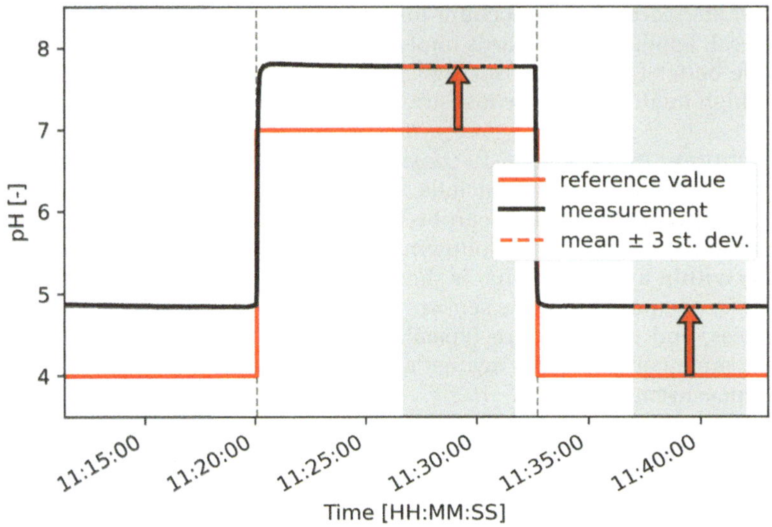

Figure 3.1 Example E3.1 – evaluation of a pH sensor's accuracy, trueness, and precision. The sensor is exposed to solutions of known pH (4 and 7), and the sensor measurements are recorded. Accuracy, trueness, and precision are evaluated under steady-state conditions by computing the measurement errors as deviations between the measured value (full black line) and reference value (full red line). The measurements used to do so are obtained by selecting a time window with a stable signal. In this particular case, the sensor is fairly inaccurate, given that the RMSE is 0.84 at pH 4 and 0.77 at pH 7. The measurement error's standard deviation is visualized with red dashed lines indicating the interval spanning six times the standard deviation around the mean value. The upper and lower values bounding this interval cannot be distinguished. Thus, the sensor precision is excellent in this case, with values for the measurement error standard deviation taking low values of 0.0018 at pH 4 and 7. Details are provided in Ohmura *et al.* (2019*a*). As a result, the accuracy (RMSE) is practically equal to the trueness (bias). Thus, the poor accuracy is mainly due to a lack of trueness in this case.

Additional remarks: Note that accuracy is an overall measure of data quality that combines trueness and precision. As explained below, trueness and precision are typically quantified as bias and standard deviation (see below). When so, the RMSE, bias, and standard deviation satisfy the following equation:

$$\text{RMSE}^2 = \text{bias}^2 + \text{standard deviation}^2 \quad (3.5)$$

Figure 3.2 illustrates this point through an analogy with archery. In panel (a), one can see the points where excellent archers may land their arrows. The landing points are close to the centre of the target while also being close to each other. This leads to high accuracy. This accuracy is reflected in a low RMSR of 0.2. In panels (b), (c), and (d), one can see the results of three archers with poorer accuracy. These three archers share the same RMSE value of 2.0 and are thus equally inaccurate. However, the average landing point (trueness) and spread (precision) of the landing points are very different between these three archers. They may need to concentrate on various aspects of their skill to improve their accuracy. Similarly, the accuracy of a sensor may not provide sufficient information to improve sensor accuracy. Inspection of the trueness and precision by decomposing the RMSE into bias and variance (see below) may be helpful to do so.

With the additional assumption that the measurement errors are distributed according to a normal density, the bias and standard deviation can be used as estimators of the mean and standard deviation for this density. This is illustrated in Figure 3.3 using the data points shown in Figure 3.2. This figure

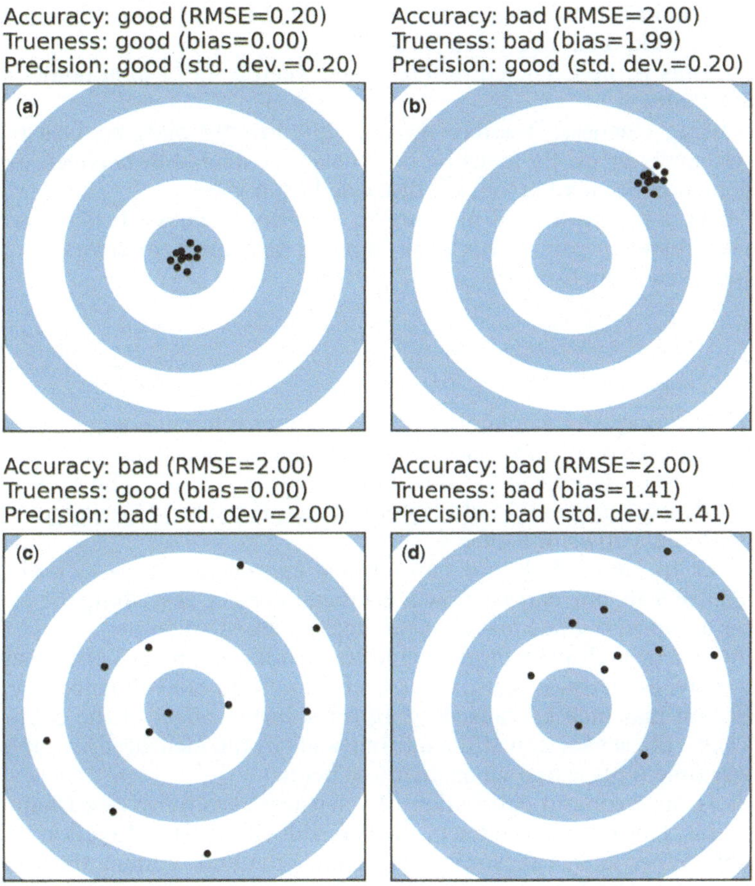

Figure 3.2 Example E3.2 – illustration of accuracy, trueness, and precision through analogy with archery. The centre of the roses is analogous to the reference value for a sensor or measurement method: (a) an excellent archer with good trueness and precision, leading to good accuracy; (b) an archer with bad trueness and good precision, leading to poor accuracy; (c) an archer with good trueness and bad precision, leading to poor accuracy; (c) an archer with bad trueness and bad precision, leading to poor accuracy. Note that the overall accuracy as measured with RMSE in panels (b), (c), and (d) is the same.

shows that measurement error distributions with distinct parameter values can produce the same accuracy.

Also note that RMSE is only one way of evaluating accuracy. Other measures, like median absolute deviation (MAD), are often preferred for their robustness against outliers. We refer to Daszykowski *et al.* (2007) and Zoubir *et al.* (2018) for introductions to robust statistics. Other measures based on the bootstrap can be found in Efron and Tibshirani (1986). Finally, measures of data accuracy could also be obtained by taking measures for model accuracy (see, e.g., Hauduc *et al.*, 2015) and replacing the model predictions with the reference values.

Trueness and bias

Next, we define and discuss trueness and the most commonly used measure for quantification, that is, bias.

Definition D3.19: **Trueness** is the degree of closeness of the mean value obtained from a large set of data points to an accepted reference value (ISO, 2006).

Definition D3.20: **Bias** is the expected difference between a set of data points and the accepted reference value (ISO, 2006).

Motivation: Assessing trueness, possibly through quantifying bias, enables the detection and quantification of systematic error. By focusing on an average deviation between measurements and the reference, random measurement errors are discarded from view.

Quantification: Bias is the most used measure to quantify trueness. It is often estimated as the mean of a set of measurement errors. Given the computed measurement errors (see Equations 3.2 or 3.3), bias can be quantified as their mean:

$$\text{bias} = \frac{1}{N}\sum_{i=1}^{N} e_i \tag{3.6}$$

As in the case of accuracy, the availability of an accurate reference value is important. See the notes above on selecting a mechanism to obtain such reference values. Figure 3.1 shows data obtained by exposing a pH sensor to solutions with known pH values of 4 and 7. The sensor appears severely biased as the mean difference is over 0.75 at the considered pH values. The sign and magnitude of the bias are indicated with red arrows in Figure 3.1.

Additional remarks: In Figure 3.2b, one can visually appreciate how a lack of trueness leads to poor accuracy. In Figure 3.3b, one can see how a Gaussian density describing the measurement error distribution now has a non-zero value for the mean due to lack of trueness.

Sensor bias is affected by multiple factors, including installation, placement, cleaning, calibration, wear-and-tear, and more. Sometimes, sensor bias is constant and not influenced by external factors. For example, a constant bias may be caused by an error in the offset of the calibration curve (see Chapter 5). In this case, the sensor can still capture trends in the measured variable. A sensor's bias may also change over time. This is known as drift (see below).

Another example is the wrongful calibration of a sensor's sensitivity (see Chapter 5), which will become visible as a bias. However, the value of the bias will change as a function of the true value of the measured variable. When such a sensor is deployed in a dynamic system, the systematic deviation leads to a measurement error whose magnitude changes with changes in the measured value. This may complicate the distinction between a calibration error and drift due to wear-and-tear. This underscores the preference to assess trueness under well-controlled conditions.

Bias may also be caused by analog-to-digital conversion (see above). Analog-to-digital conversion means that every data point is rounded up or down to the nearest multiple of the numeric resolution prior to writing it to a physical data storage device (e.g., hard disk, data server). This is also known as quantization and its consequences and mitigation have been studied in detail (see, e.g., Oppenheim and Schafer, 1975; Wang *et al.*, 2010; Widrow and Kollár, 2008). The measurement range (see above) divided by the numeric resolution (see above) equals the number of unique values (N) that can be stored for the data point (see Chapter 2, Section 2.4.1 for definition). This number (N) is often described in bits, that is, $\log_2(N)$. The number of bits can vary and can take on values between 1 bit (simple on/off signal) and 64 bits (as in most modern desktops and laptops). For continuous variables (e.g., dissolved oxygen, flow rates, pH), using 16 bits or higher is common, meaning that at least $2^{16} = 65{,}536$ unique values can be recorded within the measurement range. If the true value of the measured variable is constant, quantization will produce a constant bias. However, the sign and magnitude of this bias depends on the true value. As the true value of the measured variable is likely to change over time, the magnitude of this bias will be dynamic as well.

Bias can often be eliminated through calibration of the offset and sensitivity (see Chapter 5). However, care should be taken that factors causing bias are eliminated first to the extent that is feasible. When the cause of bias cannot be eliminated (e.g., quantization, regular wear-and-tear), this

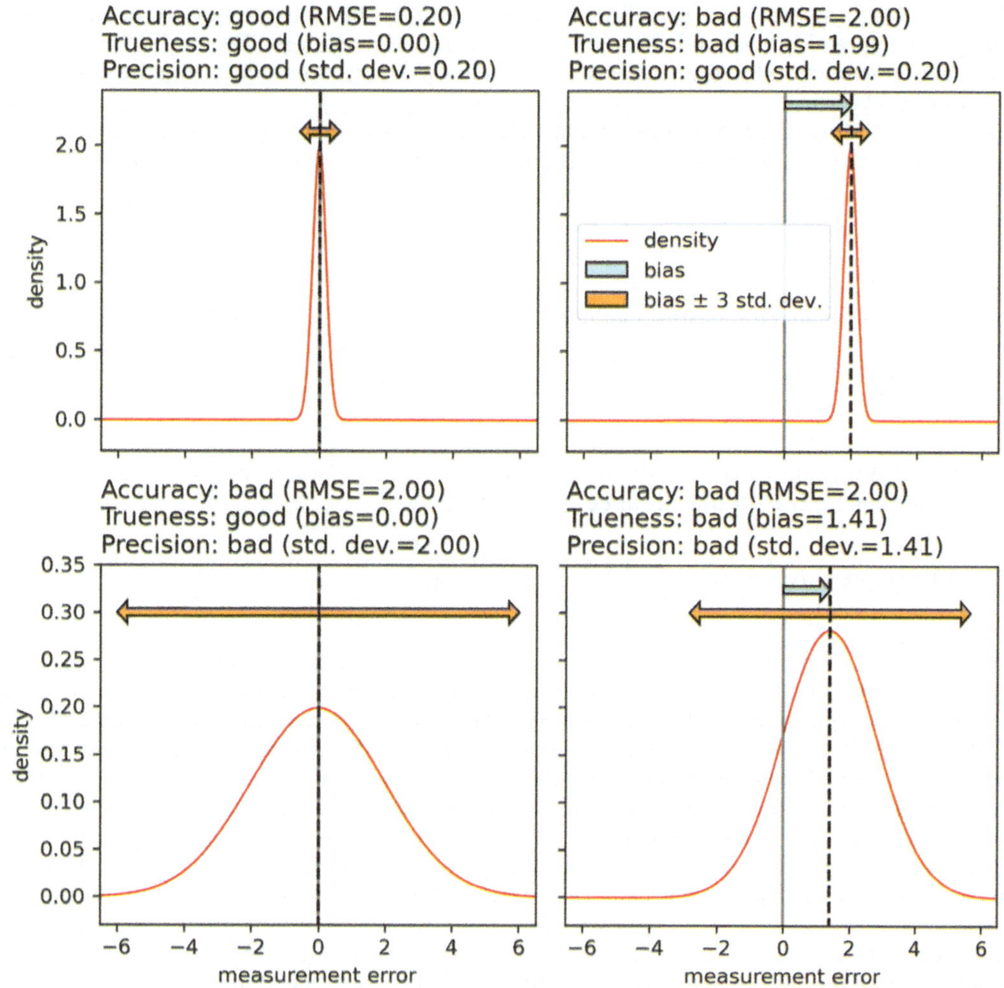

Figure 3.3 Example E3.2 – decomposition of accuracy into trueness and precision allows for describing the measurement errors with a normal distribution. This distribution has two parameters, which are the mean and standard deviation. These parameters are typically set to the bias and the measurement error standard deviation. Panels (a)–(d) show the densities matching the archery example in Figure 3.2. Importantly, poor accuracy may be a consequence of a lack of trueness (b), a lack of precision (c), or a combination of both (d).

may cause the bias to change over time (i.e., drift, see below), such that regular calibration becomes necessary.

Precision, variance, and standard deviation

Whereas trueness requires the elimination of random errors, precision requires the exact opposite. Systematic errors are removed, mainly to enable a description of the random errors. We base the following definitions on ISO (2006).

Definition D3.21: **Precision** is the degree of closeness of individual data points to each other.

Definition D3.22: The **variance** of measurement error is the expected value for the squared deviation between an individual measurement and the expected value (e.g., mean) for the measurement.

Motivation: Tracking precision, possibly through estimating the measurement error variance, gives a measure of the spread introduced through random error. Precision is affected by gross errors and random measurement errors but is not affected by systematic errors, thus facilitating the discrimination between sources of error.

Quantification: Most often, precision is quantified as measurement error variance. Given a set of measurements (as used in Equations 3.2 or 3.3), measurement error variance becomes

$$\text{variance} = \frac{1}{N-1}\sum_{i=1}^{N}(y_i - \text{average})^2 \tag{3.7}$$

where average is the mean measured value:

$$\text{average} = \frac{1}{N}\sum_{i=1}^{N} y_i \tag{3.8}$$

The standard deviation (positive square root of the variance) is a related measure of dispersion computed as the square root of the variance:

$$\text{standard deviation} = \sqrt{\text{variance}} = \sqrt{\frac{1}{N-1}\sum_{i=1}^{N}(y_i - \text{average})^2} \tag{3.9}$$

Increases in variance or standard deviation indicate a degraded precision. Unlike accuracy and trueness, the evaluation of precision does not require an accepted reference value. However, all data points should be independent measurements of the same variable under the same conditions. For this reason, the above equations require that the sensor is isolated from external variations. The bucket approach (see Chapter 5) aims to precisely do that.

Figure 3.1 illustrates how the standard deviation can be evaluated by exposing a pH sensor to solutions with known pH values of 4 and 7. The horizontal red dashed lines indicate the interval corresponding to +/-3 times the standard deviation at pH 4 and 7. The dashed lines are indistinguishable in both cases suggesting that the sensor is very precise. The computed standard deviations are both below 0.002.

As indicated, the above equations assume that the true value of the measured value is constant, that is, in a steady state. This may not always be an option. For instance, it may be impractical to move the sensor out of its measurement location to control the reference values or potential confounding factors. When so, the effect of a changing reference value on y_i should be accounted for. Indeed, the variation in the difference with the mean (y_i – average) is not due to random measurement error alone, and the above equations deliver a poor quantification of sensor precision. To solve this problem, one can opt to compute the variance or standard deviations based on Equation (3.3). Practically, one computes the bias first (Equation 3.6). Then, the precision is quantified as the following variance:

$$\text{variance} = \frac{1}{N}\sum_{i=1}^{N}(e_i - \text{bias})^2 \tag{3.10}$$

Or this standard deviation:

$$\text{standard deviation} = \sqrt{\text{variance}} = \sqrt{\frac{1}{N}\sum_{i=1}^{N}(e_i - \text{bias})^2} \tag{3.11}$$

Note that the quantification of precision now critically depends on accurate reference values. This is not the case if Equations (3.8) and (3.9) can be applied through measurement in controlled conditions.

Additional remarks: In Figure 3.2c, one can visually appreciate how a lack of precision leads to poor accuracy. In Figure 3.3c, one can see how a Gaussian density describing the measurement error distribution has a large spread due to lack of precision.

Unlike variance, the standard deviation is expressed in the same units as the original sensor signal, which facilitates interpretation. However, as computed above, the standard deviation is a biased estimate of the true standard deviation. The variance, as computed above, is thus often preferred from a statistical perspective, as it is a consistent and unbiased estimator. Alternative ways of describing precision, including robust estimates, can be found in Chen *et al.* (1997), Daszykowski *et al.* (2007), and Zoubir *et al.* (2018).

To some degree, a lack of precision can be mitigated through averaging or filtering. To illustrate this point, Figure 3.4 (top) shows a simulated true value and a noisy sensor signal for a flow rate. The sensor responds immediately but has poor precision with a standard deviation estimate under steady-state conditions of 0.2. Figure 3.4 (bottom) shows the true value and a filtered version of the sensor signal. A first-order filter is used to this end. After applying the filter, the (steady state) standard deviation is 0.05, which is better by a factor of 4. However, one must contend with a slower response of the filtered signal to a change in the measured value. This means fast changes in the measured value will not be captured accurately. In this case, the 90% response time (see below for definition) is 34 s.

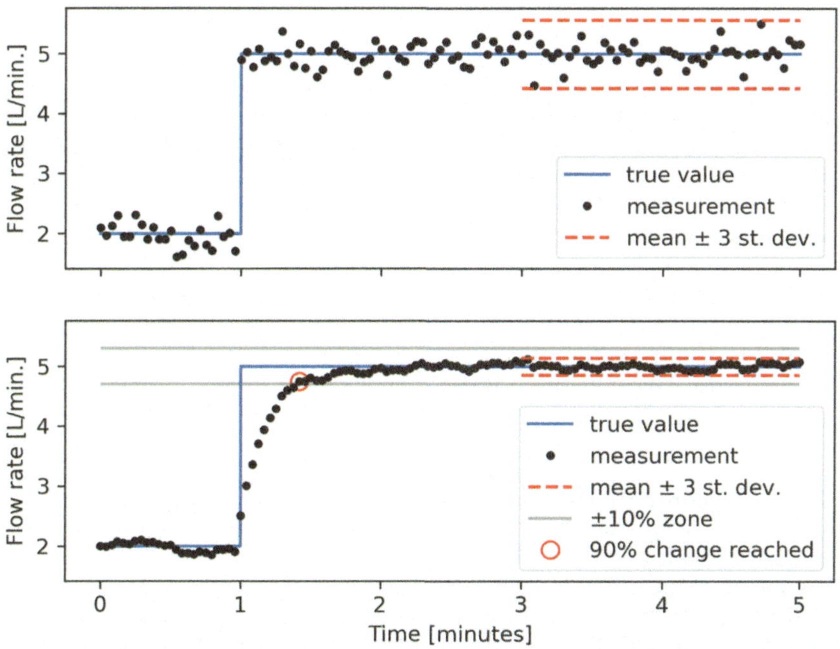

Figure 3.4 Example E3.3 – data filtering can improve precision. Top: A change in the true value of the flow rate is simulated at 1 min. The sensor responds immediately but is very noisy. The standard deviation is estimated to be 0.19 L/min between 3 and 5 min. Bottom: The precision of the sensor signal can be improved by filtering. The steady-state standard deviation estimated between 3 and 5 min is now 0.05 L/min. However, the sensor now responds slower to the applied change. The time needed to reach 90% of the imposed change in value is 34 seconds.

3.3.1.3 Static description of data quality – advanced: repeatability and reproducibility

The above concepts and measures apply to a single sensor within a short window so that the produced data quality will not change. While accuracy, trueness, and precision require precise control or measurement of the measured value, additional factors may affect the result. The following definitions help to define more precisely controlled measurement protocols to avoid such effects.

Repeatability

Definition D3.23: **Repeatability** is the precision under exactly repeated conditions, that is, obtaining independent test results using the same method in identical tests in the same laboratory, by the same operator, using the same equipment, within a short time interval.

Motivation: The evaluation of repeatability requires eliminating any factor that may influence the measurement result. A comparison with the precision evaluated without eliminating such factors can potentially pinpoint the exact cause of measurement error. See above to see how precision can be quantified.

Quantification: Repeatability is the precision obtained after carefully eliminating any potential cause of variation from the measurement process. See above to see how precision can be quantified.

Reproducibility

Definition D3.24: **Reproducibility** is the precision obtained from test results using the same method on identical test items with different laboratories, operators, and equipment (ISO, 2006).

Motivation: Whereas repeatability requires eliminating as much variation in the measurement technique as possible, reproducibility requires introducing several potential sources of error in the measurement process in a deliberate, systematic way. This can help identify systematic and random errors due to variations in sampling and measurement procedures. Because gross and systematic errors may affect individual laboratories, operators, or equipment, quantifying precision can help detect, diagnose, and mitigate measurement errors.

Quantification: Reproducibility is typically assessed through a round-robin test. In this case, a single water sample is split into multiple water samples. Each water sample is then analysed by a single laboratory, operator, or equipment. The reported measurements are used to quantify the resulting precision or variance. See above to see how precision can be quantified.

3.3.1.4 Accounting for dynamics: response time, shocks, drift, and outliers

We now introduce several terms related to dynamics. The response time describes the dynamic response of a sensor to a change in the true value of the measured variable. A non-zero response time is typical for most sensors. The remaining terms – shocks and drift – are considered symptoms of faulty sensor behavior.

Response time

Definition D3.25: The **response time** is the time it takes for an instrument to reflect a new value when subjected to a step change. It can also be defined as the time required to achieve a percentage of a step change in the true value (typically 90 or 95%).

Motivation: Whereas accuracy, bias, and variance quantify the sensor performance under static conditions (i.e., steady state), they do not quantify how well the sensor can handle dynamics in the measured variable. Most typically, the ability of a sensor to follow trends in the measured variable is quantified as the response time. A sensor's response time is critical for deploying and tuning real-time control systems.

Quantification: We refer to international standards (e.g., ISO, 2003) for detailed information on the quantification of response times and related concepts such as delay time, fall time, and rise time.

Figure 3.5 illustrates the response time concept based on experimental evaluation for a pH sensor. The true signal, based on known experimental conditions, is shown in blue, and the response that is

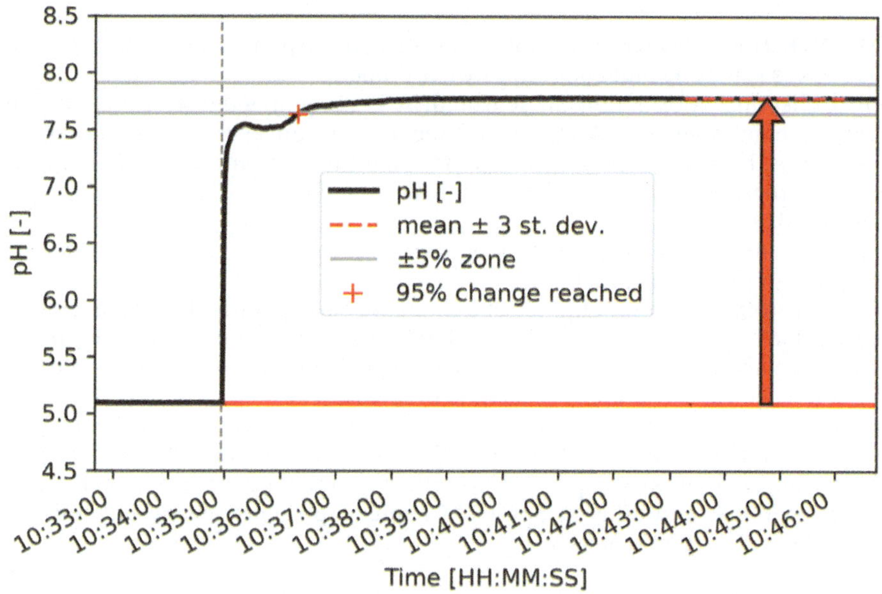

Figure 3.5 Example E3.1 – evaluation of response time for a pH sensor. The sensor is moved from a solution with pH 4 to a solution with pH 7. This causes the sensor value to change from 5.1 to 7.8. The response time is computed as the time it takes for the sensor to reach 95% of this change in value and stay within a 5%-band of the final value. In this case, the response time is 85 s. The original data are retrieved from Ohmura *et al.* (2019*b*).

obtained with the sensor is shown in red. The response time is quantified as the time it takes for the sensor signal to reflect 95% of the value change and equals 85 s.

Additional remarks: Changes in the response time of a sensor can have multiple causes, including choice of materials (e.g., permeable membranes), sensor geometry, sensor placement, mixing conditions, chemical deposition, fouling, incurred damage (membrane tearing), and normal wear-and-tear.

Step change or shock

Definition D3.26: A **step change** or **shock** is a drastic and momentary change of the sensor bias from one timestamp to the next.

Motivation: Identifying shocks in sensor signals is recommended to detect severe damage to a sensor. Moreover, shocks induced by maintenance actions (e.g., cleaning, calibration) should be quantified to enable correction of sensor signals (e.g., by bringing the measured data points before calibration closer to the measured data points after calibration).

A step change or shock is a change that happens all at once at a specific moment. This symptom is often easily detectable by the human eye as a result. A shock may result from physical damage or a maintenance action (e.g., cleaning, calibration). The deviation between the sensor value and true value (bias) may either improve or deteriorate at the time of the shock.

Quantification: The most accurate evaluation of step changes in measurement error will be obtained by regular evaluation of sensor bias through the sensor's lifetime. To accurately quantify the shock effects of a maintenance action (e.g., cleaning, calibration), bias should be quantified before and after the maintenance action. Appendix 5.A in Chapter 5 provides several methods to do so. In Chapter 6, several possible methods are presented to detect and quantify shocks that appear involuntary (i.e., not through maintenance actions).

Drift

Definition D3.27: **Drift** is a slow change of bias, variance, or response time as time progresses. It is a symptom that evolves smoothly and continuously over time.

Motivation: Collecting information on sensor drift is recommended to fine-tune the frequency by which sensors are maintained, calibrated, and replaced. Changes in sensor bias are a prioritized concern in most cases. However, changes in response time can be quantified without needing to move the sensor and are therefore useful to efficiently track the effects of fouling and scaling (see, e.g., Samuelsson *et al.*, 2018).

Quantification: Sensor drift is quantified by regular evaluation of sensor bias, variance, and response time through the sensor's lifetime.

Figure 3.6 shows an experimental example of a changing bias quantified through regular validation. It shows the pH value provided by an uncalibrated pH sensor in a medium at pH 6. In the absence of corrective calibration and maintenance actions, this sensor produces a signal that continuously moves away from the reference value (here: pH 6) as time progresses.

Additional remarks: The direction and rate by which bias, variance, or response time change is not necessarily constant. The most trustworthy approach to combat the effects of drift of the bias is regular validation and calibration. Chapter 5 lists methods to regularly collect reference measurements to quantify changes in bias and other symptoms. Chapter 6 presents several methods to detect and quantify drift through algorithmic means.

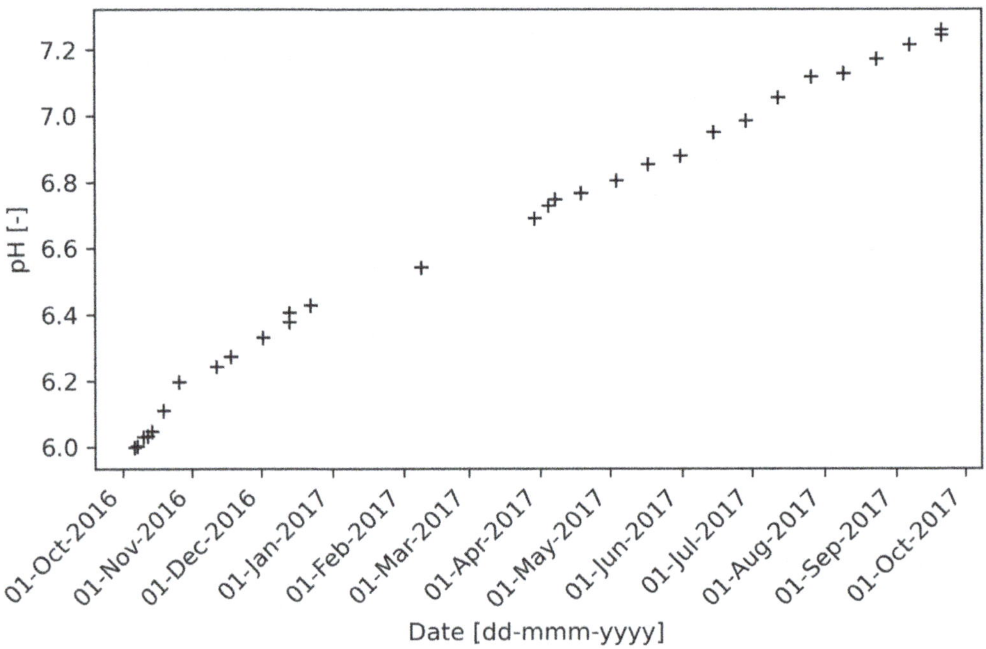

Figure 3.6 Example E3.1 – evolution of the pH sensor signal over time, obtained in a medium at pH 6. The values shown are the expected pH values if the sensor is never calibrated. These pH data points were calculated based on the measured data and the calibration values of pH sensor offset and sensitivity obtained through two-point calibrations at pH 4 and 7. Details are provided in Ohmura *et al.* (2019*a*).

Outlier

Whereas changes in bias, variance, and response time are typically infrequent (step change) or slow (drift), it is also common to observe symptoms that do not last long. In general, the corresponding data records showing short-lived symptoms are named outliers.

Definition D3.28: An **outlier** is a member of a data set that is inconsistent with the other members of that set.

Motivation: Most data analysis and automated decision-making methods are sensitive to outliers. It is, therefore, crucial to detect outliers early on. While outliers may indicate a degraded data quality, they may also indicate an accurate measurement of an anomalous process condition. Thus, it is, important to distinguish between the potential root causes of an outlier before removing or replacing individual data points.

Outliers have received significant attention because they significantly influence analytical results and models when not considered. Outlier detection and correction are fundamental for parameter estimation and data reconciliation for dynamic and steady-state systems since they can induce heavy biases in parameter estimates and reconciled data (Albuquerque and Biegler, 1996).

Outliers may be caused due to variability inherent in that measurement or the monitored process. This may or may not require corrective action, that is, the recorded value may be very odd but still correct. In other cases, an outlier may be caused by deviations from the experimental procedure, calculations, or data-acquisition errors. When so, outliers should be investigated and rejected when confirmed as an erroneous measurement (Grubbs, 1969).

Outliers are typically assigned to one of three categories:

- Point outlier or global outlier: A data point whose value is well outside the range of the complete data set. Typically, such outliers can be detected automatically by comparing them to sensor-specific metadata such as the measurement range or the variable range (see Sections 3.2.2.6 and 3.2.2.7). This is illustrated with dissolved oxygen measurements in Figure 3.7 (top). The variable range in this example is 0.5–1.5 mg/L, and two data points are labelled as point outliers.
- Contextual outlier: A data point whose value deviates significantly from the rest of the data points in a specific context or condition, even though the data point could be possible in another context. Figure 3.7 (top) shows one such example with a rectangle. The data point value is within the variable range but obviously different from measurements taken before and after. In this specific temporal context, it is thus considered an outlier.
- Collective outliers: This case does not refer to a single data point but rather a subset of the complete data set. This subset of data is a collective outlier if the individual data points do not fit the previous definitions of point or contextual outliers but still deviate significantly from the complete set. Often, this means certain patterns occur that may be considered normal in one circumstance but not in the considered case. Figure 3.7 (top) shows an example in grey shading. The data points in the identified time are slightly higher than the data points before and after. There are also similar data collected about 20 min later. However, this bump in the concentration may be considered abnormal because an operator judges that it occurred at an abnormal time. Naturally, the identification of collective outliers is often more challenging and subjective than point outliers or contextual outliers.

Additional remarks: The consequences of unmitigated outliers are also shown in Figure 3.7. In this example, the dissolved oxygen (DO) is controlled with a proportional integral (PI) controller and the variable used for control is the aeration modified with a mass flowmeter. For example, around 13:18, the measured DO increases from 0.99 to 2.03 mg/L, probably due to an electrical glitch. The PI controller reacts to the sudden DO change and decreases aeration. However, after 20 s, the DO returns to the correct value (1.0 mg/L), and the DO control reacts in the opposite way. If not accounted for, this type of disturbance may cause process upsets or cause excessive wear-and-tear in actuating equipment (e.g., valves, pumps).

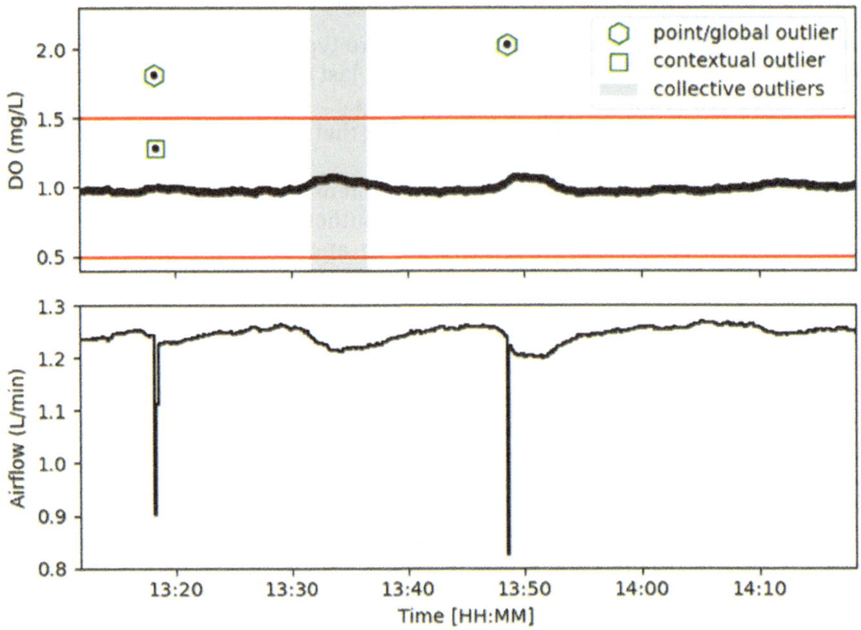

Figure 3.7 Example E3.4 – illustration of outlier types. Top: dissolved oxygen measurements as a function of time with outliers indicated: hexagons indicate point outliers identified based on the variable range (see Section 2.2.7); The square indicates a contextual outlier; The grey area indicates data points that are labelled as collective outliers. Bottom: airflow setpoint as a function of time. The presence of point and contextual outliers causes large disruptions in the airflow. Original data collected for the study was reported by Zhang *et al.* (2021).

3.3.2 Describing measurement errors through their root causes

3.3.2.1 Overview

Errors can have many different causes. Identifying the source of these possible errors will help propose strategies to reduce their occurrence or, if that is not possible, filter them out of the dataset with minimal loss of information. In the above, we discussed how sensor measurement errors can be quantified with measures like accuracy, bias, variance, and response time. The following paragraphs discuss measurement errors according to typical causes or sources.

We largely follow the definitions provided by McGhee (2005). In this work, errors are assigned to three main groups based on their causes: systematic errors, gross errors, and random errors. These three categories are described below. An overview is given in Table 3.2. Note, however, that

Table 3.2 Classification of measurement errors according to the nature of their root cause.

Presence in Time	Typical Effect	
	The root cause is expected to produce constant or slowly varying error values	*The root cause is expected to produce fluctuations of the measurement error*
The root cause is persistent, that is, present continuously or over long periods	Systematic error	Random error
The root cause is momentary, that is, at a unique point in time		Gross error

distinguishing between these categories is not trivial. Because random errors are mainly described by quantifying variance, and non-random errors are primarily described by quantifying bias, it is common to consider systematic errors and gross errors as a single category (see, e.g., Narasimhan and Jordache, 1999).

3.3.2.2 Systematic errors

Definition D3.29: A **systematic error** is persistent in time and is likely to produce a measurement error that is constant or changes slowly.

Systematic errors are often caused by persistent instrument deficiencies, incorrect calibration, and structural defects in data acquisition or processing. Due to the persistence, it is possible to quantify the symptoms through repeated measurements. Multiple data points are expected to be affected the same way, that is, with the same magnitude and direction of the error.

We recognize the following frequent causes leading to systematic errors:

- **Calibration error.** Calibration errors can occur due to many factors, for example, incorrect selection or preparation of standard solutions, lack of proper cleaning of the sensor before calibration, temperature differences between the standard solution used for calibration and the temperature in the measuring location, blockage problems in tubing or valves in automatic calibration systems, and lack of a proper response of the sensor in the calibration range. Calibration error may produce an incorrect characterization of the sensor's calibration curve, including offset and sensitivity (see Chapter 5). In turn, the symptoms of calibration error may include increased bias and variance. Calibration error can be minimized by adopting standard operating procedures for sensor calibration, possibly including precise control of the medium during calibration (see Chapter 5).
- **Confounding factors.** Any variable or factor that distorts the measure of an instrument is called a confounding variable, confounding factor, or confounder. Its presence can lead to an increase in measurement (positive confounding) or a decrease in measurement (negative confounding). Confounding is typically mitigated by measurement of the known confounding factors. For example, typical ammonium ion selective electrodes (ISE) are also sensitive to potassium. An increase in potassium could lead to an increased electrode response which could be erroneously assigned to ammonium. To avoid this problem, some commercially available ISE sensors for ammonium also include an additional potassium-sensitive electrode. The most common symptom of confounding factors is the presence of bias. Note that this bias may change over time. In the above example, the potassium concentration may be influenced by the application of de-icing salts during the winter, thus leading to a time-varying bias. In turn, this may be described as drift. Controlling or measuring confounding factors is especially important during sensor calibration.
- **Fouling.** The accumulation of materials on the sensor surface is a common phenomenon. In many cases, the accumulated material can easily be removed by gentle mechanical cleaning (water and/or cloth). When so, the accumulation process is known as fouling. Biofouling is the specific condition when fouling is caused by the attachment and growth of microorganisms. Fouling typically results in reduced mass transfer between the bulk liquid and the sensor surface. The most common symptom of this is an increase in response time. This may still allow proper sensor calibration if sufficient wait times are applied. A side effect of mass transfer limitations is that random variations in the bulk liquid will have a smaller effect on the measurements, which may manifest as an apparent decrease in variance. In the case of biofouling, the accumulation of microorganisms as a biofilm can change the concentration of the compound of interest (e.g., oxygen, nitrate) near the sensing element, thus introducing a change in offset and/or sensitivity (see Chapter 5) and impeding sensor calibration. The most recognizable symptom of this phenomenon is bias. Errors caused by fouling are typically mitigated through regular inspection and mechanical cleaning.

- **Nonlinearity.** Sensor calibration often assumes a linear, instantaneous response to changes in the measured value. However, nonlinear response curves are commonly observed, for example, ammonia ion-selective electrodes are known to have a nonlinear response to the true ammonium concentration below 1 mgN/L. Sensors may also exhibit hysteresis, meaning that upward trend in the measured variable leads to a negative bias (measurement too low) and vice versa. The symptoms of these types of nonlinear behaviours include bias and changes in variance.
- **Imperfect mixing.** Most control systems assume tank reactors are perfectly mixed, so the measurements at a specific point represent the whole reactor. In practice, the water may not be mixed equally throughout the tank. This can lead to dead zones (zones with no or limited mixing) and bypassing. It also may cause fluctuations in the measured variable. The symptoms of imperfect mixing may include bias, variance, and changes in response time. Proper placement and installation can help avoid these phenomena. Computational fluid dynamics modelling can help to reveal some of these challenges and allows tuning sensor placement.
- **Instrument error.** The symptoms of instrument error may include increased bias, increased variance, drift, and changes in response time. Problems in this area are diverse and include, among others, defective instrument construction, limitations of the instrument due to friction or hysteresis, use of the instrument outside its recommended range, hardware abuse or misuse, and lack of proper maintenance.
- **Scaling.** Scaling is the chemical deposition of materials on the surface of the sensor. In contrast to fouling, the deposited material cannot be removed easily with mechanical cleaning alone. The symptoms of scaling include an increase in the response time, as in the case of fouling. A bias is less likely until the level of scaling limits mass transfer to the sensor entirely. Errors caused by scaling are typically mitigated through regular inspection, chemical cleaning, and, eventually, part replacement.
- **Wear-and-tear.** Many sensors are not robust to prolonged exposure to the harsh environment in which they are deployed. Many sensors will produce degraded data quality as time progresses. This can lead to symptoms like increased bias, increased variance, drift, and changes in response time, or a complete breakdown and failure to produce any data point. Wear-and-tear can be minimized through sensor placement and installation, sensor maintenance, and part replacement.

3.3.2.3 Gross errors

Definition D3.30: A **gross error** is caused by a phenomenon that occurs rarely and during short periods, often affecting a single data point only.

Due to low frequency, it is generally difficult to describe and quantify the symptoms through repeated measurements alone. Similarly, the exact condition under which the error is produced may be challenging to reproduce. For this reason, this type of error is considered non-repeatable.

Quite often, gross errors are of human origin. This may be caused during water sample collection or preparation (e.g., wrong dilution factor) or analysis (e.g., incorrect instrument reading). They may also be caused by momentary sensor faults. We distinguish the following generic causes:

- **Wrongful reading or recording.** The measurement is executed correctly, but the produced value is read or recorded wrongly. This is most common for laboratory measurements.
- **Wrongful water sample preparation.** The water sample is improperly prepared before it is measured. This may include changes to the expected procedure, including chemical dosing, dilution, filtering, and mixing steps. These errors are relevant for laboratory measurements and online analysers (e.g., ammonium analyser).
- **Computational error.** An erroneous measurement is produced due to miscalculation. Such errors can take place in a laboratory setting (e.g., application dilution factor) and in online sensors (e.g., due to electrical glitch or software error).

3.3.2.4 Random errors

Definition D3.31: A **random error** is caused by unpredictable yet persistent variations in the data collection process and affects all measurements by the same sensor in the same or similar way.

Random errors are caused by unpredictable, always-present variations in water sample preparation, measurement device, and other experimental conditions. Random errors differ from systematic and gross errors because they cannot be reproduced exactly. The primary symptom of random error is an always-present variation, often described as noise. Random errors have multiple causes that are often difficult to distinguish. They include:

- Uncontrollable variations in water sample preparation.
- Unpredictable variations in the measured environment, including temperature and pressure.
- Electric and magnetic field noise caused by variations in electric voltage, current, and/or magnetic fields. This type of error is particularly relevant when analog measurements are communicated over unshielded cables.
- Power quality variations. The power supply, whether the electrical grid, a battery, or an alternative energy source (solar, biomass recovery), can cause unpredictable variations in power quality. In turn, this may affect data points in poorly shielded devices.

3.3.3 Diagnosis: identifying the root causes that explain symptoms

Most algorithmic sensor signal quality tools are focused on the detection of symptoms and the removal of data of deteriorated quality. However, detecting a symptom and removing the affected data is often insufficient to ensure good or improved data quality. Removing the root cause of the low data quality will be more effective.

Several studies report on the development and testing of methods for fault diagnosis, that is, identifying the root cause that can best explain the observed symptoms of degraded data quality (see, e.g., Venkatasubramanian, 2003*a*, *b*, *c*)). This is challenging as no generally applicable mapping exists between root causes and symptoms. Table 3.3 shows reasonable matches between the symptoms and root causes described separately above.

Unfortunately, many symptoms are not unique to a single root cause, so multiple potential root causes must be investigated. For example, drift may be caused by normal wear-and-tear and confounding factors. Indeed, an ammonia ISE may exhibit drift due to a change in concentration at the reference electrode. This can be addressed through calibration. However, drift may also occur due to a change in a confounding factor. In the case of an ammonia ISE, drift can be caused by a change in potassium concentration. Importantly, this kind of effect cannot be mitigated effectively by calibration. This example illustrates how a symptomatic description of measurement error may not suffice to identify appropriate corrective action.

In addition to the fact that there is often not a single root cause that explains observed symptoms, the assessment of measurement errors is further challenged by feedback loops. When sensors are used for feedback control, it can be hard to detect changes in data quality, as the feedback mechanism will effectively mask many symptoms or reduce their magnitude. In such cases, it is recommended to jointly analyse both the sensor data and the control commands used to manipulate the controlled variable. Chapter 6 offers several multivariate techniques that can be used for this purpose.

When detected, the diagnosis of degraded sensor data quality is best based on a combination of knowledge of the monitored process, knowledge of the sensor, and analysis of the recorded data. Chapter 5 provides several experimental methods useful for sensor diagnosis, particularly by controlling confounding factors and other circumstances. Chapter 6 provides algorithmic methods that can be used to detect and sometimes help identify the root cause of sensor data quality degradation. Naturally, the costs associated with using these tools should be considered and weighed against improved data quality measures.

Table 3.3 Possible root causes for typical symptoms of degraded sensor data quality.

Root Causes	Symptoms				
	Bias	Variance	Inadequate Response Time	Drift	Outlier
Systematic Errors					
Calibration error	X	X			
Confounding factors	X			X	
Fouling	X	X	X	X	
Nonlinearity	X	X			
Imperfect mixing	X	X	X		
Instrument error	X	X	X		X
Scaling	X	X	X	X	
Wear-and-tear	X	X	X	X	
Gross Errors					
Wrongful reading or recording					X
Wrongful water sample preparation					X
Computational error					X
Random Errors					
Uncontrollable variations in water sample preparation		X			X
Unpredictable variations in the measured environment		X			X
Electric and magnetic field noise		X			X
Power quality variations		X			X

Reasonable matches between root causes and affected metrics/symptoms for sensor data quality are marked with *X*. Unfortunately, classical measures to quantify the symptoms are often insufficient to isolate a unique root cause. Both experimental methods (Chapter 5) and algorithmic tools (Chapter 6) are proposed to eliminate root causes and refine the diagnosis.

3.4 METADATA PROVIDING CONTEXTUAL INFORMATION: ANNOTATION

It is common practice to annotate data in projects relying on online sensor signals to provide contextual information for the data user. Through visual cues (arrows, colour, shading, boxes) and structured or unstructured annotations (definitions below), one can alter how data records are interpreted. In turn, this interpretation can affect data analysis, identification of mechanistic and empirical models, process control, and decision making. Annotations are, therefore, a form of metadata (see definition in Chapter 2). Data annotations can describe one or more data records in one or more sensor signals. This is often achieved through a manual process executed by a human expert (pointing, clicking, dragging, typing).

Due to varying scopes of attention between experts, available annotations typically consist of a mix of observations of a symptom (e.g., *outlier*) and explanations of their causes (e.g., *electric surge*). In some cases, a more automated approach can be deployed to facilitate a more consistent annotation process. Either way, data annotation, when successful, enables effective communication between experts with various affiliations, skillsets, and responsibilities.

To our knowledge, data annotation has not been researched in depth, and the way it is implemented is highly variable. The existing literature on data management offers only broad statements and recommendations regarding the data annotation process. Akkiraju *et al.* (2020) recommend the use of labels (see Definition D3.33) and suggest three areas for improvement:

- Remove ambiguity by creating and continuously refining definitions for terms and annotation guidelines.
- Bring annotation to scale by combining an internal team of annotators and external crowd work.
- Speed up the annotation process using machine learning tools (see Chapter 6).

Of these three, we suspect the first and last are the easiest to adopt in wastewater utilities. Crowd work would be more challenging to deploy since (a) data needs to be shared at least temporarily with (a portion of) the public, and (b) annotation may require some awareness of the field or the specific context of the data. The contents of this chapter should help with the first item. The third item is primarily addressed in Chapter 6. In what follows, we aim to

- Describe the current variety in annotation practices
- Offer suggestions on how current annotation practices could be improved.

3.4.1 Purpose

The main purpose of data annotation is to modify how data are interpreted or used in an ongoing or future project. There are two primary ways this is achieved:

(1) Data annotation is often used to remove data points from the user's view for certain purposes, for example, when outlying points are annotated in such a way that they are ignored during model calibration. A typical example consists of marking a data point as an outlier, thus implying that the data point is of low quality.
(2) The provided markers or comments may draw attention to an annotated data point, for example, to point the data end-user to events of interest.

Both kinds of annotation can highlight important information for future analysis, thus changing how these data are interpreted. Since data annotation can be used to both divert and attract the user's attention, special care must be taken during data annotation and use to ensure the intended effect is created.

3.4.2 Basic definitions

Definition D3.32: **Annotation of sensor signal data** is the process of providing additional information to data records or timestamps to inform future use of the sensor signal data.

Importantly, the process of annotation never removes data from a data set. Note also that a single label or comment can be assigned to any number of data records, timestamps, or time windows. As a result, one can use annotation to describe any of the following:

- Individual data records. For example, to mark an outlying value. See Chapter 2 for the definition of a data record.
- Single timestamps. For example, to indicate an event that may not coincide with any data record (e.g., sensor installation event). See Chapter 2 for a distinction between timestamps and data points.
- Multiple data records, possibly produced by several sensors. For example, to mark a calibration issue or coinciding outliers appearing simultaneously in multiple sensor signals.
- Time windows, which are marked by a start and end timestamp. For example, to indicate a period of disruption (e.g., rain event) or a change in operation (e.g., ammonia controller switched off).

A single timestamp, data record, or time window could be annotated with multiple labels or comments, either by different experts or by the same expert with another purpose. This is due to several factors, such as:

- Multiple experts differing in opinion and providing different annotations to the same data records.
- Different annotations being provided for different purposes or at different times.

As a result of the above potential multiplicity, storing annotated sensor signals typically requires the use of many-to-many relationships between the provided annotations and the available data records. Section 4.4.5.1 in Chapter 4 shows how this can be accounted for during the design of a relational database.

Definition D3.33: A **label** is a structured and pre-determined type of annotation with the following properties:

- It is used repeatedly to annotate similar events or phenomena
- It features at least two elements:
  - A unique identifier (alphanumeric string) describing the type of event or phenomenon.
  - A description of a type of action, causal event, or symptom.

- In addition, the label may also include the following:
  - Identification of the relevant instruments.
  - Identification of the relevant data records described by the label.
  - Identification of the relevant time point or window.
  - Identification of the expert providing the description.
  - Time the label was created.
  - Purpose of annotation.
  - Software and version used for annotation.

Note that the term *label* is considered synonymous with the term *flag* in this work. Only the term label is used in what follows. The major advantage of labels is that they are structured and lead to an efficient data annotation strategy. On the downside, they assume a common language and understanding among data annotators and users and require careful management of the labels available for data annotation.

Definition D3.34: A **comment** is a piece of free-form text used for annotation purposes. A single comment can be assigned to any number of data records, any number of timestamps, or any number of time windows.

Unlike labels, comments are completely flexible and allow annotation in an unstructured, free-form manner. This allows new phenomena to be recognized and annotated easily during data annotation. The downside is that comments, when unmanaged, can lead to cluttered data annotations, thereby preventing a systematic approach to interpreting the data at a later stage.

3.4.3 Addressing challenges in data annotation

Effective data annotation proves challenging in many cases. In what follows, we provide some general recommendations that could be adopted.

3.4.3.1 Cross-linking annotations with other metadata and signals

As indicated above, Akkiraju *et al.* (2020) recommend using labels (see Definition D3.33) for annotation and suggest implementing a quick feedback loop between data annotators and data users. Clearly, the challenge in creating and managing effective data annotations is not unique to wastewater treatment alone. Today, data annotations are often generated for a specific task or project and reside in isolated systems, for example, on a computer managed by a modelling expert. Where considered attractive, data annotation efforts should be geared towards generating reusable data annotations, thus preventing repetition and fostering a shared knowledge base across data annotators and users.

3.4.3.2 Flexible annotation methods lead to heterogenous metadata

Data annotation is usually a flexible process through which a process expert provides metadata that is helpful for a later user. Unfortunately, the flexibility available during annotation poses a challenge during data use, as interpreting large sets of comments in free-text form can be extremely difficult.

Annotation is often used to highlight the occurrence of events that are rare, possibly completely unexpected, or never seen before. As the range of events and conditions one may want to highlight through annotation is extremely diverse, it is typical to allow the provision of comments (freeform text) for annotation purposes. For example, a sensor technician may indicate a unique type of damage to a sensor, or an operator may indicate a remarkable deviation from normal operating conditions. Moreover, the terms used to describe events may vary over time or from one person to the next, even if applied to the same system. This vastly complicates the efficient use of annotations.

In addition to the diversity mentioned above, annotations can be used to highlight the occurrence of causal events (e.g., rain events), the execution of deliberate actions (e.g., change in operation, calibration event), or a description of human observations or sensor signals. A clear distinction between causes (actions, events) and effects (observations, symptoms, sensor signals) is rarely made today when designing systems for annotation based on human inspection.

The heterogeneous nature of flexible annotations makes their use challenging. This is often referred to as one of the four V's of big data, namely *Variety*. This type of variety requires a trade-off between ease-of-entry during annotation and ease-of-use later. To our knowledge, there is no universal way to strike this balance, as also noted below.

3.4.3.3 Communication across specializations, teams, and time

Quite often, the person providing the annotations and the person using the annotated data are different. For example, a service technician might perform maintenance work on a sensor and subsequently generate an annotation for the duration of the work. However, the process engineer is the one who uses the annotated data to interpret sensor performance and process behaviour. Moreover, the person annotating the data and the person receiving the annotations may never meet, especially if a project uses a long historical data set, including annotations provided by a former staff member. This means that it may be hard to understand the original mindset or intention of the person that annotated the data. As with the challenges mentioned above, the effects of this could be reduced by using standardized workflows and user interfaces for data annotation and structured annotation formats.

3.4.3.4 Standardizing annotations

It is common to advocate for a structured approach to annotation to maximize the efficiency by which annotated data can be navigated. Standardization is predominantly executed by providing specific categories and terms for annotation purposes. It is thus considered useful to reduce acceptable annotations to a restricted set of labels, each with a unique and standardized description using commonly accepted terms and phrases only while limiting the use of free-form comments.

In some cases, free-form text may be abandoned entirely in favour of structured labels, specifically by selecting a limited number of descriptions available to the data annotator (e.g., dropdown menus, radio buttons, check marks). However, this kind of extreme approach may lead to the loss of valuable metadata, for example, when the person annotating data cannot find a suitable option to describe a new type of event. Thus, one must be mindful not to restrict the expert annotator so much that discouragement leads them to discard valuable information.

In practice, a balance must be struck between the convenience of the data user (more structure) and the convenience of annotating expert (more flexibility). This balance requires a trade-off between the stakeholders involved in the provision and the use of annotated data. The exact trade-off, if identified, is likely unique to the system or the team. It is, therefore, crucial to involve all persons sharing an interest in the data annotation process. In Section 3.4.4, we describe several ways annotations can be categorized.

3.4.3.5 Automating annotations

In addition to carefully choosing a level of standardization of data annotations, it may be of interest to incorporate algorithmic tools for data annotation. Several methods are available to automatically detect symptoms of a degraded sensor data quality (see Section 3.3.1) or the presence of process anomalies. Algorithmic methods permit more efficient and more objective generation of annotation metadata. Chapter 6 lists several methods that can be deployed to this end.

3.4.4 Categorization of data annotations

Annotations can be categorized according to multiple aspects, including temporal and spatial scope, explanatory power, source of information, and the annotation process. These aspects are explained next. Careful consideration of the categories below can facilitate a systematic approach to data annotation, thus decreasing the messiness of the produced metadata and, in turn, increasing the value of annotated data.

3.4.4.1 *Categorization according to temporal and spatial scope*

A single label or comment can annotate single or multiple data records. As a result, we distinguish between the following labels and comments:

Definition D3.35: A **point annotation** is an annotation assigned to one data record only.

Definition D3.36: A **collective annotation** is an annotation assigned to multiple data records at once. The annotated data records may have the same or distinct timestamps and may belong to one signal only or multiple signals. Collective annotations are typically used for events that start and end at a different time.

Figure 3.8 (top) shows a point annotation while Figure 3.8 (bottom) shows a collective annotation.

3.4.4.2 *Categorization according to explanatory power*

In many cases, annotation is focused on providing a symptomatic description of data records, for example, to identify outliers, signal drift, high noise levels, and more. In some cases, it is also feasible to identify the causal mechanism that led to the observed symptoms. Accordingly, labels and comments can be classified as follows:

Definition D3.37: A **symptom annotation** is an annotation that describes observed symptoms only does not provide any causal explanation.

Figure 3.9 (top) shows a symptom annotation.

Definition D3.38: A **causal annotation** is an annotation that includes information about the root cause for the observed symptom.

Figure 3.9 (top) shows a symptom annotation while Figure 3.9 (bottom) shows a causal annotation for the same phenomenon.

3.4.4.3 *Categorization according to the location of information*

Annotations can also be classified according to the source of information for the provided label or comment. For example, a label could be provided based on the level of satisfaction of a known relationship between sensor signals (e.g., a hydraulic balance). In this case, all information leading to the label is available within the same data set. In other cases, the label could be based on information outside the available data set. For example, the expert annotator may know of special weather-related events (e.g., storms) or changes in an upstream infrastructure (e.g., the sewer) that are not captured as measurements in the available data set. Accordingly, one can classify labels and comments as follows:

Definition D3.39: A **direct annotation** is an annotation based only on inspection or analysis of the data records being annotated.

Definition D3.40: A **circumstantial** or **indirect annotation** is an annotation based on information residing outside of the data set subject to annotation.

Figure 3.10 (top) shows a direct annotation. There is no information apart from what the expert could derive from the data itself. Figure 3.10 (bottom) shows a circumstantial annotation. The provided annotation is based on information that is not obvious by inspecting the annotated data points only.

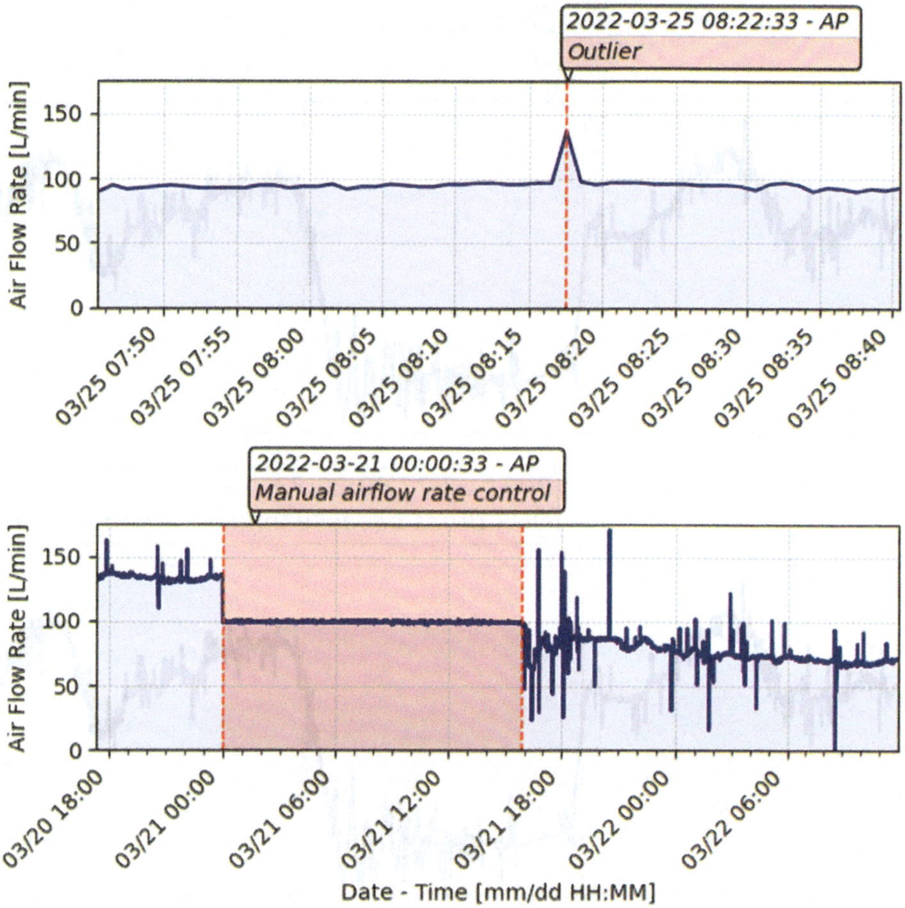

Figure 3.8 Example E3.5 – Data annotations according to temporal and spatial scope for an airflow rate time series in a conventional activated sludge bioreactor. An author acronym, timestamp, and the annotation text are shown. Top – point annotation; bottom – collective annotation.

3.4.4.4 Categorization according to the annotation process

As annotating can be a cumbersome, subjective, and tedious process, it is not surprising that many research studies have focused on the automated generation of labels or comments to describe data records. This research mainly focuses on anomaly detection, fault detection, and fault diagnosis. Chapter 6 provides an overview of the algorithmic tools available for this purpose. Depending on the use of such tools, one can distinguish between the following categories:

Definition D3.41: An **expert annotation** is an annotation that is assigned following the inspection by a human expert of data records without assistance of data analytics or algorithmic tools.

Definition D3.42: A **computer-aided annotation** is an annotation that is assigned following an inspection by a human expert of data records selected for inspection through data analytics or algorithmic tools.

Definition D3.43: An **automated annotation** is an annotation that has been assigned through data analytics or algorithmic tools without any human intervention.

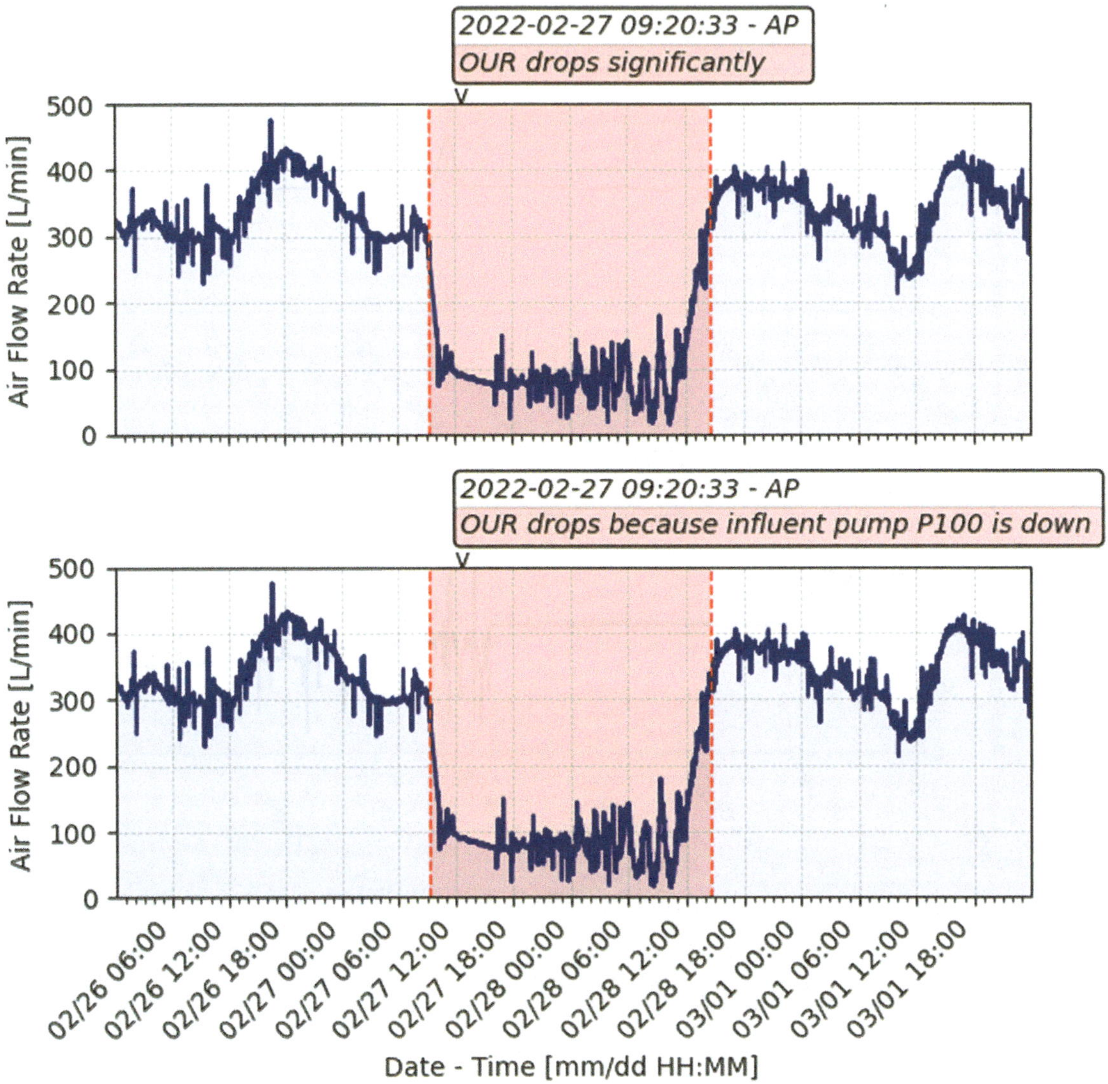

Figure 3.9 Example E3.6 – data annotations according to their explanatory power for an airflow rate time series. Top – symptom annotation; bottom – causal annotation. An author acronym, timestamp, and the annotation text are shown.

3.5 TAKE-HOME MESSAGE

- Metadata comes in a wide variety of forms and can be focused on the following:
  - Information about the equipment and processes involved in data collection (data-generating process).
  - Information about the quality of the produced signals (data quality).
  - Information about the context within which the data were produced (e.g., operational changes, weather).

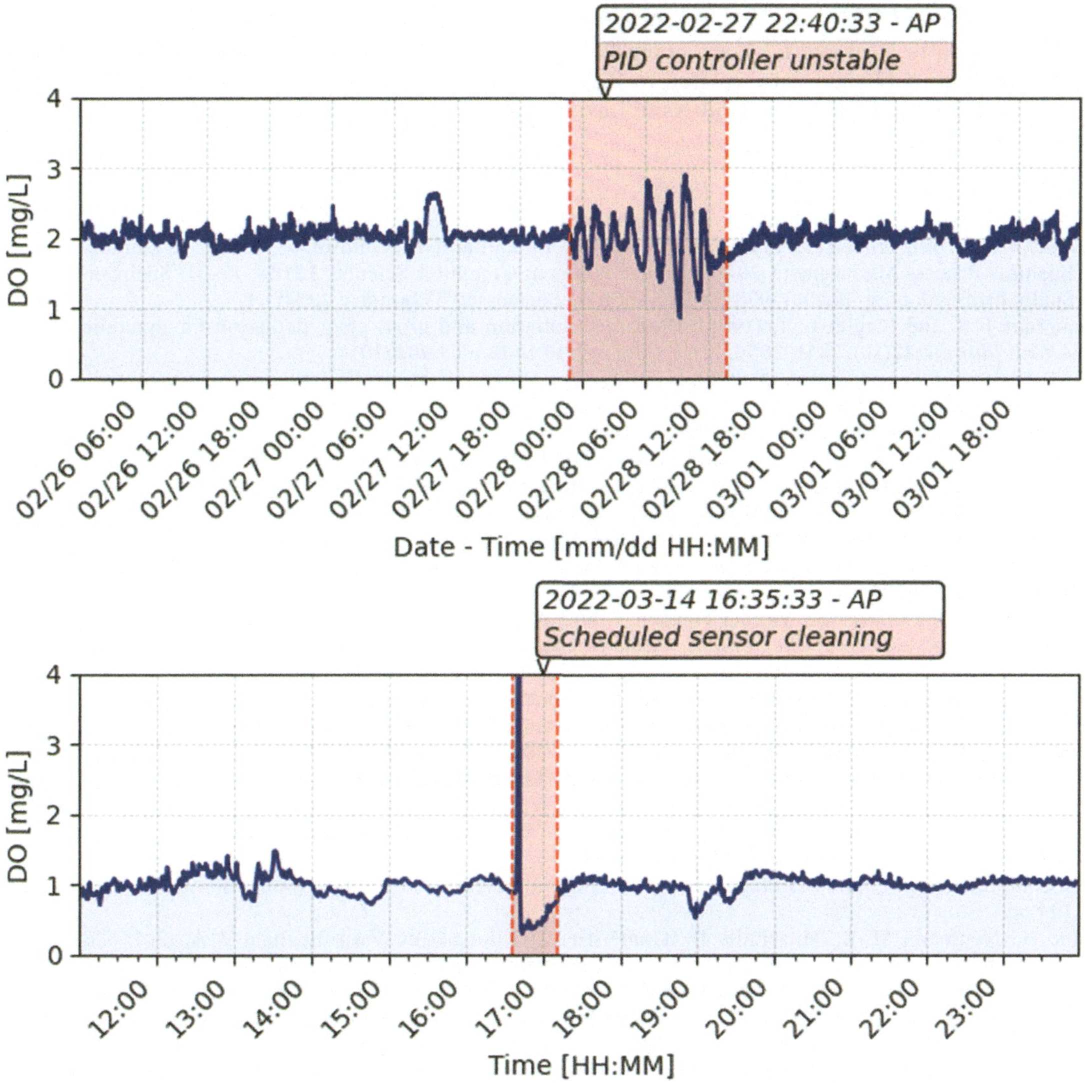

Figure 3.10 Example E3.7 – example of data annotations according to the location of information for an airflow rate time series. Top – direct annotation; bottom – indirect annotation. An author acronym, timestamp, and the annotation text are shown.

- A comprehensive approach to metadata collection starts with listing the metadata required for current or future projects or tasks. This chapter gives an overview of the most relevant types of metadata.
- Careful data management practices will include the collection of metadata to:
  - Know with confidence where, when, how, and why sensor signals were produced.
  - Assess the data quality produced by a sensor throughout its lifetime.
  - Adequately frame sensor signals in the context in which they were produced.

ACKNOWLEDGEMENTS

The data relating to Example E3.1 are open data (Ohmura *et al.*, 2019*b*). Example E3.4 is provided by Juan Baeza (unpublished data, related to Zhang *et al.*, 2021). Examples E3.5 through E3.7 were provided by the model*EAU* team (Université Laval, Québec, QC, Canada).

REFERENCES

Akkiraju R., Sinha V., Xu A., Mahmud J., Gundecha P., Liu Z., Liu X. and Schumacher J. (2020). *Characterizing machine learning processes: A maturity framework*. In: Fahland, D., Ghidini, C., Becker, J., Dumas, M. (eds) Business Process Management 2020. Lecture Notes in Computer Science, **12168**, 17–31. Springer, Sevilla, Spain. https://doi.org/10.1007/978-3-030-58666-9_2 (accessed 31 January 2024).

Albuquerque J. S. and Biegler L. T. (1996). Data reconciliation and gross error detection for dynamic system. *AIChE Journal*, **42**(10), 2841–2856. https://doi.org/10.1002/aic.690421014

Alavi N., Warland J. S. and Berg A. A. (2006). Filling gaps in evapotranspiration measurements for water budget studies: evaluation of a Kalman filtering approach. *Agricultural and Forest Meteorology*, **141**(1), 57–66. https://doi.org/10.1016/j.agrformet.2006.09.011

Ba-Alawi A. H., Loy-Benitez J., Kim S. and Yoo C. (2022). Missing data imputation and sensor self-validation towards a sustainable operation of wastewater treatment plants via deep variational residual autoencoders. *Chemosphere*, **288**(3), 132647. https://doi.org/10.1016/j.chemosphere.2021.132647

Chen J., Bandoni A. and Romagnoli J. A. (1997). Robust estimation of measurement error variance/covariance from process sampling data. *Computers & Chemical Engineering*, **21**(6), 593–600. https://doi.org/10.1016/S0098-1354(96)00295-5

Daszykowski M., Kaczmarek K., Vander Heyden Y. and Walczak B. (2007). Robust statistics in data analysis – a review: basic concepts. *Chemometrics and Intelligent Laboratory Systems*, **85**(2), 203–219. https://doi.org/10.1016/j.chemolab.2006.06.016

De Mulder C., Flameling T., Weijers S., Amerlinck Y. and Nopens I. (2018). An open software package for data reconciliation and gap filling in preparation of water and resource recovery facility modeling. *Environmental Modelling & Software*, **107,** 186–198. https://doi.org/10.1016/j.envsoft.2018.05.015

Efron B. and Tibshirani R. (1986). Bootstrap methods for standard errors, confidence intervals, and other measures of statistical accuracy. *Statistical Science*, **1**(1), 54–75. https://www.jstor.org/stable/2245500

Grubbs F. E. (1969). Procedures for Detecting Outlying Observations in Samples. *Technometrics*, **11**(1), 1–21. https://doi.org/10.1080/00401706.1969.10490657

HART Communication Foundation (2013). Hart Field Communications Protocol Application Guide Revision 7.1. 194 pp.

Hauduc H., Neumann M. B., Muschalla D., Gamerith V., Gillot S. and Vanrolleghem P. A. (2015). Efficiency criteria for environmental model quality assessment: a review and its application to wastewater treatment. *Environmental Modelling & Software*, **68,** 196–204. https://doi.org/10.1016/j.envsoft.2015.02.004

Isermann R. and Ballé P. (1997). Trends in the application of model-based fault detection and diagnosis of technical processes. *Control Engineering Practice*, **5**(5), 709–719. https://doi.org/10.1016/S0967-0661(97)00053-1

ISO (2003). ISO Water quality – on-line sensors/analysing equipment for water – specifications and performance tests. Standard ISO 15839:2003, International Organization for Standardization, Geneva, CH. https://www.iso.org/standard/28740.html (accessed 31 January 2024).

ISO (2006). Statistics–Vocabulary and symbols – Part 2: Applied Statistics. Standard ISO 3534-2: 2006, International Organization for Standardization, Geneva, CH. https://www.iso.org/standard/40147.html (accessed 31 January 2024).

ISO (2019*a*). Date and time – representations for information interchange – Part 1: Basic rules. Standard ISO 8601-1:2019, International Organization for Standardization, Geneva, CH. https://www.iso.org/standard/70907.html (accessed 31 January 2024).

ISO (2019*b*). Date and time – representations for information interchange – Part 2: Extensions. Standard ISO 8601-2:2019, International Organization for Standardization, Geneva, CH. https://www.iso.org/standard/70908.html (accessed 31 January 2024).

McGhee J. (2005). Calculation and Treatment of Errors. In: *Handbook of Measuring System Design*, https://doi.org/10.1002/0471497398

Narasimhan S. and Jordache C. (1999). Data Reconciliation and Gross Error Detection: An Intelligent Use of Process Data. Gulf Professional Publishing, Houston, Texas, 350pp. https://doi.org/10.1016/B978-0-88415-255-2.X5000-9

Ohmura K, Thürlimann C. M., Kipf M., Carbajal J. P. and Villez K. (2019*a*). Characterizing long-term wear and tear of ion-selective pH sensors. *Water Science and Technology*, **80**(3), 541–550. https://doi.org/10.2166/wst.2019.301 (accessed 31 January 2024).

Ohmura K., Thuerlimann C. M., Kipf M., Carbajal J. P. and Villez K. (2019*b*). pH Sensor Ageing Experiment (SoDAN-dataset1). https://doi.org/10.5281/zenodo.3229195 (accessed 31 January 2024).

Oppenheim A. V. and Schafer R. W. (1975). Digital signal processing. Englewood Cliffs, NJ: Prentice-Hall, 1975. New Jersey.

Osman M. S., Abu-Mahfouz A. M. and Page P. R. (2018). A survey on data imputation techniques: water distribution system as a use case. *IEEE Access*, **6,** 63279–63291. https://doi.org/10.1109/ACCESS.2018.2877269

Samuelsson O., Björk A., Zambrano J. and Carlsson B. (2018). Fault signatures and bias progression in dissolved oxygen sensors. *Water Science and Technology*, **78**(5), 1034–1044. https://doi.org/10.2166/wst.2018.350

Udugama I. A., Öner M., Lopez P. C., Beenfeldt C., Bayer C., Huusom J. K., Gernaey K. V. and Sin G. (2021). Towards Digitalization in Bio-Manufacturing Operations: A Survey on Application of Big Data and Digital Twin Concepts in Denmark. *Front. Chem. Eng.*, **3**, 727152. https://doi.org/10.3389/fceng.2021.727152

Venkatasubramanian V., Rengaswamy R., Yin K. and Kavuri S. N. (2003*a*). A review of process fault detection and diagnosis – part I: quantitative model-based methods. *Computers & Chemical Engineering*, **27**(3), 293–311. https://doi.org/10.1016/S0098-1354(02)00160-6

Venkatasubramanian V., Rengaswamy R. and Kavuri S. N. (2003*b*). A review of process fault detection and diagnosis – part II: qualitative models and search strategies. *Computers & Chemical Engineering*, **27**(3), 313–326. https://doi-org.ornl.idm.oclc.org/10.1016/S0098-1354(02)00161-8

Venkatasubramanian V., Rengaswamy R., Kavuri S. N. and Yin K. (2003*c*). A review of process fault detection and diagnosis – part III: process history based methods. *Computers & Chemical Engineering*, **27**(3), 327–346. https://doi-org.ornl.idm.oclc.org/10.1016/S0098-1354(02)00162-X

Wang L. Y., Yin G., Zhang J. F. and Zhao Y. (Eds.) (2010). System Identification with Quantized Observations, Birkhäser, Basel, Switzerland.

Widrow B. and Kollar I. (2008). Quantization Noise: Round off Error in Digital Computation, Signal Processing, Control, and Communications. Cambridge University Press, Cambridge.

Zhang C., Guisasola A. and Baeza J. A. (2021). Achieving simultaneous biological COD and phosphorus removal in a continuous anaerobic/aerobic A-stage system. *Water Research*, **190,** 116703. https://doi.org/10.1016/j.watres.2020.116703

Zoubir A. M., Koivunen V., Ollila E. and Muma M. (2018). Robust Statistics for Signal Processing. Cambridge University Press, United Kingdom, 312pp. https://doi.org/10.1017/9781139084291 (accessed 31 January 2024).

doi: 10.2166/9781789061154_0069

Chapter 4

Structuring and storing signals with their metadata: practical considerations

N. Nicolaï[1], *K. Villez*[2] *and Q. Plana*[1,3]

[1]*modelEAU, Université Laval, Québec, QC, Canada*
[2]*Oak Ridge National Laboratory, Oak Ridge, TN, USA*
[3]*Eurecat – Technology Center of Catalonia, Barcelona, Spain*

Keywords: data life cycle, data storage, database technology, flat-file, relational database, time-series database, process historian, data warehouse, data pipeline, SQL

This chapter aims to provide a comprehensive overview on structuring signal data and their metadata, highlighting key considerations for optimal management and storage. Specifically, it focuses on: (a) the relevance of data organization; (b) what data to store and what to keep; and (c) data management methods. Thus, this chapter answers *where* and *how* to store metadata efficiently. Chapter 3 explains *what* is considered metadata. Chapters 5 and 6 present *how to collect* certain metadata through dedicated sensor validation tests (Chapter 5) or algorithmic analysis (Chapter 6).

4.1 HOW TO READ THIS CHAPTER

The main purpose of this chapter is to motivate the need for systems to structure, store and document collected data, also known as data management. In addition, we describe the main types of systems available to address this need, together with their main properties, utility, and examples. The scheme below indicates the structure of this chapter and the dependencies between sections. Section marked in blue or white are considered basic and essential to understand the data structuring and existing technologies. The sections marked in grey are complementary and helpful sections to learn how these technologies can be implemented and documented with industry examples.

4.2 WHY DATA ORGANIZATION IS IMPORTANT

In previous chapters, common types of data and metadata were described in detail. In this chapter, we motivate the need to prioritize, structure, and organize data and illustrate how this can be achieved in practice by means of different data management technologies. We define data storage technologies as follows:

Definition D4.1: A **data management technology** is the combined set of hardware and software used for the systematic collection, description, exploration, modification, storage, and retrieval of sensor signals and their metadata.

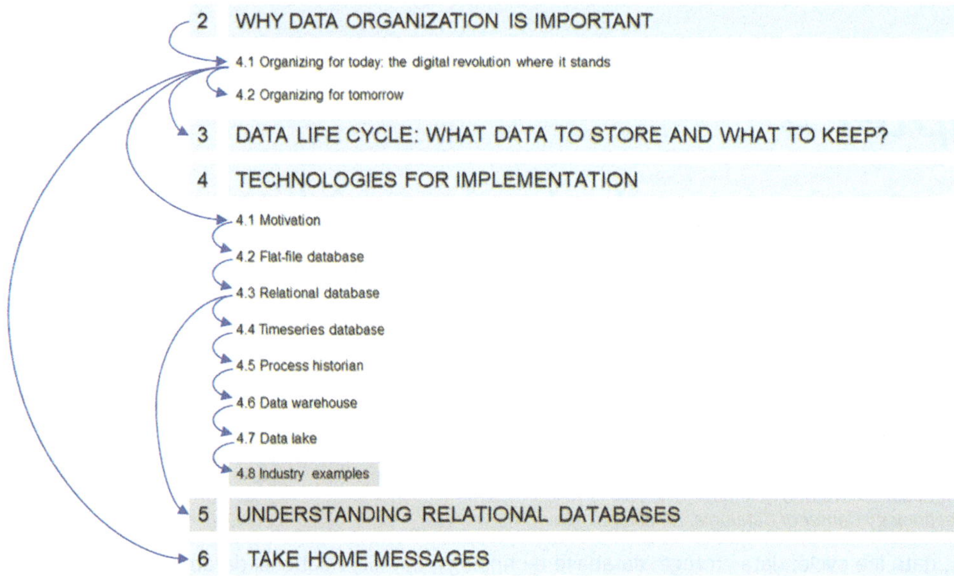

There are several benefits to having access to data that is easy to store, inspect, navigate, and use. In an ideal setting, the collection, storage, and retrieval of data is highly automated and leads to convenient use of the data for all data users. In practice, trade-offs may need to be made between the ease of data registration and ease of data retrieval. In what follows we highlight the major benefits that can be expected when applying a systematic approach to data management. We first discuss the immediate benefits that follow from improved data management practices. After that, we discuss benefits that can be expected in the long term.

Definition D4.2: **Data management** consists of methods, architectural techniques, and tools for gaining access to and managing delivery of data in a consistent way across different data types in a company. The purpose of data management is to fulfil all data requirements for use cases, applications, and business processes.

4.2.1 Organizing for today: the digital revolution where it stands

4.2.1.1 Data as a resource for information

Sensor signals without context are near useless (see Chapter 3). Indeed, without critical information such as unit of measurement, location, and information on sensor maintenance, potentially informative time series can quickly turn stale and useless. Today, most wastewater treatment plants collect sensor data that is primarily used for human or automated supervision and feedback controls. Only the most recent historical data records are typically needed to satisfy this need. In addition, online sensor data are also aggregated over longer periods (e.g., hours to years) for reporting to stakeholders and regulatory bodies. The low resolution that is needed to satisfy this need often does not require detailed contextual information. Current practice often assumes that manual inspection is sufficient to ensure the required level of accuracy.

The advent of the digital revolution, including increasing number of sensors, the use of physics-based, empirical, and hybrid models as digital twins for the plant require a new approach. The data history used for model building and validation extends the historical horizon from days to years so that one cannot rely anymore on human memory for contextual information. In addition, this historical horizon will now typically include multiple sensor maintenance events and fault events, all of which may be important to account for during model identification and deployment. Manual inspection of data records becomes untenable for this. For this reason, data users should also have access to a digitized record of such contextual information in the form of structured and accessible metadata. Only by providing access to both data and the metadata necessary to interpret them, can the data user extract the information required for any decision process.

4.2.1.2 Better decision making and cost saving

Well-managed metadata, including aspects of data generation, quality, and context, enables the data user to interpret the available data with confidence. Reducing uncertainty during data interpretation will reduce scepticism concerning decisions made based on data and allow decision-makers to proceed with more confidence when optimizing their process and plants and drive operations closer to their physico-chemical or safety limits in view of increased effectiveness or efficiency. This argument holds especially when decisions are made long after data collection, which typically is the case for applications that include models.

The most urgent barrier to achieving this is that utilities must be able to pool several sources of information together to arrive at a decision-making process that can account for all data available. Often, this means that different disparate data sources (e.g., process historian, laboratory information management systems, written records, asset management systems, computerized maintenance management systems, IoT platforms, enterprise resource planning, etc.) need to be effectively connected for efficient data retrieval.

4.2.2 Organizing for tomorrow

In the above, we briefly described the expected benefits that can be expected immediately after adopting improved data management practices. In what follows, we present potential benefits that arise from long-term commitments to improved data management. Some of the benefits discussed next are necessarily speculative at this time.

4.2.2.1 Preparing for unknown unknowns

We highly recommend reviewing current data management practices not only in view of current needs and projects but also in view of future projects, which may greatly benefit from a comprehensive approach to data and metadata management.

More specifically, data management practices today often only reflect the immediate need for information, including process supervision, control, and reporting to stakeholders and regulatory agencies. It is a frequent occurrence that the use of the data for other tasks, such as plant design, retrofitting, process control design and tuning, are not considered. As a result, the large volumes of data managed by wastewater utilities may prove useless for such tasks. In turn, this often results in executing additional, costly measurement campaigns to cover the gap in the data (see, e.g., Hauduc *et al.*, 2009; Rieger *et al.*, 2010). Such costs and loss of time can be avoided by improving routine data collection and management practices with such tasks in mind. By doing so, future questions that are currently undefined, could be answered without expensive measurement campaigns or data reconciliation efforts.

4.2.2.2 Integration across the fence: data brokerage

Today's water industry is highly fractionated. In practice, data are largely produced and used within the fence of water and wastewater utilities. Only a small fraction of the available data is communicated to interested parties outside of the fence. Similarly, the fraction of data collected outside of the fence is small in most cases. Examples include weather data or signals produced by sensors in upstream utilities (e.g., sewers) or connected systems (e.g., construction, demographics, industrial projects or activities, traffic etc.). However, there is an increasing push to approach the urban water cycle in a more holistic manner. For this goal to be achieved, utilities will necessarily need to share data among themselves and others. For the shared data to be most useful to all stakeholders, an integrated data management will be needed. Where feasible, common practices and standards across connected components of the urban water infrastructures should be established. A nice example of such a holistic collaboration across the urban water cycle is the work performed during the recent Covid-19 pandemic. It was illustrated how wastewater data can provide a wealth of information about epidemiology at urban and state-level scales. Utilities should start considering themselves as a producer of valuable, and possibly marketable, data for decision-making across many aspects of society. Leveraging this potential may require engagement with data brokerage systems with additional data quality and metadata requirements (Torfs *et al.*, 2022).

4.3 DATA LIFE CYCLE: WHAT DATA TO STORE AND WHAT TO KEEP?

Data, once collected, can become stale and useless over time. A concept referred to as the time-value of data. The only way to counter this natural tendency, is to also collect relevant metadata along the way (Michener, 2006). Having this in mind, data sets are best considered as resources with an expiration date. To maximize utility of available data, we recommend that data are subject to multiple phases along their life span. The different phases are listed below, specifically for sensor measurement data.

(1) *Planning*. This is the phase where new sensors are considered, purchased, installed, calibrated, and connected to the data-collecting infrastructure. During this phase, one should attempt to collect and structure metadata concerning the data-generating process (see Chapter 3, Section 3.2.2). The intended storage location of the data and how long they can and should be stored should also be defined at this time. The next phases and associated activities, to the extent that is feasible, should be defined as well.

(2) *Acquisition*. Rather trivially, this is the phase during which data are actively being collected. This phase may have a defined length (e.g., the length of an intense measurement campaign), or may be undefined in length (e.g., a sensor installed for monitoring and control purposes).

(3) *Maintenance and pre-processing*. Depending on the purpose of the data collection, data derivatives may be needed, instead of the original raw data. Typical derivatives may include (a) modified signals that are improved in data quality (e.g., noise reduction, removal of inaccurate measurements) and (b) aggregated values derived from the raw sensor signals for a specific purpose (e.g., monthly averages for reporting). This phase may include the definition, implementation, and updating of data processing steps that are considered necessary to attain and maintain the desired data quality. Inspiration may be taken from software versioning tools (definition below) to make this easier.

(4) *Usage*. Sometime after sensor deployment, one expects to use the data for the intended purpose. Depending on the task at hand, the data may be used long after data collection (e.g., model-based process design or retrofitting), directly after collection (e.g., real-time process control), or somewhere in-between (e.g., data analysis for operational troubleshooting and exploration). Critically, the data use phase should be defined early on, ideally in the planning phase.

(5) *Accessing/Sharing/Re-using*. It is quite likely that the data being collected will be re-used or re-purposed after the original data use phase has started or even finished. New, previously unintended use of the data should be described well and lead to a redefinition of phase 4 (data use). Ideally, new types of data usage should be checked against data availability, data fit-for-purpose, including a check on whether the data quality is adequate for the newly proposed use.

(6) *Archiving*. Once the data are collected and not modified or used anymore, one should consider archiving for future use. Documentation should be added to ensure the future user can use and interpret the data in a reliable manner.

(7) *Purging*. Data that have met their expiry date will degrade the quality of other data present in the same store and should therefore be removed in a timely manner (e.g., sometime after the shutdown of a unit operation). If available, the organizations data retention policy could provide guidance on how to do this correctly.

Definition D4.3: **Software versioning** is the process of documenting incremental changes in software and assigning unique version names or unique version numbers to refer to unique, historical states of the software. The goals of software versioning include (a) effective collaboration between programmers, (b) managing compatibility with other software, and (c) guaranteeing reproducibility of results obtained with the software.

To address this phase-based approach in a structured and organized way, a data governance strategy is highly recommended. This plan should cover aspects such as metadata generation, data preservation methods, purpose definition, expected analyses, and is ideally defined before a new project or data collection effort begins. Ultimately this may lead to data being well-managed in the present and prepared for preservation in the future. Special attention should also go to adequate back-up systems, user access policies, decisions about the types of data that can/should be stored in third-party applications (e.g., cloud storage vs. on-premise storage). Note that although the benefits of proper data governance are numerous, detailed cost–benefit analysis have yet to be reported for applications in the water industry. In the future, this knowledge gap might be filled if data assets become a marketable good available to industry stakeholders (Section 4.2.2.2).

Definition D4.4: Data **governance** is the exercise of authority and control (planning, monitoring, and enforcement) over the management of data assets.

Definition D4.5: A data **retention policy** clarifies what data should be stored or archived, where that should happen and for how long before it is disposed of.

The collection, usage, archiving, and purging phases may also be affected by a data retention policy that answers to multiple stakeholders and covers different aspects. It is therefore important that such a policy is well defined up-front. Some guidance is provided here. First and foremost, it is important to consider local regulatory requirements that specify how long certain data should be available for future compliance, whose reasons may not be identified yet. The policy might allow some data to be stored in a reduced or compressed form (e.g., lower temporal resolution), whereas other data should be stored in a raw format. It may also define adequate means of storage, thereby recommending or requiring specific storage media and excluding other devices for data storage (e.g., backup servers, external hard disks, etc.). Note that with the advent of cloud service providers, cheap long-term storage is available with acceptable data retrieval speeds and flexibility. It should also be clear from the start (i.e., before restoring archived data) whether the dataset is stored as original data, which metadata it includes, whether it's comprised of aggregated data, and whether it contains documentation. Data retention is also important when defining adequate backup plans. Backup plans often make use of a cascaded approach with respect to retention time, as to prevent unnecessary backups sticking around forever. For example, a data governance policy may require daily backups to be retained for 7 days, weekly backups for 4 weeks, monthly backups for 12 months, and annual backups for 7 years.

4.4 TECHNOLOGIES FOR IMPLEMENTATION

4.4.1 Motivation

When requesting data for analysis, chances are that you end up interacting with one or more databases. Sometimes this interaction is facilitated by a user-friendly front-end application, other times you manipulate data stores directly using a querying language. In any case, understanding how the data and metadata are structured in a database might greatly benefit the final analysis. Database technologies are not new. For a relatively long time in the history of information technology, flat-file data stores and relational databases were the go-to solutions for the structured storage of an organization's datasets. With the arrival of the big data era, new database technologies are needed to effectively handle the ever-increasing data volume, variety, and velocity in a reliable and efficient way.

The remainder of this chapter will introduce key technologies involved in the storage and structuring of data. The potential use of different database types will be explored, including the strengths and weaknesses of each technology. As the focus of this report is specifically on sensor data and metadata, relevant database technologies will only be discussed. Throughout this section it will be shown how metadata can come into use, and which database technologies are best suited to structure and make use of contextual information.

Special attention is provided to process historians, although these are essentially software applications build around more fundamental database technologies such as a relational or time-series databases. However, historians provide special features for the storage of sensor data and metadata which are worth exploring.

Since sensor data and metadata can ultimately be used to support business decisions, modern database technologies that are increasingly used in data analytics and data science use cases will also be discussed (i.e., data warehouse and data lake). The heterogeneity in different database technologies will be illustrated with relevant examples drawn from industrial applications.

4.4.2 Flat-file database

A flat-file database is commonly composed of a text file with little to no markup. It uses a simple structure, often a table, to store data. Common to this type of data storage is that records have a uniform format and are stored sequentially in the database file. Because of this simplicity, structured relationships between data records are typically not present. Flat-file databases are mostly self-contained, meaning that a single file is used to store data while it does not depend on other files or software applications to reveal structure and interpret the data.

Table 4.1 summarizes the major benefits and drawbacks of a flat-file database. Flat-file databases are often chosen as the first solution to store tabular data, including time series. Its simple data structure with little overhead requires only a minimal number of skills to save and query the data it stores. Indeed, knowledge on how to use spreadsheet software (e.g., MS Excel, Google Sheets, LibreOffice Calc, etc.) is one of the basic skills for modern professionals and therefore widely available. When spreadsheets grow large, however, the speed of retrieving data decreases and thus also the ability to extract valuable information. Indeed, flat-file databases rely on a linear search method to query the data, in which every record of the file must be checked for a match. This method works fine for a small list but does not scale to larger datasets where a lot of irrelevant data needs to be parsed before the correct record can be located. Moreover, because of the lack of structured relations between data, flat-file databases typically do not include extensive contextualization in the form of metadata which rely heavily on data relationships.

Example: A comma separated values (CSV) file is a prime example of a row-oriented flat-file database. Each line of text in such files represents a data record described by one or more fields separated by a comma (,) (see Figure 4.1). In the case that the data values contain commas, alternative delimiters are used such as a tab (\t) (tab separated values (TSV) file format) or fixed widths (positional file format where the start of a field is specified as an offset from the beginning of a line). Manipulation of the records in these flat-file databases is typically performed using basic text editors, custom-coded applications or using commercially available applications. In the latter case, the data are stored and compressed by default in the file extension native to the application software (e.g., xlsx, GSHEET, ods, etc.) but can easily be converted to other file formats that also support flat-file databases.

Table 4.1 Advantages and disadvantages of flat-file databases.

Advantages	Disadvantages
• Simple data structure without overhead makes it easy to understand the data.	• Not meant to have structured relationships or hierarchies between records, which complicates metadata storage.
• Familiar software applications for data manipulation are known to a broad range of professionals (text editors, spreadsheets, etc.).	• Data are not indexed which makes data retrieval slow for large datasets.
• Requires little implementation effort, nor schema design.	• Fixed maximum file sizes. Larger datasets need to be partitioned over multiple files.
• Fast querying for small datasets.	• Files are often stored locally and therefore not widely accessible to manipulate data by multiple concurrent users.

```
Date/Time,NH4+-N,NO3--N,PO4+-P,TOC,Comment
2018-11-02 11:00:00,,9.3,0.60,2.90,
2018-11-06 14:00:00,,,,3.40,
2018-11-13 11:30:00,<LOD,5.9,1.30,3.12,
2018-11-15 09:30:00,<LOD,5.3,0.40,2.36,
2018-11-19 12:20:00,0.041,11.0,1.20,5.87,
2018-11-21 15:00:00,3.050,11.5,1.20,8.40,
2018-11-26 14:15:00,<LOD,5.2,0.56,3.40,
2018-11-26 16:00:00,,,,4.70,
2018-11-28 14:30:00,<LOD,6.6,1.00,3.00,
2018-12-03 14:00:00,<LOD,5.8,1.10,6.80,
2018-12-06 15:30:00,0.038,8.2,0.74,4.50,
2018-12-06 12:00:00,,,,2.20,Batch 1.2.a.1
2018-12-06 12:00:00,,,,2.56,Batch 1.2.a.2
2018-12-06 12:00:00,,,,3.91,Batch 1.2.b.1
2018-12-06 12:00:00,,,,3.13,Batch 1.2.b.2
2018-12-06 12:00:00,,,,2.08,Batch 1.2.c.1
2018-12-06 12:00:00,,,,2.51,Batch 1.2.c.2
2018-12-06 12:00:00,,,,2.59,Batch 6.1
2018-12-06 12:00:00,,,,2.96,Batch 6.2
2018-12-08 10:00:00,<LOD,5.4,0.36,2.75,
2018-12-10 09:50:00,<LOD,5.0,0.44,3.44,
2018-12-12 11:30:00,<LOD,8.0,1.00,2.90,
```

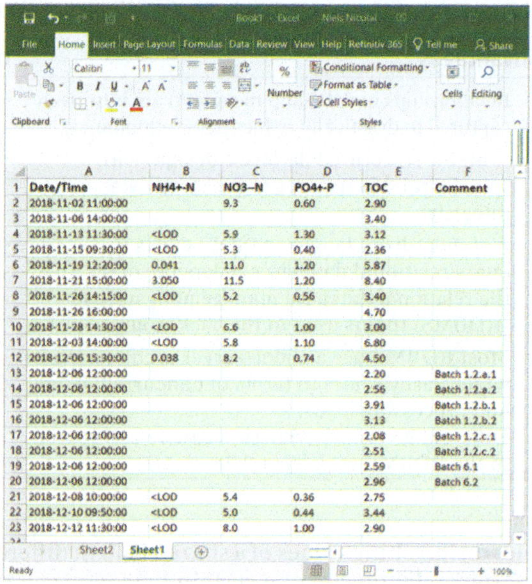

	A	B	C	D	E	F
1	Date/Time	NH4+-N	NO3-N	PO4+-P	TOC	Comment
2	2018-11-02 11:00:00		9.3	0.60	2.90	
3	2018-11-06 14:00:00				3.40	
4	2018-11-13 11:30:00	<LOD	5.9	1.30	3.12	
5	2018-11-15 09:30:00	<LOD	5.3	0.40	2.36	
6	2018-11-19 12:20:00	0.041	11.0	1.20	5.87	
7	2018-11-21 15:00:00	3.050	11.5	1.20	8.40	
8	2018-11-26 14:15:00	<LOD	5.2	0.56	3.40	
9	2018-11-26 16:00:00				4.70	
10	2018-11-28 14:30:00	<LOD	6.6	1.00	3.00	
11	2018-12-03 14:00:00	<LOD	5.8	1.10	6.80	
12	2018-12-06 15:30:00	0.038	8.2	0.74	4.50	
13	2018-12-06 12:00:00				2.20	Batch 1.2.a.1
14	2018-12-06 12:00:00				2.56	Batch 1.2.a.2
15	2018-12-06 12:00:00				3.91	Batch 1.2.b.1
16	2018-12-06 12:00:00				3.13	Batch 1.2.b.2
17	2018-12-06 12:00:00				2.08	Batch 1.2.c.1
18	2018-12-06 12:00:00				2.51	Batch 1.2.c.2
19	2018-12-06 12:00:00				2.59	Batch 6.1
20	2018-12-06 12:00:00				2.96	Batch 6.2
21	2018-12-08 10:00:00	<LOD	5.4	0.36	2.75	
22	2018-12-10 09:50:00	<LOD	5.0	0.44	3.44	
23	2018-12-12 11:30:00	<LOD	8.0	1.00	2.90	

Figure 4.1 Data presented in Table 2.3 (Chapter 2) expressed as a flat file. Left: CSV file format. Right:.xlsx Spreadsheet in the Microsoft Excel editor.

Additional remarks: Although not designed for this purposes, flat-file database can be used to represent relations between pieces of information. Usually, this is achieved by means of an additional textual note field. For example, the following annotation *recordID Y is related to the event captured by recordID X* could be manually added as a comment to record X to denote its relation to record Y.

Some argue that flat-file databases should not be considered as real database technologies (Dunning & Friedman, 2014). In the context of data management, the use of flat-file databases is indeed often limited to sharing raw data and aggregated results between applications and end users. In that sense, the flat-file acts as an intermediate data store with the sole purpose of transferring data into a more refined database technology. Flat-file storage may seem outdated because it does not bring all the bells and whistles of more comprehensive data storage technologies. However, most organization still have a lot of raw data stored as flat files. Think of spreadsheets that contain measurement data stemming from experimental trials, CSV files generated by a stand-alone measurement instrument, and so on.

4.4.3 Relational database

Relational databases are a proven technology to store large data sets together with structured metadata. A high-level overview of this type of database is provided here. A more detailed description, as well as some relevant examples is provided in Section 4.4. Table 4.2 summarizes the major benefits and drawbacks of relational databases.

Relational databases store data in relations, which the user perceives as tables. In contrast to flat-file data storage, a relational database intuitively captures the associations that exist among those tables. Each row in a table represents a data record which is characterized by a unique identifier (i.e., the primary key). Whereas each column in a table represents an attribute, characteristic, or fact that is relevant to the table subject. Rather than storing all data in one wide table with many attributes, the data are separated into multiple tables according to a specific subject, that is, some real-world understanding of the data. In this way, data redundancy is reduced to ensure data integrity and ease of maintenance. The latter is referred to as database normalization. To make this work, the database

Table 4.2 Advantages and disadvantages of relational databases.

Advantages	Disadvantages
• Relationships in data (and metadata) are intuitively captured to drastically reduce data redundancy. • A highly structured database design, with specific rules is in place to preserve data integrity. • A standardized language (SQL) exists to query relational databases independent of the relational database management system (RDBMS) that is used in the background. • Most RDBMS use a client–server architecture which easily scales in terms of concurrent users and processing loads. • Different levels of permission can be configured in the RDBMS to protect data access.	• A well-designed data model is essential to for a well-performing database. • Querying the database requires a basic understanding of SQL and the logical data model underlying the database. • Ideally, relational databases are actively administrated and fine-tuned (e.g., backup plan, retention policy, security updates, resource allocation, design changes, etc.) • Developed for transactional data, and therefore not originally intended to efficiently handle time-series data (e.g., slower storage and retrieve speeds, complex SQL queries for time-series alignment and subsampling).

relies on matching values of a shared field in different tables (i.e., primary key-foreign key relationships) to form a relationship of a certain type between the data (i.e., one-to-one, one-to-many, and many-to-many). The way data are separated into different tables, as well as the definition of the relationships amongst those tables, is determined by the logical data model used to structure the database. Data models are highly specific for each application. However, generic principles exist to design logically structured, reliable, and flexible relational databases independent of the application (Hernandez, 2021). Also note that besides typical alpha-numeric data (e.g., measurement values, text, datetime, GPS coordinates, etc.), most relational databases nowadays support querying of semi-structured data (e.g., flat files like XML, JSON, etc.). Figure 4.2 shows an example of such a relational structure where each box represents a table and the arrows the identified relationships between them.

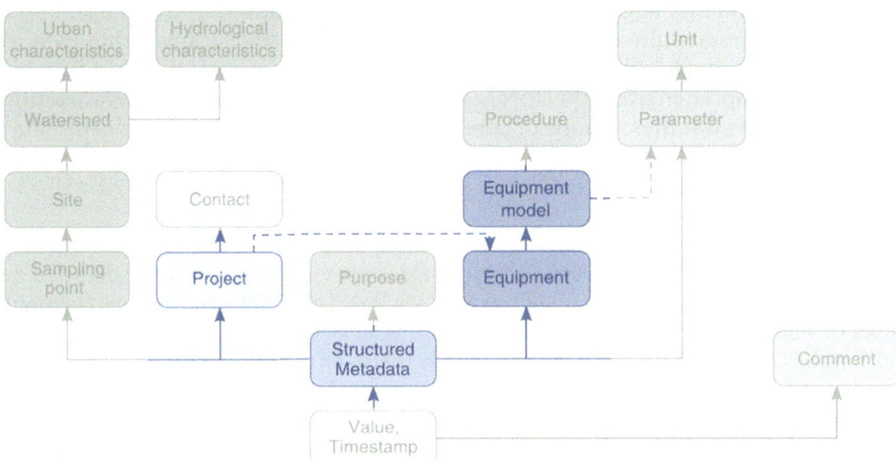

Figure 4.2 Logical data model relevant for the storage of water quality data in a relational database. Highlighted in blue are the relations relevant to the example in Figure 4.3 and Table 4.3. Adapted from Plana *et al.* (2019) and Therrien *et al.* (2020).

Example: A major difference between flat-file data storage and relational databases is the way data are accessed and modified. Searching a flat-file requires one to understand the physical lay-out of the file. In contrast, a relational database can be queried using conditional statements that take the form of an instruction that is readable by a human user. This requires the user is familiar with the relationships among the tables in the database. The formal language used to accomplish these manipulations in relational databases is called structured query language (SQL) which originated from relational algebra. Typical SQL statements combine relational operations such as *select*, *project*, and *join* to manipulate one or more tables to result in another table visible to the user (i.e., the result set) (see Figure 4.3 and Table 4.3). It is important to understand that this resulting table is not stored in the database as such, but instead it is the output of a query. If desired, the output table can still be stored as a virtual table based on the query, rather than being defined in the data model itself (i.e., a *view*).

Additional remarks: Besides the logic used to define a data model, it can be important to also consider how information is stored for the database to access it (i.e., physical design). This is especially relevant when working with huge databases composed of millions of data records and relations. One term that often pops-up in discussions on database performance management is *indexing*. In the same way that an index can be used to locate specific terms in scientific publications (e.g., technical books), a database can use indexes on specific columns to quickly locate rows in tables. An index can be thought of as a special table which contains columns to locate rows in another table, along with information describing where the rows are physically stored. The default structure applied for database indexing is typically based on B-trees (Beaulieu, 2020). The fact that relational databases can make use of high-performant indexes, gives them a tremendous advantage over flat-file databases which are stuck to linear searches.

Although the relational model and SQL are non-proprietary concepts, a dedicated software program is needed to manipulate the data files and the physical hardware on which it is stored. This, and much more, is handled by a relational database management system (RDBMS). A wide variety of open and closed source, as well as commercial and free RDBMS exists. Among the most popular enterprise solutions are MySQL, SQL Server, Oracle database, DB2, SQLite, and PostgreSQL. Each

```
SELECT DISTINCT dbo.equipment.Equipment_ID AS 'Equipment ID',
    dbo.equipment.Equipment_identifier AS 'Name',
    dbo.equipment_model.Manufacturer AS 'Manufacturer',
    dbo.equipment_model.Equipment_model AS 'Model'
FROM dbo.metadata
    LEFT OUTER JOIN dbo.project ON dbo.metadata.Project_ID = dbo.project.Project_ID
    LEFT OUTER JOIN dbo.equipment ON dbo.metadata.Equipment_ID = dbo.equipment.Equipment_ID
    LEFT OUTER JOIN dbo.equipment_model ON dbo.equipment.Equipment_model_ID = dbo.equipment_model.Equipment_model_ID
WHERE dbo.project.Project_name='pilEAUte'
```

Figure 4.3 Example of a basic SQL query that combines select, project, and join operations to retrieve all active equipment that is part of the pilEAUte project. Adapted from Plana *et al.* (2019) and Therrien *et al.* (2020).

Table 4.3 Result set of the query in Figure 4.3.

Equipment ID	Name	Manufacturer	Model
6	pH_006	WTW	Sensolyt 700 IQ
15	pH_015	HACH	Hach pHD sc Online Process pH Sensor
26	Spectro_002	s::can	spectro::lyser
38	Ammo_004	s::can	ammo::lyser
41	Varion_001	WTW	VARION PlusAN/A
...	...	...	...

Only the first five entries are presented here. Adapted from Plana et al. (2019) and Therrien et al. (2020).

of them has specific advantages and disadvantages, as well as highly specific niche features. Yet all of them have the same basic functionalities: (1) to define structural rules according to a relational data model, (2) to allow for concurrent data entry and modification using the SQL standard, and (3) to protect and secure your database based on user-defined roles with corresponding permission levels. Although the RDBMS software is used to manage a relational database, interacting with the data is not necessarily restricted to the RDBMS only. Thanks to the widespread availability of software interfaces, it is possible to use general programming languages (e.g., C++, Java, Python, and R) to directly query most relational databases.

4.4.4 Time-series database

Time-series data are characterized by the repeated recording of a variable (e.g., a sensor signal value) over time, together with the time instance at which the recording was made (i.e., timestamp). Thanks to the decreasing cost of data storage devices, time series stemming from business assets, manufacturing plants and industrial processes are being generated at unprecedent volumes. Hence, the need for solutions that efficiently store and retrieve data from one or more time series in a particular time range. Although flat-file and conventional relational databases can be/are being used for this purpose, dedicated time-series databases exist that exploit the peculiarities of time-series data for faster and more intuitive data storage and retrieval (Dunning & Friedman, 2014; Nielsen, 2019). To this end, time-series databases usually take advantage of one or more of the following characteristics:

(1) The data records composing a time series are chronologically ordered. Moreover, this order needs to be preserved for the data to make sense.
(2) Time-series data often represent records in a sensor signal. Interpolation between data records is therefore acceptable.
(3) Many time series are composed of regularly spaced data records, that is, with a constant time interval between subsequent records.
(4) For a lot of applications, the more recent data records in a time series are more likely to be of interest in comparison to records that were stored a long time ago.
(5) Time series from steady-state operations are rarely most significant. More interesting are periods with dynamic changes such as trends, transients, outliers, and so on.
(6) Time series that have been recorded are rarely updated afterwards.
(7) Individual data records are often less informative as compared to short sequences or aggregated statistics of a time series.
(8) Time-series data are usually characterized by very few independent relationships in comparison to relational data.

Table 4.4 summarizes the major benefits and drawbacks of time-series databases.

Considering the specific nature of time series, out-of-the-box relational databases will not result in the high performance and scalability often needed for time-series analysis. An important reason is that relational databases do not naturally consider continuity in the time-series data which makes queries based on the alignment of multiple time series, often recorded at different frequencies, a complicated exercise. Also, standard SQL does not easily allow queries that require a modified temporal resolution (i.e., time-bucketing) (TimescaleDB, 2017).

As a general guideline, when time range queries represent most data retrieval requests it is a good idea to consider time-series databases as they are purposely built for it. The catch, however, is that time-series database technologies are not (yet) as mature as relational databases. Time-series databases can be self-contained solutions or can be built on top of other existing database technologies. An example of the former technology is InfluxDB, whereas TimescaleDB and OpenTSDB are built on a relational database (PostgreSQL) and a NoSQL database (Apache HBase or Google Bigtable), respectively. Each of them claiming advantages over the other. The way to interact with each of those time-series databases is strongly dependent on their implementation.

Table 4.4 Advantages and disadvantages of time-series databases.

Advantages	Disadvantages
• Purposely built for high performance, scalable storage, and retrieval of large volumes of time-series data.	• No standards among time-series database solution providers, which makes query language and functionalities software dependant.
• A variety of time centric functions are usually provided for time-series analysis.	• Collectors to interface time-series databases with industrial equipment (PLC, SCADA, DCS, etc.) must be custom developed.
• Time-series data can easily be joined with metadata stored in a relational database.	• Graphical applications for the user to interact with the database are limited in functionality. Custom applications have to be developed if non-specialists need to perform analysis on the data.
• A few of the most popular time-series databases are license free and open source.	

Additional remarks: The exact details on how to improve performance for the storage of time series are highly technical and not covered in-depth here. Yet, there are some key points that are interesting to understand. For example, the fact that time-series data tend to be visualized as long lists (i.e., one row for each data record in the series) whereas a wide format (i.e., multiple data records, of the same time series, are spread over different columns in a clever way) is more adequate, as it can drastically reduce the number of rows that need to be scanned if one also takes into account the chronological nature of the series. The lossless compression (definition below) of subsequent timestamps and values using the least number of bits can also reduce the volume of data stored (Chiarot & Silvestri, 2021). By having an automated retention policy in a time-series database, older data records can be stored at lower temporal resolution or aggregated over time. Finally, by caching more recent data records in memory rather than storing them directly on disk, latency can be substantially reduced.

Because each time-series database uses a different way of modelling and storing data, there currently exist no standard language to query time-series databases. Some commercial solutions use a language that completely inherits standard SQL and add custom functions to handle time series, while others have introduced an entirely new and independent querying language (Figure 4.4). Again, both

```
-- Microsoft SQL Server
SELECT timestamp, value
FROM pilEAUte_values
WHERE timestamp >= 1662909036 AND timestamp <= 1663513836
AND signal_name = 'temperature'

-- TimescaleDB
SELECT time, temperature
FROM pilEAUte_values
WHERE time > now() - '7 days'

-- InfluxDB
from(bucket: "pilEAUte_values")
    |> range(start: -7d)
    |> filter(fn: (r) => r._measurement == "temperature")
```

Figure 4.4 Querying the last 7 days of a temperature signal stored in a relational database (top) (i.e., SQL compliant) and two different time-series databases (middle and bottom) (i.e., custom query language). Note how querying a time series for a specific range in the relational database is complicated by the fact that an explicit time range is expected. Moreover, Unix epoch time is used to store, and therefore also retrieve, the time series.

approaches claim to have advantages over the other. Fortunately, most time-series databases provide the same functionalities to the users. For example, off-the-shelve time-centric functions for filtering, selection, aggregation, transformation, down sampling, interpolation, and so on.

Querying time series often requires that the user pre-defines a time range for further analysis. Deciding on the desired range for analysis upfront may result in multiple iterations before the actual period of interest is pinpointed. This is especially true when dealing with time series recorded over long periods of time. To efficiently query for interesting periods in time-series data, innovative methods for time series querying are actively being researched. By using motif discovery algorithms, patterns of interest can be identified in the same or different time series. Based on the initial pattern provided by the user (e.g., an initially selected period of interest, or based on a free-form sketch) a query can be constructed to find similar patterns and rank them according to their match (Correll & Gleicher, 2016; Torkamani & Lohweg, 2017). Thereby greatly improving the way time series can be queried.

Given that time-series databases are highly specific for the storage of time series, they cannot directly be used to store contextual information that relates to the recorded time series (e.g., the type and ID of the sensor that was used to generate the time series). Because such metadata is mostly static information with multiple relationships, it is better captured in a more structured database. That is why time-series databases allow dynamic time series to be joined with static information stored in a flat-file or relational database. As a result, contextual metadata can be used in front-end applications to construct a query relating to the correct time series of interest.

4.4.5 Process historian

4.4.5.1 General description

Process historians are commercial software packages that are available for the efficient acquisition, storage, organization, and retrieval of online measurement data stemming from industrial processes. Because these measurements are recorded over time, they are in fact time series. Note however that in the world of process automation, the term *tag* is often used to refer to a specific time series. Each tag of the historian is characterized by a unique and structured name for identification purposes and has some defining attributes. Despite the slight difference in terminology and usage, deep down historians are in fact time-series databases. Likewise, they can be self-contained solutions or built on top of another database technology (e.g., a relational database). Yet, historians can do much more when compared to generic time-series databases (Middleton, 2022).

The main purpose of a historian is to collect real-time measurement data and to store it efficiently for later interpretation. A lot of historians also store additional data to further clarify operational incidents and business questions. For example, historians can store alarm and event data important to operational troubleshooting activities. Also, aggregate features can be computed automatically and stored to provide summary statistics for certain variables. This is especially relevant when solving big-picture questions and when defining business strategies. Another type of data that is stored by most historians is the configuration data. These data describe the configuration of the software and hardware and is mostly used for administrative purposes of the historian infrastructure.

Table 4.5 summarizes the major benefits and drawbacks of process historians.

Additional remarks: There exist a few features that make a historian particularly tailored to the needs and desires of the process and manufacturing industries. First, historians are highly compatible with other software and hardware used for industrial automation (e.g., DCS, SCADA, MES, ERP, PLC, etc.). This means that users do not need to develop automatic programs for the low-level data transfer between those data sources and the historian (i.e., data collectors). Bear in mind that in the past, each vendor had a proprietary protocol for its ecosystem of software and hardware products. This meant that industrial software solutions such as historians were often evaluated based on the availability of specific interfaces. With the adoption of standardized communication protocols (e.g., ODBC, OPC UA, MQTT) the importance of this selling point has somewhat decreased. While interoperability has improved over the years, it is important to highlight that just like most other industrial automation

Table 4.5 Advantages and disadvantages of process historians.

Advantages	Disadvantages
• Purposely built for the high performant, scalable storage, and retrieval of large volumes of time-series data.	• High price. Typically, the user must pay per 100–1000 tags on an annual basis.
• The software package is heavily tailored to industrial processes or even specific sectors (e.g., water and wastewater industry).	• Setting up a historian takes a lot of manhours which can significantly increase project cost.
• Historian vendors provide a whole ecosystem of additional software components that help in organizing and analysing the stored data (i.e., the historian suite/stack).	• Black-box systems which reduces flexibility and configuration options. Support from vendor needed for custom extensions.
• Data collectors are readily available following standard industry protocols.	• Its dependency on other software products (Windows Server, Microsoft SQL Server, etc.) requires additional licenses, which brings additional costs.
• Graphical-user interfaces make it easy for the user to query for the correct data without writing any SQL code. • Robust solution to guarantee 100% uptime. • Product integrators and support are familiar with the needs of the process industry.	• Because of the (1) significant investment in both cost and time required to install a historian, (2) the dependency on the vendor for future updates and extensions, and (3) the dependency on legacy systems. Vendor lock-in can occur, where the user is financially forced to remain with the same vendor despite a desire for change.

software, historians are still very much optimized for the Microsoft stack of software technologies (Siemens SIMATIC Process Historian, GE Proficy Historian, AVEVA Historian, OSIsoft Pi, Rockwell Automation FactoryTalk Historian, Honeywell Process History Database, ABB S+ Historian, Emerson DeltaV Historian).

Historians are being used to store data for safety critical systems with large financial turnovers. This requires a robust software and hardware infrastructure with sufficient redundancies, such that 100% uptime can be guaranteed. A fall-back strategy is often included when remote data sources are not able to access the historian database because of communication issues. In that case, the data will be stored temporarily in a local buffer until the historian is back online (i.e., store and forward cache). Another feature is multi-tiering where multiple historians can be setup to collect data at different frequencies for different purposes. For example, a tier 1 historian is often located on site and collects raw measurement data at relatively high frequencies useful for operations. On top of it sits a tier 2 historian, this one is located outside the secure internal network infrastructure of the plant, and acts as a storage for aggregate data or measurement data at low frequency, like a data warehouse specific to all process operational data (Section 4.4.6). A tier 2 historian can be used to answer high-level business questions relevant for determining a long-term strategy, and because of that it might be connected to the internet for remote access. Integrating the historian in the enterprise network fits within the recent trend to tightly incorporate systems for operational technology (OT) with information technology (IT) systems, that is, IT/OT convergence.

Originally, data historians were only aimed at efficiently storing and retrieving measurement data by a select number of users. Today, modern historians offer a lot more functionalities that are particularly useful for the process industries. In many cases these additional software components were developed by the suppliers and end-users themselves to solve specific problems, over the time they were gradually integrated and fine-tuned as part of the package available to other users. For example, most historians have dedicated highly specialized graphical-user interface or web application to visualize and analyse the data by many concurrent users. Out-of-the-box analyses packages can be used for calculating process efficiency, mass balancing, determining power consumption, investigating alarm logs, monitoring safety systems, and so on. Moreover, extensions can be configured for automatic reporting of summary data, sending out notifications based on user-defined conditions or anomalies, managing user and their permissions, and much more. Depending on the vendor, asset management software and data contextualization tools are available for predictive maintenance and for the storage of metadata relevant to the user of the process historian (e.g., OSIsoft AssetFramework).

4.4.5.2 Compression techniques deployed in historians

Large process plants can have millions of tags that need to be recorded by the historian. To handle these large volumes of real-time data, a diverse set of data-handling techniques is included in most historians. As in generic time-series databases, lossless compression techniques are used to store the data using the least number of bits as possible without loss of information. In addition, several methods are typically available to get even higher compress ratios by excluding redundant intermediate data records from storage (see also Section 3.2.4.2 in Chapter 3). This is acceptable because a lot of process variables measured in real time do not change significantly for long periods of time. The motivation to deploy lossy compression is two-fold. On the one hand it significantly reduces the physical space needed to store the data. On the other hand, it limits the bandwidth required for data transmission towards other applications or databases. We apply the following definitions:

Definition D4.6: Data **compression** is the process of encoding information using fewer bits than the original representation. This reduces the volume required on the physical storage device.

Definition D4.7: **Lossless compression** is data compression that ensures that the original data can be reconstructed perfectly from the compressed data, that is, storage without loss of information. Compression is achieved by eliminating redundant parts in the data.

Definition D4.8: **Lossy compression** is data compression that allows certain information to be lost. As a result, one can only construct an approximation of the original data based on the compressed data. While information will be lost in the process, high levels of compression can often be achieved with small losses in information.

The most common approach to compress measurement data in real time is to only store data records of a tag that significantly differ from the previously recorded data point. This is known as *delta storage* or *report on exception*. One way to define this significance is by setting fixed user-defined limits on the allowed change of a value before being recorded (i.e., dead-band compression, or Box-Car compression). Alternatively, the magnitude of change between two consecutive recorded data points can be used as a basis for deciding which data points to discard (i.e., backward-slope compression, or rate of change compression). A combination of the two is also possible (see Figure 4.5). Another popular compression method is the swinging-door algorithm. It is important to keep in mind that with the application of these compression algorithms, the data records that are not registered are lost forever. The time interval between data records that are kept will also vary from one record to another. A higher-resolution time series can be reconstructed from the compressed time series using simple linear interpolation (Choudhury *et al.*, 2008).

Interestingly, the previously described lossy compression techniques are relatively straightforward compared to the complex algorithms also applicable to image, audio, and video compression (e.g., wavelet compression used for the JPEG standard). A crucial difference between the two lies in the way the compression algorithms are applied. The simple, and therefore very intuitive, methods used by historians operate in real-time on a point-by-point basis (i.e., they decide to keep or discard a new measurement whenever it becomes available). In contrast, more complex compression algorithms usually require a larger sequence, that is, with a defined start and endpoint, before compression can be executed. Also note that the discussed compression methods do not guarantee high compression ratios, this is because they heavily depend on the variance of the time series. Tags that exhibit low-signal-to-noise ratio will have low compression ratios. Moreover, lossy compression methods can

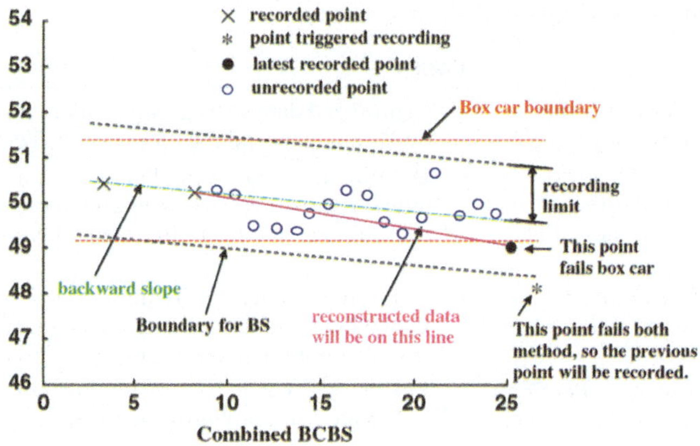

Figure 4.5 Illustration of the combined Box-Car-backward-slope (BCBS) compression algorithm used by some process historians. When the present data point fails to meet the criteria of both Box-Car and backward-slope method (*), the point prior to the present measurement (•) is stored (Choudhury *et al.*, 2008).

degrade data quality by altering summary statistics, introducing non-existent time delays as well as nonlinearities in the dynamic profile (Thornhill *et al.*, 2004). That is why it can be important to understand and document the functioning of the algorithms used by the historian as supplementary metadata. It might in fact be worth evaluating the impact of these compression algorithms for specific signals, before introducing them.

Note that a distinction can be made between lossy compression for the short term, and aggregation for the long term. In case of the latter, average values taken over extensive periods of time can be calculated, stored, and reported. However, for plant operations and analysis, aggregation tends to compress the data so much that it becomes uninformative for operational staff. The use of lossy compression techniques to capture short-term dynamics is considered more popular as a practice.

4.4.6 Data warehouse

A data warehouse is a central data repository that stores mostly aggregated data and metadata coming from one or more disparate data stores. The focus of a data warehouse is primarily on data processing for business intelligence, for that reason it is categorized as an analytics database as opposed to the operational database technologies discussed before. Table 4.6 summarizes the major benefits and drawbacks of data warehouses.

The importance of a data warehouse is best understood by considering the data-driven decision-making hierarchy in large organizations. Located at the lowest level of an organization are the users of the operational systems and the associated low-level management. Their role is to keep the wheels of the business spinning. Examples for a water utility are the operators and engineers that report on process conditions, document the progress of maintenance activities, log events and anomalies in logbooks, keep track of assets status, and so on. To perform their job, they interact with operational data stored in different isolated operational systems which are often applications connected to a relational database in the backend. Examples of operational systems include the process historian, laboratory information system, asset management software, computerized maintenance management software, spreadsheets, logbooks, IoT platforms, and so on. The focus of these systems is very much on execution and solving short-term issues related to temporary events. One level above in the hierarchy is the organization's middle management which integrates, analyses, and summarizes the information stored in the databases of the operational systems. Based on the insights generated at this level, tactical decisions are made on the medium-to-long term. At a water utility this could, for example, be the

Table 4.6 Advantages and disadvantages of data warehouses.

Pros	Cons
• Centralizes disparate data such that it can be used by non-specialist users.	• Centralized data storage can result in monoliths which are overly complex and therefore not easily maintained nor queried.
• Intuitively stores both the business measurement data as well as the associated context (i.e., business and technical metadata).	• Defining the dimensional model is a monumental task which requires expert input and agreement on metadata terminology amongst different stakeholders.
• Data collectors for integration are readily available following standard database protocols.	• Enforcing a schema will almost certainly result in some of the business data not being included. Initially this might be irrelevant, but sooner or later this can change.
• Available as managed service (PaaS) on public clouds with optimized ecosystem integrations for data ingestion, processing, and BI analytics	• Businesses evolve. As a result, business questions change over time, which requires continuous revisions of the logical schema.
• Complete transparency and flexibility on how data structured in the dimensional model.	• Ad-hoc data analysis on large datasets is hampered due to long retrieval times, limited granularity, and the predefined database schema.
• Structured data schema allows for generating targeted queries that aid rapid data-driven decision making.	• Data should be highly structured. Moreover, this assumption does not seem to be possible. Unstructured data come from many sources such as IoT sensors, text documents, video/audio files, and can be of various formats (JSON, PNG, PDF etc.). In those cases, a strict structure cannot always be enforced.

usage of different operational datasets by the process engineers for short-term process performance evaluation as well as for the definition of corrective process actions. Ultimately, the top management of an organization makes use of these analysis results together with external information, to make long-term strategic business decisions. An example relevant to a water utility would be an economic feasibility analysis for the expansion of treatment capacity considering trends in the historic influent quantity and quality, process and equipment efficiency, population forecasts, climate change, and so on.

The hierarchy presented above clearly illustrates how high-level decisions rely on accurate yet summarized *de-siloed* operational data. Because business intelligence has different needs, rhythms, and structures, as compared to short-term operational decision making, a different system is required for middle and high-level management to store and interact with the organization's analytical data. Democratizing disparate data for non-specialist business users is exactly what a data warehouse was designed for. An important point that cannot be forgotten when making business decisions based on data, is the context in which the data were collected, that is, metadata. The power of a data warehouse is that it intuitively stores both relevant operational data as well as the metadata required for thorough analysis. The exact functioning of a data warehouse is best understood considering the broader concept of the enterprise data pipeline.

Definition D4.9: A **data pipeline** is the combined sequence of software processes and tools used to transfer and transform data between one or more source systems and a target repository. Along its way, the data are structured in such a way that its analysis is facilitated such that new insights can readily be obtained.

Every data pipeline starts with a source, usually operational databases which provide the backbone of most organizations are the primary sources of raw information (see Figure 4.6). Next comes the extraction of the useful data from these sources. Note that this is often a complicated process as data must come from software applications that might be developed by different vendors, coming in various

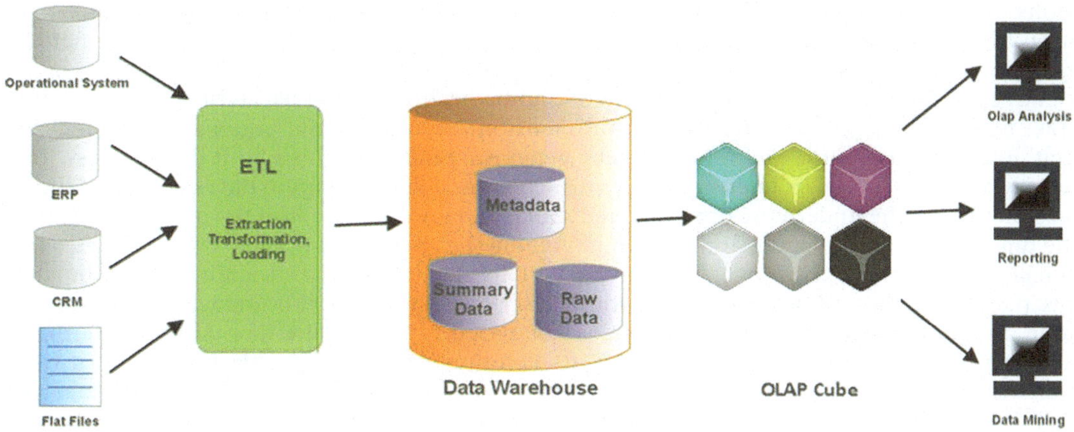

Figure 4.6 Schematic representation of the data pipeline used for data warehouses. Obtained from Panoply (2023).

formats, hosted on isolated computational infrastructure, and operated by different stakeholders. After this, the raw data need to be refined by transforming it into a result of interest in an agreed upon standard ready for efficient querying and analysis. For example, while business decision makers require detailed data to respond to changing questions, they are almost never interested in individual entries stored in the raw datasets. Instead, numerous operational database entries will be aggregated into summary statistics, such as medium-to-long-term performance results, to define high-level answers. Once the raw data are transformed, the resulting information is loaded into the data warehouse. Note that this extract-transform-load (ETL) process is typically configured in a batch-like fashion, for which a certain amount of raw data must be acquired before execution. However, streaming ETL processes can be configured to do this as soon as new data are available. The latter is useful when interested in analysing massive inflows of data in real time (Figure 4.6).

Like relational databases, the structure of a data warehouse must be designed in such a way that it serves its usage optimally. Whereas operational systems built upon relational database technology emphasize on data integrity and ease of maintenance (i.e., normalized databases without redundancies), data warehouses aim at maximizing the business information that can be extracted with minimal querying effort. In other words, data warehouses are designed to make data accessible and simple for its business users. This often results in data warehouses that purposely contain redundant information which simplifies querying, improves performance, and is resilient to change (i.e., denormalized databases).

The methodology used for the design of the underlying data warehouse structure is referred to as dimensional modelling. Depending on the physical implementation of the dimensional model, a distinction can be made between the star schema and the online analytical processing (OLAP) cube. The former is the implementation in a general-purpose relational database, whereas the latter concept is used when the data warehouse is implemented in a purposely fit multidimensional database (MDB). Although different in their implementation, the underlying logic is the same (Kimball & Ross, 2013).

Dimensional models are built around the concept of a fact table. A fact table contains performance measurements or business metrics that are generated as part of organizations ongoing operational business processes, that is, summary statistics. The data on each row of this table is typically a continuously valued numbers representing either a transaction (e.g., items sold per day) or periodic snapshot (e.g., daily average sensor value). As business processes continuously generate such metrics, the underlying ETL processes ensure that the fact table is constantly growing with the most recent data. It is therefore not uncommon for an organization to have fact tables with millions of rows. Associated with a fact table are one- or multi-dimensional tables. Dimensions contain the context in which the measurement in the fact table was generated, that is, metadata. These dimensions are aimed describing the who, what, where, when, how, and why of a fact. To this end, each dimension contains a variety of characterizing attributes that capture part of this context (see Chapter 3). The attributes of a dimension are used by the organization's analysts to search, filter, or classify the fact table, thereby generating improved insight in the business. In the same way as the fact tables, dimension tables are populated and kept up to date because of the ETL process acting on the underlying operational databases. However, because the metadata contained in the dimension is rather static and limited in volume (i.e., a handful of table rows), updates to the dimension tables do not occur on a frequent basis. This also justifies denormalization of dimension tables (i.e., undo the effects of normalization described above) as the impact of their minimal volume on query performance is negligible as compared to that of the fact tables. Note that the descriptive nature of dimensions makes that its columns can be very wide with a lot of textual content. Eventually, the association between fact and dimensions is established using primary key–foreign key relationships (i.e., like data storage in relational databases). The storage of metadata in a data warehouse is not only restricted to the business context in which the data were collected (e.g., type of equipment used for a measurement, operator responsible for the data collection, etc.) but can also include technicalities about the physical technologies used to store and manage data (e.g., structure of the data warehouse, data lineage, information on algorithms for summarization, etc.). In the field of data modelling, a separation is therefore made between business metadata and technical metadata. In fact, some big data stores even add meta-metadata which is used to understand and manage the metadata itself (Hay, 2006). In this report, the emphasis is on business metadata, more specifically the kind that helps understanding sensor measurement data. Two examples of star-like schemas can be found in Section 4.5.5. In Example E4.1, the *signal* table acts as the facts table, whereas in Example 2 the dimension tables are linked to the *Metadata* table which is directly used by the *values* table.

Because of the tabular nature of data warehouses, the type of data it accepts is restricted. Originally only highly structured data (i.e., data that can be structured directly in rows and columns) was applicable to data warehouses. Over the last decades database technology has greatly improved, and semi-structured data formats like JSON and XML are now also accepted and available for querying. This increased the flexibility of modern data warehouses as the dimensional attributes (i.e., column headers of the dimension tables) do not necessarily need to be modified when additional data fields are included (Epstein & Roberts, 2022). Despite its increased flexibility, it is important to understand that the fundamental dimensions of a data warehouse are created based on pre-defined business questions. If some data are considered irrelevant during design, it might not be possible to simply add it later without significantly revising the data warehouse. Such a revision, when applied, is known as *schema-on-write*.

The final stage of a data pipeline is where the user interacts with the data for analysis and visualization. As mentioned before, most data warehouses are based on relational database technology, therefore they can be queried directly using SQL, or by using a general programming language (e.g., C++, Java, Python, and R) that is able to construct and execute SQL queries. Yet, high-level decision makers might not have the time or ability to translate their business questions into complex SQL constructs. Nowadays, a great number of intuitive, flexible, and easy-to-use business intelligence tools is available that seamlessly integrate with data warehouses without the user having to write a single line of code. Given the pre-defined and structured nature of a data warehouse, the focus of these tools is typically on descriptive analytics (i.e., what happened), to answer recurring business intelligence questions in the form of reports and dashboards (comprehensive historic data analysis and real-time data interpretation, respectively). While it is possible to perform ad-hoc exploratory data analysis on a data warehouse, the technology is best suited for running routine and production ready queries to gain insight on information pertaining to a previously established topic.

There exist two main reasons for this limitation in its usage: (1) querying for large subsets of data for exploratory analysis is very demanding for a data warehouse and might end-up taking significant amounts of time, (2) data warehouses typically do not store raw or high-resolution data. Because of this, data warehouses are mostly the playground of business analysts rather than data scientists. This is because the former type of users typically performs routine business analysis using pre-defined types of questions, whereas data scientists tend to operate iteratively without pre-defined recipes (Gopalan, 2022).

4.4.7 Data lake

The concept of a data lake was introduced in the early 2010s as a solution to the growing number of challenges faced when administrating and processing the data of massive enterprise data warehouses. At the time, data collection and processing requirements ramped up because of the growing digital footprint of global businesses. Simultaneously, the systematic analysis of large volumes of semi-/unstructured data, stored as binary files or located in NoSQL databases, took off (e.g., audio, images, videos, document collections, social media posts, click streams, log files). Although data warehouses are very successful at centralizing structured operational data for business intelligence, they were not able to scale affordably for these big data applications. Moreover, data warehouses did not natively support semi-/unstructured data storage and analysis, nor were they designed to provide rapid answers to ad-hoc queries fired by curious data scientists.

The answer to this technological gap came by introducing yet another data storage paradigm, coined the data lake. Instead of providing mostly business analysts with pre-digested data following a carefully crafted dimensional model, the data lake stores raw data (structured and unstructured) in its original form and format without enforcing an overarching data structure. The result is that various users with different needs can find answers in the data on a self-service basis without having to first understand the intricacies of a previously defined data model. Or as the original inventor of the term elegantly describes: 'If you think of a data warehouse as a store of bottled water – cleansed, packaged and structured for easy consumption – the data lake is a large body of water in a more natural state. The contents of the data lake stream in from a source to fill the lake, and various users of the lake can come to examine, dive in, or take samples' (Dixon, 2014).

Since data are stored as-is in a data lake, there is no need to justify storage upfront. All data can be made available, even when there is no direct use case in-mind. This makes the data lake suited for much more than operational reporting and business analysis. A prime example is exploratory data analysis, which often takes-off without a completely defined goal in mind. In this case, a data scientist or process engineer can reveal potential insights by incrementally querying the data lake (using an SQL-like language) without knowing upfront what data might end up being useful. Something that was difficult to achieve in data warehouse. On the downside, having massive amounts of data automatically being dumped into a storage location can quickly become unwieldy if not managed correctly, that is, a data swamp. To govern storage and consumption by specific users, data lakes are typically composed of an entire ecosystem of components besides the actual data store.

Data integration services are an integral part of data lakes. They allow for the orchestrated ingestion of raw data coming from different isolated sources. In contrast to the ETL process of data warehouses, a slightly different approach is often used when considering data lakes. Since the data lake accepts any kind of structured or unstructured data in their natural form, there is no need to transform the data before it is loaded. Once the end-user is ready to start analysing part of the data, they can be transformed as needed. The biggest advantage of such an extract-load-transform (ELT) procedure is the flexibility and speed it offers at a lower cost, as there is no need to transform the data into a structure before it can be used for analysis. A concept referred to as *schema-on-read*.

While the data consumer of a data lake perceives a single central repository of data, the reality is that these data are distributed amongst many underlying (cloud) computers that efficiently work together. Logically centralized, yet physically decentralized. By using such an architecture, the time needed to process massive amounts of data can significantly be reduced because of distributed and parallel work. Modern data lakes are therefore built on cheap object storage in the cloud (e.g., AWS Simple Storage Service (S3), Azure Blob Storage, Google Cloud Storage) combined with analytics engines for large-scale data processing, analytics and/or data science (e.g., Apache Spark, Upsolver, Presto).

Because of the apparent lawless nature of the data lake, users might end up with a system that is difficult to navigate. This is where a data catalogue comes in. By providing intuitive tools to tag datasets and attributes with additional metadata, a catalogue of all data stored in the lake can be constructed such that non-expert users can quickly search for what they need using conventional business terminology. This is very similar to the usage of a catalogue in a library. Note that cataloguing data requires a common understanding of terminology and naming, which is where business glossaries, taxonomies, ontologies, and naming conventions come in handy. Once a common language is available, cataloguing can be performed manually with the help of data stewards or automatically with the help of artificial intelligence (Gorelik, 2019).

Another way to bring order in the chaos of the data lake, is by constructing layers. These are zones that contain data that meet certain requirements. For example: a bronze layer that contains all the originally ingested immutable data and is not accessible to most users; a silver layer consisting of a subset of the raw data that has been verified but remains in original format useful for technical data scientists and data engineers; a golden layer were the most refined and valuable data lives and is accessible to business analysts and high-level business decision makers. Layering the data not only refines the information content, but it also increasingly adds structure to the data. This is why mid-level layers in data lakes often use relational concepts to store the data, whereas top-level layers contain consumption ready datasets following de-normalized and read-optimized data models (Simon, 2021).

Like data warehouses, data lakes can be coupled with a plethora of tools for analysis and visualization. Because of the unstructured nature of data lakes, dedicated tools for big data exploration and analytics are of particular interest. This includes tools for the development, deployment, and tracking of artificial intelligence models. For example, tools for machine learning operations (MLOps) aimed at version controlling code and data used to train data-driven models. Hence, with the adoption of data lakes, the focus shifted from solely descriptive (i.e., what happened) and diagnostic analysis (i.e., why did it happen) of business data, to more predictive (i.e., what could happen) and ultimately prescriptive analytics (i.e., what should happen) on big data.

Pros	Cons
• Business data and metadata are centrally stored in raw format, providing unlimited flexibility to future usage. • Low storage and maintenance cost. • All data formats are accepted, that is, structured/semi-structured/un-structured. • Data collectors for integration are readily available following standard database protocols • Available as managed service (PaaS) on public clouds with optimized ecosystem integrations for data ingestion, processing, and analytics. • Supports both well-planned business intelligence and ad-hoc data science.	• Centralized data storage can result in monoliths which are overly complex and therefore not easily maintained nor queried. • Data quality is not enforced because there is no schema. • Does not support data transactions. • Poor concurrency control. • Accurate querying is complicated by the lack of consistent data structure. • Non-standard data storage formats might need manual reformatting.

4.4.8 Industry examples

The remainder of this section is devoted to two examples that illustrate the concept of a data pipeline and how raw data and their metadata collected in operational databases find their way up the business decision hierarchy for their transformation in valuable insights and actions.

As indicated in Section 4.3 on the data life cycle and in Section 4.4.6 on data warehouses, data flows through different stages in its lifecycle. The flow from one step to the other can be thought of as following the flow through a pipeline. Therrien *et al.* (2020) presents the different conceptual stages of such a data pipeline specifically for WRRFs. Note that although it is highly desirable to have an autonomous data pipeline based on automatic data flow, routine manual manipulations are not uncommon.

4.4.8.1 Example 1: pilEAUte wastewater treatment pilot plant

pil*EAU*te is a highly instrumented and controlled pilot wastewater treatment plant for both carbon and nitrogen removal operated by model*EAU*, Université Laval, QC, Canada. It consists of two parallel biological treatment lanes to compare and optimize control strategies for nutrient removal. The pilot was configured to represent mainstream nitrification/pre-denitrification plant design. However, it is prepared for possible extension towards enhanced biological P-removal and addition of carriers to study biofilm-enhanced treatment.

There exist multiple data sources in the pilot plant that can be used to assist operational decision making (Figure 4.7). Signals such as sensor measurement values, actuator states, setpoint values, and

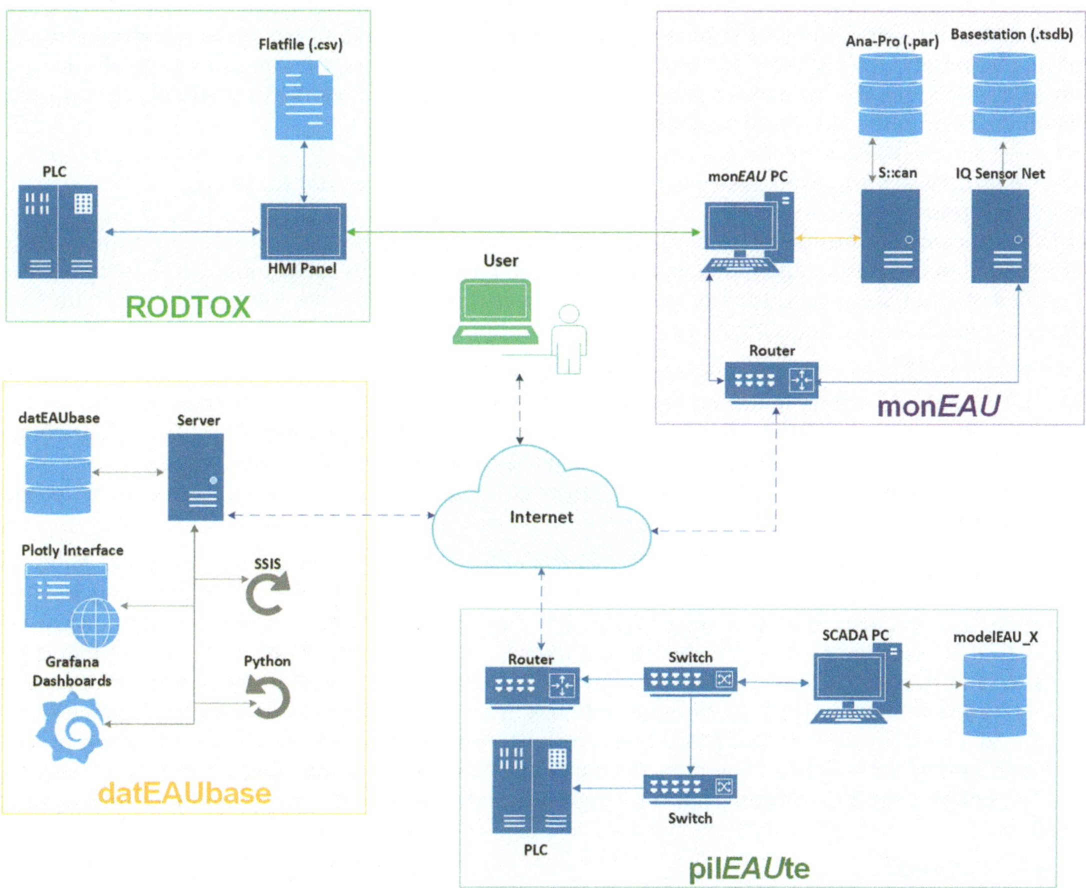

Figure 4.7 Data flow diagram of the pilEAUte wastewater treatment pilot.

so on are obtained directly from the process control and automation system (*PLC/SCADA*). These signals are stored as time series in a relational database without any contextual information by the signal TAG name (*modelEAU_X*). Additional stand-alone sensor systems are also made use of for the detailed characterization of influent and effluent composition (*monEAU* online sensors and *RODTOX* at-line batch respirometry). These systems are, however, not integrated physically in the core process control and automation structure. Instead, the data are stored on the local PCs in flat-file databases (.par and.csv file formats) and a dedicated time-series database (.tsdb file). Moreover, logbooks for maintenance activities (e.g., calibration, validation, cleaning, annotation) also contain valuable operator observations. The latter are stored as flat files in Excel spreadsheets.

On their own, each of the data sources of the pil*EAU*te plant are isolated data silos which complicates comprehensive data analysis. To solve this issue, most data sources of the pil*EAU*te were integrated over secure internet tunnels (*VPN*) into a centrally available data store which acts like a process historian (dat*EAU*base). Integration services (*MS SQL Server Integration Solutions – SSIS* and custom *python* scripts) automatically extract new data on a periodic basis, transform them in the expected format (e.g., SI engineering units) and load them in the desired database (*ETL process*). To assist the operator, user-friendly interfaces were also developed to intuitively visualize sensor data and metadata (*plotly interface* and *Grafana dashboards*). In the back, the central data store runs a relational database (*MS SQL Server*) according to a purposely designed data model (Figure 4.2). Note that the data model makes extensive use of descriptive metadata (related to instrumentation, locations, contact persons, etc.) to structure all data. In that aspect it operates very similar to a business data warehouse. As a result, operators have all information centrally available for improved decision making on the short and medium term.

4.4.8.2 Example 2: simplified process data flow diagram of Hampton Roads Sanitation District (HRSD) facilities

HRSD is a wastewater utility which manages multiple large WRRFs in southeastern Virginia, United States. Data flow, structure, and management across all large facilities are standardized as represented in Figure 4.8. Data are produced and managed to meet the requirements for permit compliance and operations monitoring. Data relevant to process operation are produced by four major sources:

(1) Plant operators (e.g., manual sludge blanket measurement).
(2) Electrical and instrumentation equipment (E&I) (e.g., pump rotational speed).
(3) Distributed control system (DCS) in charge of process automation and supervisory control (e.g., process command or state, setpoint values provided by plant operators).
(4) Laboratory analysis (e.g., final effluent ammonia concentration obtained from composite/grab water samples).

Process data are subsequently stored in six different locations depending on its purpose. Note that the HRSD network is partitioned between a more secure control side and a general usage business side. HRSD's IT security policy requires data that can only be transferred out of the control side network to the business side (i.e., unidirectional data flow) to protect operations from possible cyber-attacks and network outages. This means that modifications to the automatic process operation can only be made from within the control network. This policy was found invaluable after HRSD experienced a ransomware attack in 2020 which shuttered the business network but did not interfere with plant operational control. The data infrastructure thereby satisfies HRSD's needs for security, permit compliance and operational monitoring and analysis for both plant operators and business management.

Control network

- Local control historian – short-term process data storage for critical monitoring points (up to 30 days).

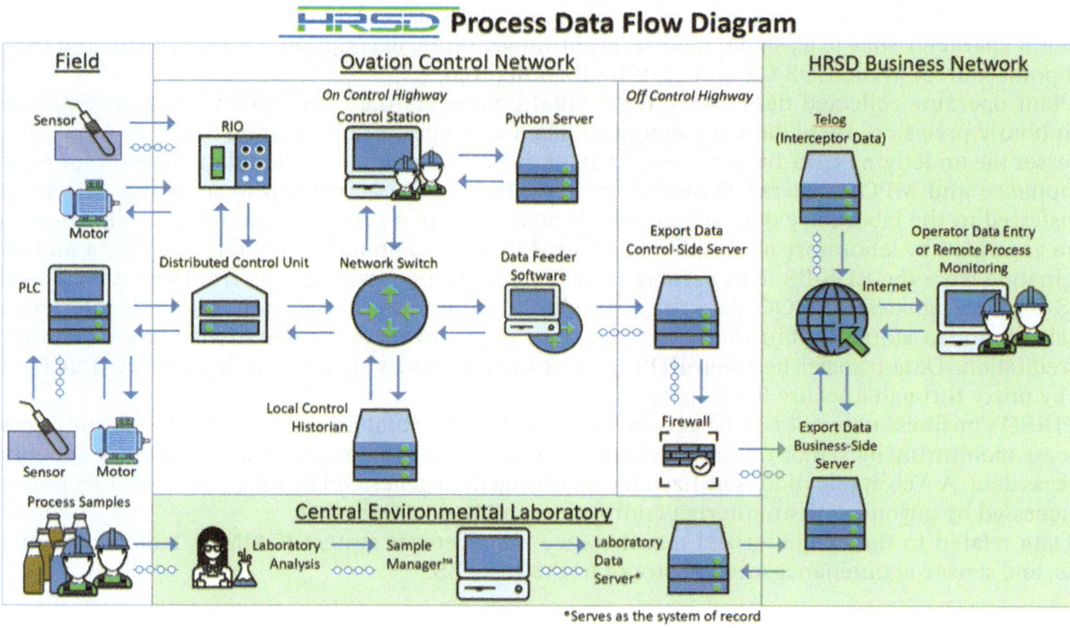

Figure 4.8 Data flow diagram of the HRSD wastewater utility.

- Export data control-side server (EDS-CS) – clone of DCS tags and graphics. Includes all E&I field-generated data.
- Python server – at two of HRSD's facilities, OPC communication and custom Python scripts are configured to connect to specific tags for research purposes (i.e., data-driven process operations and digital twins).

Business network

- Export data business-side server (EDS-BS) – clone of DCS tags and graphics. Includes all E&I field-generated data, some laboratory-generated data, and plant operator-generated data. Interface allows for process monitoring via time-series trending and data export.
- Laboratory data server – all data generated by the laboratory analysts and relevant to permit compliance as generated by the DCS or plant operators. Serves as system of record for permit compliance and the generation of monthly plant operating reports (MPOR), that is, spreadsheet for high-level operations monitoring.
- Telog (interceptor) – data generated by pump stations, tides, or rainfall.

All data are stored as time series. Data are continuously communicating from the E&I/DCS system to the local historian using ethernet-based TCP/IP protocol (if designated) or to the EDS-CS before being passed to the EDS-BS. Depending on the signal the associated timestamp precision can range from seconds (e.g., E&I and DCS) to days (e.g., plant operators). To minimize excessive data storage from data generated at high-frequency, signals are configured to only store a new value after a certain change in value has been observed (i.e., Box-Car compression, Section 4.4.5). The threshold has been standardized to 1% of the range for analog signals. For example, a disssolved oxygen signal designated in a range from 0 to 10 mg/L will only be stored if the measurement has changed by at least 0.1 mg/L as compared to the previous record. For digital signals only the change in state is recorded. Both the historian and the EDS servers are based on relational databases. Additional metadata can be stored as point characteristics (e.g., signal quality, point range, units, description, alarm priority and limits, and point source) in the EDS-CS and EDS-BS if configured.

Plant operator collected data and short textual comments (i.e., annotations) have a timestamp with hourly precision, these data are submitted via a web application running in the browser which accesses the underlying EDS-BS database. At an early hour, the previous days data relevant for permit compliance and MPOR generation stored in the EDS-BS are compressed to a flat file (.CSV) and transferred to the laboratory data server, which makes use of a relational database in the back end. Data generated by laboratory analysis or calculated using a combination of laboratory data and data originating from the EDS-BS (e.g., aerobic solids retention time) will also be transferred back to the EDS-BS if designated. QA/QC data (e.g., standards, calibration dates, matrix spikes, interference checks) are also stored within the laboratory data server to meet the requirements for laboratory accreditation. Data transfer between EDS-BS and the laboratory data server is completed using flat file by proxy through a secure file transfer site.

HRSD's business network can be accessed externally by virtual private network (VPN) for remote process monitoring by senior operations staff or access to sample manager for laboratory generated process data. A web application is utilized for recording data generated by the plant operators and can be accessed by anyone with an internet connection and login credentials.

Data related to the computerized maintenance management system (CMMS), biological activity tests, and sensor maintenance, are not stored to the EDS-BS.

4.5 UNDERSTANDING RELATIONAL DATABASES

Relational databases are immensely popular for the structured storage of business data in the back end of various operational software applications. As discussed before, relational databases can also be used for the storage of time-series data and metadata stemming from process instrumentation. Moreover, the highly efficient time-series storage and retrieval aspects of data historians are often built on top of a relational technology. For this reason, the structures as well as the fundamental concepts of relational databases are discussed here in more detail. Special attention is given to the inclusion of metadata and data quality descriptors.

4.5.1 Definition

A relational database is based on the relational model of data. In other words, it is a database whose logical structure is made up of a collection of relations. In this context, a relation is understood as a table with columns (a.k.a. attributes or fields) and rows (a.k.a. tuples or records) (see Figure 4.9).

A column is the smallest organizational structure of a relational database and represents the different aspects that define the records in a table. The name of a column within the table is unique but it can be the same in other tables in the same relational database. In addition, the values in a

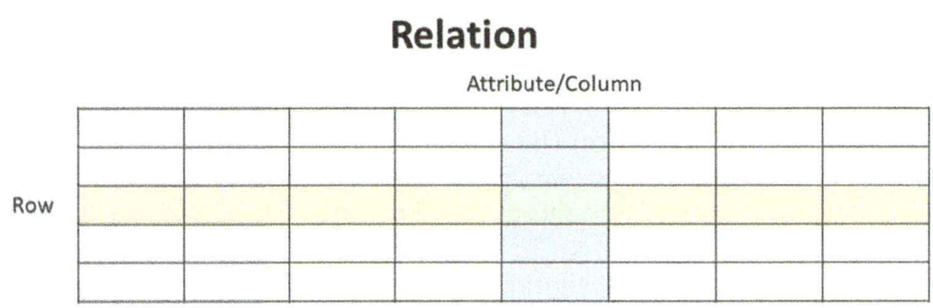

Figure 4.9 A relation (or table) in a relational database.

column must have the same data type and can be viewed in any order without affecting the meaning of the data.

Regarding the rows, each record is a unique instance of whatever type of data (e.g., signal value, type of sensor, analytical method, data post-processing method, etc.). In other words, there are no duplicate rows in a relation. Each row is identified with a unique value, named primary key, containing one or a combination of columns. Ensuring an exclusive primary key, makes each row unique. A concept referred to as entity integrity. As presented in Table 4.7, each row corresponds to a value measured at one time. Automatically, a key is given to each row as an identification, and it makes it unique.

The general structure of a relational database contains two types of tables: primary tables and lookup tables. Primary tables are the core tables in a database. It is possible to identify them because they contain the most relationships to other tables. A primary table typically contains a primary key which is an integer number adjusted automatically (i.e., to ensure its uniqueness) and at least one text field containing or grouping the most important information of the database.

Lookup tables are meant to reference tables built mainly with a primary key and text fields containing categories, a category name, and a description of this category. This type of tables generally has a single relationship with the primary table or to another lookup table. Once the relationship is established, the primary key of a lookup table becomes a foreign key on the primary or the other lookup table.

In addition, the link between primary or foreign keys from different tables can be one-to-one, one-to-many, or many-to-many (Figure 4.10). Consider a table describing specific instruments (i.e.,

Table 4.7 Example of relation.

Key	Timestamp	Value	Parameter	Measurement Unit	Sensor	Location
1	03:45:10	7.123	pH	–	PH18065	Primary clarifier outlet
2	03:45:15	7.324	pH	–	PH18065	Primary clarifier outlet
3	03:45:20	7.163	pH	–	PH18065	Primary clarifier outlet

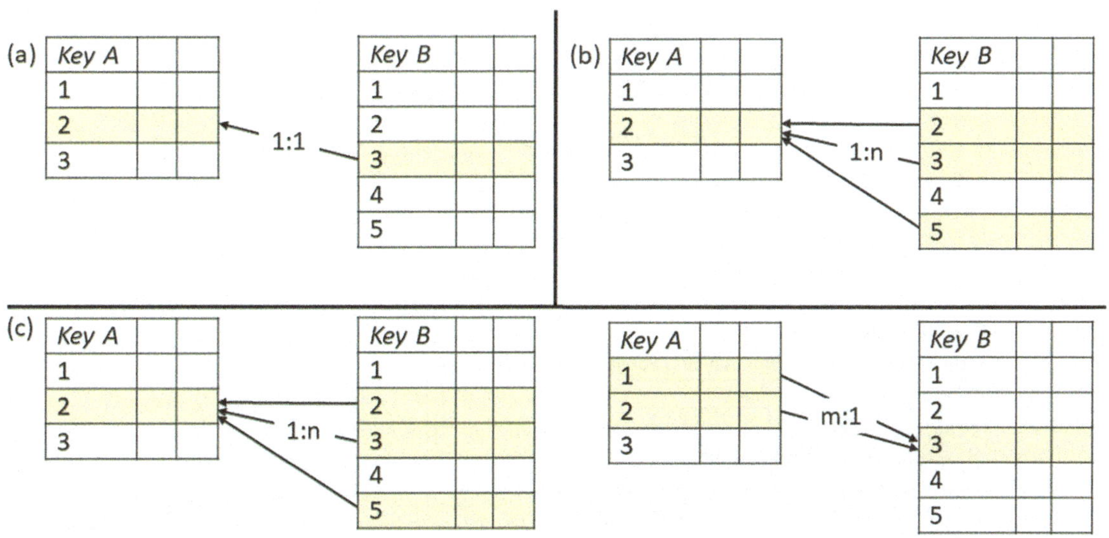

Figure 4.10 Different types of relationships found in a relational database: (a) one-to-one, (b) one-to-many and (c) many-to-many relationships.

rows) and the key characteristics (i.e., columns), and a table describing the analysed water samples (i.e., rows) and associated results (i.e., columns). A common one-to-one relation occurs when a single lab-analytical result is obtained from a single water sample, due to the destructive nature of the analytical method. Similarly, a one-to-many relationship can be that a concentration in a water sample is measured multiple times with the same instrument. Considering the many-to-many relationship, a relevant example would be for example that one instrument can measure multiple parameters, but the same parameter can also be measured by different instruments.

4.5.2 The most common data relationships

As presented earlier (Chapter 3 and Section 4.2), signals used to monitor and control treatment processes should be documented, that is, contextualized, to make them useful and readily available for further analysis and interpretation of different purposes. In the field of wastewater treatment, this documentation is understood as providing information about when, where, and how a data point was collected, who collected it, for what reason it was collected, and what is the quality of that data point. Below the main relationships used to tie metadata to the signal data are described in detail.

4.5.2.1 [data point] produced by [device]

This is one of the most established relationships in the wastewater sector. Linking the collected data points together with the device that produced them. This allows to identify and provide information on how a data point was obtained. In Figure 4.11, a simplified example is provided on what type of information can be stored for a device and for data points in tables, and how both tables can be linked.

4.5.2.2 [data point] produced at [location]

Another important relationship can be established to capture where the data points were produced. This piece of information provides more background on the precise physical location and therefore environment that data points have been obtained. Figure 4.12 shows a simple example of what type of location table and possible information to describe this specific place where a data point is collected.

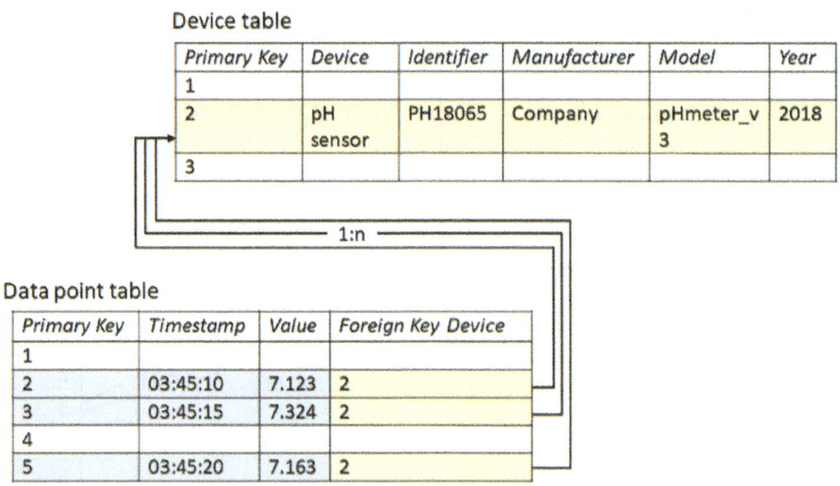

Figure 4.11 Scheme of how a data point can be linked with information of the device that has produced it.

Location table

Primary Key	*Site name*	*Sampling location*	*Latitude*	*Longitude*	*Address*
1					
2					
3	Primary settler 3	Outlet channel	42.460367	1.367154	2324 Water St, Water City

1:n

Data point table

Primary Key	*Timestamp*	*Value*	*Foreign Key Location*
1			
2	03:45:10	7.123	3
3	03:45:15	7.324	3
4			
5	03:45:20	7.163	3

Figure 4.12 Example of information included in location and data point tables.

4.5.2.3 [quality] based on [data points]

Other relevant information that can be associated with data points, is the quality level. This can be any information related to the techniques or methods implemented for a quality check of the measured value. This last relationship is described in more detail in Section 4.5.3.

4.5.3 Relationships describing data quality

As presented earlier (Chapters 2 and 3, and Section 4.5.2), providing context related to data modification and quality checks increases the reliability, reproducibility, and robustness the data. This will affect the trust the data analyst will have in the dataset. Information on data quality can be provided directly by the devices that produces the data points or by the end-user itself. Quality checks can be included in the same table as the data point table (e.g., see Figure 4.13) or separated in an independent table if more extensive descriptions are needed (i.e., multiple fields and combinations) (e.g., Figure 4.14). Both examples, Figures 4.13 and 4.14, show the possibilities on tagging raw and processed data after an average moving filter with noise removed and without insignificant digits. Details related to post-processing and/or validation methods applied to a data point (e.g., version of the data point or simply a quality flag) can be part of the relational structure. By keeping track of the origin of data, it is easier to understand and consider the effect of certain transformations after they were applied.

Definition D4.10: **Data provenance** or **data lineage** refers to all metadata describing the processes that influenced the data set of interest, thereby providing a historical record of the data and their origins. It includes context on data origin, transformations, and storage. As a result, additional transparency is provided which simplifies root-cause analysis in the data analytical process.

For example, a bias was detected in a pH signal and automatically compensated for. The database, however, only stores the corrected values. This complicates analysis later. If further details on the correction method are available, it is possible to reconstruct an unbiased data series and study effects of the modifications applied after the original raw data were collected.

The complexity of the desired information to be included in a relational database can be simplified using a hierarchical approach. In other words, any information related to data quality can be stored as a set of tables into a tree-like structure.

Analytical methods table

Primary Key	*Method Name*	*Method type*	*Code version*	*Description*
1	Raw data	Measurement	N/A	N/A
2	Moving average	Univariate	3.2.1	Specific details of the code version
3	Principal component analysis	Multivariate	0.1.9	Specific details of the code version

1:n

Data point table

Primary Key	*Timestamp*	*Value*	*Foreign Key*
1	03:45:10	7.123	1
2	03:45:15	7.324	1
3	03:45:10	7.12	2
4	03:45:15	7.18	2
5	03:45:20	7.158	1
6	03:45:20	7.16	2

Figure 4.13 Example of analytical methods table and tagged to the data point table.

Analytical methods table

Primary Key	*Method Name*	*Method type*	*Code version*	*Description*
1	Raw data	Measurement	N/A	N/A
2	Moving average	Univariate	3.2.1	Specific details of the code version
3	Principal component analysis	Multivariate	0.1.9	Specific details of the code version

1:n

1:n

Raw data point table (link to the key 1 in analytical methods)

Primary Key	*Timestamp*	*Value*	*Foreign Key*
1	03:45:10	7.123	1
2	03:45:15	7.324	1
3	03:45:20	7.158	1

Data point table after moving average (link to the key 2 in analytical methods)

Primary Key	*Timestamp*	*Value*	*Foreign Key*
1	03:45:10	7.12	2
2	03:45:15	7.18	2
3	03:45:20	7.16	2

Figure 4.14 Example of analytical methods table and separate tables depending on the data version and analytical method.

4.5.4 Defining a relational structure

To successfully build a relational database it must be considered that the collected raw data have a highly variable format; the database needs to be flexible and adaptable since it is growing continuously, and monitoring programs progress and vary overtime; and personnel collecting and managing the data also varies overtime (Camhy *et al.*, 2012; Horsburgh *et al.*, 2008). Thus, a relational database should offer:

- Keep data accurate by grouping multiple database operations that will occur to complete a database update.
- Consistency to maintain the data points within the defined rules.
- Isolation for each transaction and keep it separate and independent on others.
- Durability to recover data from a failed transaction ensuring that data changes are permanent.

According to Harrington (2016), there are several steps to follow during the structure definition of the database:

(1) Define the objective of the relational database for water quality data.
(2) List the data to be stored (i.e., information as presented in Chapter 3). For example, related to a device, it can be listed as the name of the device, the manufacturer, the year of fabrication, the measurement principle, and so on.
(3) Map the links between the data to be stored. For example, a sensor is linked to a location and a parameter, or a value is linked to a sensor, a parameter, a location, and an analytical method.
(4) Feasibility assessment of implementing a new or a modified database.
(5) Design the database according to the links mapped at point 3.
(6) Develop and test the database.
(7) Implement the database and link the SCADA system to it.
(8) Evaluate the database.

As examples, hydrological and water quality databases have been published. For example, the Consortium of Universities for the Advancement of Hydrologic Science, Inc (CUAHSI) has presented the observations data model (ODM) for hydrological data storage (Horsburgh *et al.*, 2008), or the Water Quality Portal by the US Environmental Protection Agency (EPA) for water quality data (EPA, 2023). However, information storage has been limited on providing information on the location or the device of the measurement rather than on the data quality. Recently, new water quality relational databases have been presented with emphasis on metadata promoting data quality information, for example, the WaterHubDataPool (WHDP) by Eawag (ETH Zurich, 2019) and dat*EAU*base by model*EAU* research group from Université Laval (Plana *et al.*, 2019).

4.5.5 Industry examples

4.5.5.1 Example E4.1: WaterHubDataPool

The WHDP is a database designed to collect and store data produced within the WaterHub at Eawag. The WaterHub is an experimental facility to study the collection, treatment, and distribution of water resources within the NEST building. The facility handles a set of source-separated water streams, including urine, light-grey water (e.g., rain), dark-grey water (e.g., bathwater), and black-water (toilet flush water). The WaterHub hosts a changing set of technologies to treat these water streams and recover usable water or other resources from them. The WHDP was designed to collect and store data from these processes in a way that is simultaneously comprehensive, future-proof, and easy-to-use. In the following, the data model used in the relational database will be discussed in more detail. This data model is visualized in Figure 4.15. Additional information about this database is publicly available at ETH Zurich (2019).

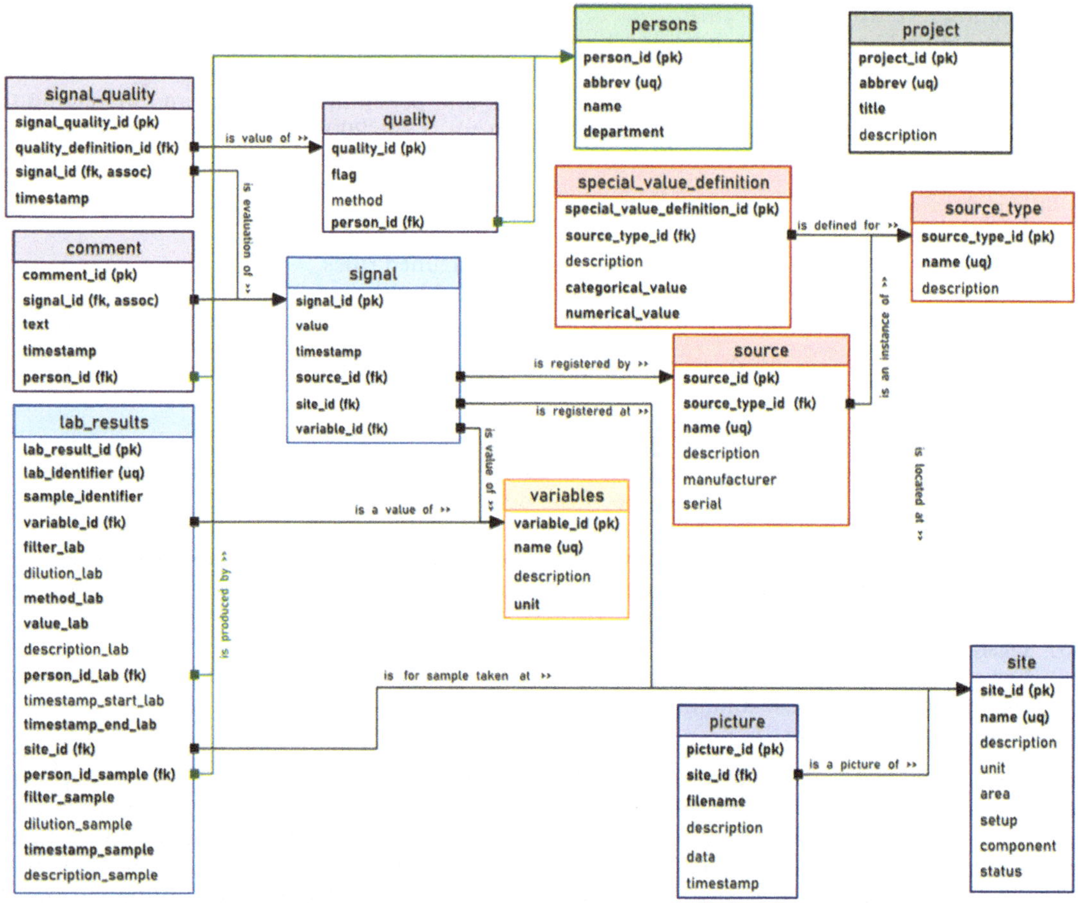

Figure 4.15 Example E4.1: Layout of a relational database for the WaterHubDataPool (WHDP) at Eawag.

The database layout includes elements for high-frequency signals (i.e., actuators and sensors) as well as low-frequency signals (i.e., laboratory analysis). We discuss the most important tables relevant for storage of high-frequency signals and corresponding metadata. Each table includes one primary key (*pk*) which is assigned automatically and usually not visible to the data user. These are not discussed in further detail.

The measurement records are stored in the *signal* table, which is located centrally in the database layout. This table includes all actuator and sensor data records. To complete a data record, it is necessary to know the data point (stored as *value*), the *timestamp*, the data-producing device (stored as *source_id*), the location of the measured variable (stored as *site_id*) and the variable being measured (stored as *variable_id*). The entries for *source_id, site_id, variable_id* are foreign keys (*fk*) and are used to point to entries in other tables, namely the tables *source*, *site*, and *variables*. These tables are mainly used to describe the data-generating process (see Chapter 3, Section 3.2.2) and are discussed next:

- Table *source*: Each entry in this table describes one actuator or sensor. An actuator or sensor is described with the device type (stored as *source_type_id*), a unique *name*, a free-text *description*, the *manufacturer*, and a serial number (*serial*). The *source_type_id* points to the table *source_type*, which lists and describes all known device types with a *name* and a *description*. By means of this structure, information that is shared between devices of the same type only needs to be entered once. For example, there may be multiple pH electrodes in use that share the same information about make and design.

- Table *site:* each entry in this table describes a measurement location with a unique *name*, a *description* of the location, a room (stored as *unit*), an area within the room (stored as *area*), an experimental setup (stored as *setup*), a *component* of the setup, and an indicator of the site status (e.g., planned, installed, active, inactive).
- Table *variables*: Each entry in this table describes one variable with a unique *name*, *description*, and *unit*.

The table *picture* is used to link to filenames of pictures stored elsewhere to specific entries in the *site* table. The table *special_value_definition* is used to document values for a data point that have a unique meaning for a given device type. This kind of information is used to automatically infer and document missing and out-of-range data (see Chapter 2, Section 2.4.1.3 and Chapter 3, Sections 3.2.2.6 and 3.2.2.7).

The WHDP layout provides a flexible way to store metadata describing signal quality and contextual information through the *comment* table. This enables the storage of free-form text to annotate the data. A single comment is described as the unique data record (stored as *signal_id*), a free-form comment (stored as *text*), the time the comment was written (stored as *timestamp*), and the person who provided the comment (stored as *person_id*). The values for *signal_id* point to the entry in the *signal* table. Similarly, *person_id* points to the table *persons*, which contains information about the human experts available to annotate the data.

The WHDP layout also provides a structured way to store context as metadata. It is mainly aimed at storing metadata describing signal quality (see Chapter 3, Sections 3.3 and 3.4 for relevant metadata). To this end, the table *quality* lists all labels available to describe known anomaly and fault types. Each label is described with a name (e.g., *outlier*, stored as *flag*), a name for the method used to create this label (e.g., *EWMA*, *visual inspection*, stored as *method*), and the person who provided the label (stored as *person_id*). A dedicated table *signal quality* is used to link the *signal* and *quality* tables. This enables assigning a single label in *quality* to multiple data records in the *signal* table while allowing multiple labels in *quality* to be linked to the same data record in the *signal* table.

The remaining tables (*lab_results*, *project*) are not relevant to describe signal data and are therefore not discussed in detail here.

4.5.5.2 Example E4.2: dateEAUbase

The dat*EAU*base was developed to store all data collected as part of research projects by the model*EAU* research group located at Université Laval, Québec, Canada. The main objective of this database is to centralize and store data from on-line measurements and laboratory analysis results coming from measurement campaigns conducted at various water bodies (e.g., rivers, sewer systems, WRRFs, etc.) (Plana *et al.*, 2019). The database was also designed in such a way that it intuitively stores relevant metadata needed to document and track the stored data along its lifecycle (see also Figure 4.2 and Section 4.4.8.1).

The database structure presented in Figure 4.16, can be divided in two main parts:

(1) The core tables comprised of the *Values*, *Metadata* and *Comment* tables. The *Values* table can be thought of as a fact table in a data warehouse, as it contains on-line measurements, results of laboratory analysis, timestamps, and the corresponding foreign key to the descriptive *Metadata* table. The *Metadata* table groups the foreign key numbers of the rest of the dimensions, thereby acting as an intermediate that stores all existing dimensional configurations. Annotations specific to one or more values can also be added (e.g., sensor failure, cleaning period, outlier, etc.) using the *Comment* table.

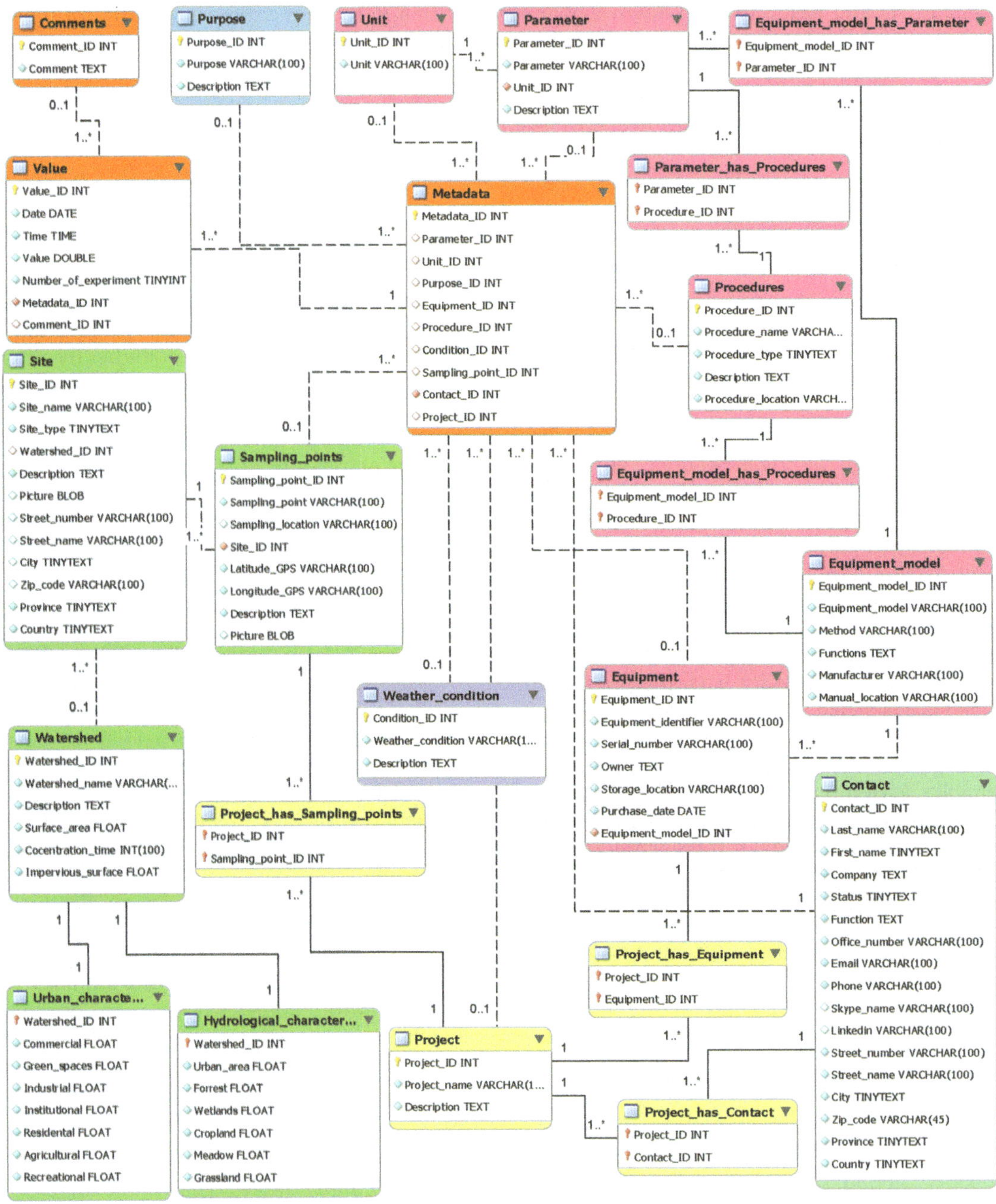

Figure 4.16 Example E4.2: Scheme of the datEAUbase by modelEAU (Plana *et al.*, 2019).

(2) Multiple dimensions containing descriptive measurement metadata. Including information about the precise location where the measurements were taken, the device or method used for the measurement, the purpose of the measurement, the person responsible of the measurement, the weather conditions at the time of measurement, and the project linked to the measurement.

The most relevant metadata tables are detailed below:

- Location information: the main table about the location information is *Sampling_points* table which allows to specify the precise location of the measurement including the GPS coordinates (e.g., inlet of the primary clarifier number 2; 46°41004″N 72°42059″W). The *Sampling_points* table is linked to the *Site* table containing broader information of the site such as the address, a picture, and a description. Related to this table, further information on the watershed, urban and hydrological characteristics is also provided and linked to the *Site* table.
- Device or method information: this part of the database is built by several tables, for example, *Equipment*, *Equipment_model*, *Procedure*, *Parameter*, and *Unit* tables. These tables provide information about the devices used for the on-line measurements or the equipment for the laboratory analysis, the procedure used to obtain a value, as well as the parameter and the units of measurement.

4.6 TAKE-HOME MESSAGE

- There is a wide variety of database and data management technologies that enable storing signal data together with the metadata to describe them.
- Relational databases are among the most mature technologies and facilitate a straightforward registration of signals with descriptive metadata in a structured manner.
- During the setup of a database or any other data management technology a trade-off is made between the effort associated with data collection and storage (data ingestion) and the effort required to use the data (data retrieval).

ACKNOWLEDGMENTS

Writing a chapter on the structured storage of data, relevant to the water and wastewater industries, covers a lot of domain expertise. To make this story as comprehensive as possible, the authors relied on input from multiple field experts. We thereby wish to thank Dr Stijn Van Hoey (Fluves, Ghent, Belgium); Frederick Van Leeckwyck (Factry BV, Ghent, Belgium); Chris Hidalgo (Signature Automation LLC, TX, USA); Annika Schönfeld (Ruhrverband, Essen, Germany); and Manoj Yegnaraman (International Society of Automation – ISA Water/Wastewater Division, NC, USA) for the enlightening discussions that contributed to this work.

REFERENCES

Beaulieu A. (2020). Learning SQL: Generate, Manipulate, and Retrieve Data. 3rd edn. O'Reilly Media Inc., Sebastopol, CA, USA.

Camhy D., Gamerith V., Steffelbauer D., Muschalla D. and Gruber G. (2012). Scientific data management with open source tools – an urban drainage example. Proceedings IWA/IAHR 9th International Conference on Urban Drainage Modelling, September 4–6, 2012, Belgrade, Serbia. URL: http://hikom.grf.bg.ac.rs/stari-sajt/9UDM/ShortAbstracts/158-ShortAbstract.pdf

Chiarot G. and Silvestri C. (2021). Time series compression survey. *ACM Computing Surveys*, **55**(10), 1–32, accepted. https://doi.org/10.1145/3560814

Choudhury A. A. S., Thornhill N. F. and Shah S. L. (2008). Diagnosis of Process Nonlinearities and Valve Stiction: Data Driven Approaches. Springer-Verlag Berlin, Heidelberg, Germany.

Correll M. and Gleicher M. (2016). The semantics of sketch: Flexibility in visual query systems for time series data. IEEE Conference on Visual Analytics Science and Technology (VAST), Baltimore, MD, USA, pp. 131–140, https://doi.org/10.1109/VAST.2016.7883519

Dixon J. (2014). Pentaho, Hadoop, and data lakes. URL: https://jamesdixon.wordpress.com/2010/10/14/pentaho-hadoop-and-data-lakes/ (accessed 27 January 2023)

Dunning T. and Friedman E. (2014). Time Series Databases: New Ways to Store and Access Data. O'Reilly Media Inc., Sebastopol, CA, USA.

EPA. (2023). What is the Water Quality Portal (WQP). URL: https://www.waterqualitydata.us/wqp_description/ (accessed 27 January 2023)

Epstein B. and Roberts P. (2022). Accelerate Machine Learning with a Unified Analytics Architecture: Deploy Machine Learning Models in Minutes, not Months. O'Reilly Media, Inc., Sebastopol, CA, USA.

ETH Zurich. (2019). WHDP documentation. URL: https://whdp.readthedocs.io/en/latest/ (accessed 27 January 2023)

Gopalan R. (2022). The Cloud Data Lake. O'Reilly Media, Inc., Sebastopol, CA, USA.

Gorelik A. (2019). The Enterprise Big Data Lake. O'Reilly Media Inc., Sebastopol, CA, USA.

Harrington J. L. (2016). Relational Database Design and Implementation: Clearly Explained. 4th edn. Elsevier, Cambridge, MA, USA.

Hauduc H., Gillot S., Rieger L., Ohtsuki T., Shaw A., Takács I. and Winkler S. (2009). Activated sludge modelling in practice: an international survey. *Water Science and Technology*, **60**(8), 1943–1951, https://doi.org/10.2166/wst.2009.223

Hay D. C. (2006). Data Model Patterns: A Metadata Map. Elsevier, San Francisco, CA, USA, https://doi.org/10.1016/B978-0-12-088798-9.X5000-1

Hernandez M. J. (2021). Database Design for Mere Mortals: 25th Anniversary Edition. 4th edn. O'Reilly Media, Inc., New York, NY, USA.

Horsburgh J. S., Tarboton D. G., Maidment D. R. and Zaslavsky I. (2008). A relational model for environmental and water resources data. *Water Resources Research*, **44**, W05406, https://doi.org/10.1029/2007WR006392

Kimball R. and Ross M. (2013). The Data Warehouse Toolkit: the Definitive Guide to Dimensional Modeling. 3rd edn. John Wiley & Sons, Inc., Indianapolis, Indiana.

Michener W. K. (2006). Meta-information concepts for ecological data management. *Ecological Informatics*, **1**(1), 3–7, https://doi.org/10.1016/j.ecoinf.2005.08.004

Middleton E. (2022). Myths about historians. AVEVA. URL: https://www.aveva.com/content/dam/aveva/documents/white-papers/WhitePaper_AVEVA_HistorianMyths_09-19.pdf

Nielsen A. (2019). Practical Time Series Analysis. O'Reilly Media Inc., Sebastopol, CA, USA.

Panoply. (2023). Data Warehouse Testing 101. URL: https://panoply.io/data-warehouse-guide/data-warehouse-testing-101/

Plana Q., Alferes J., Fuks K., Kraft T., Maruéjouls T., Torfs E. and Vanrolleghem P. A. (2019). Towards a water quality database for raw and validated data with emphasis on structured meta-data. *Water Quality Research Journal*, **54**(1), 1–9, https://doi.org/10.2166/wqrj.2018.013

Rieger L., Takács I., Villez K., Siegrist H., Lessard P., Vanrolleghem P. A. and Comeau Y. (2010). Data reconciliation for wastewater treatment plant simulation studies – planning for high-quality data and typical sources of errors. *Water Environment Research*, **82**(5), 426–433, https://doi.org/10.2175/106143009X12529484815511

Simon A. R. (2021). Data Lakes for Dummies. O'Reilly Media, Inc., Sebastopol, CA, USA.

Therrien J. D., Nicolaï N. and Vanrolleghem P. A. (2020). A critical review of the data pipeline: how wastewater system operation flows from data to intelligence. *Water Science and Technology*, **82**(12), 2613–2634, https://doi.org/10.2166/wst.2020.393

Thornhill N. F., Choudhury A. A. S. and Shah S. L. (2004). The impact of compression on data-driven process analyses. *Journal of Process Control*, **14**(4), 389–398, https://doi.org/10.1016/j.jprocont.2003.06.003

TimescaleDB. (2017). TimescaleDB vs. PostgreSQL for time-series: 20x higher inserts, 2000x faster deletes, 1.2x-14 000x faster queries. URL: https://www.timescale.com/blog/timescaledb-vs-6a696248104e/

Torfs E., Nicolaï N., Daneshgar S., Copp J. B., Haimi H., Ikumi D., Johnson B., Plosz B. B., Snowling S., Townley L. R., Valverde-Pérez B., Vanrolleghem P. A., Vezzaro L. and Nopens I. (2022). The transition of WRRF models to digital twin applications. *Water Science and Technology*, **85**(10), 2840–2853, https://doi.org/10.2166/wst.2022.107

Torkamani S. and Lohweg V. (2017). Survey on time series motif discovery. *WIREs Data Mining Knowledge Discovery*, **7**, e1199, https://doi.org/10.1002/widm.1199

doi: 10.2166/9781789061154_0103

Chapter 5

Practical procedures for sensor quality assessment

J. Alferes[1], O. Grievson[2], C. Heffernan[3], D. Lumley[4], T. Massiri[5], A. Menniti[6], O. Samuelsson[7], K. Villez[8] and E. Zegers[9]

[1]*VITO – WaterClimateHub, Ostend, Belgium*
[2]*AtkinsRéalis/visiting Professor at the University of Exeter*
[3]*Metropolitan Water Reclamation District of Greater Chicago, Chicago, IL, USA*
[4]*formerly DHI Group and Gryaab AB, Gothenburg, Sweden*
[5]*Käppala Association, Stockholm, Sweden*
[6]*Clean Water Services, Tigard, OR, USA*
[7]*IVL Swedish Environmental Research Institute, Stockholm, Sweden*
[8]*Oak Ridge National Laboratory, Oak Ridge, TN, USA*
[9]*City of Toronto, Toronto, ON, Canada*

Keywords: bias, drift, data quality, sensor maintenance, sensor validation, quality assessment, quality control.

Sensors are increasingly deployed for process monitoring and control. These produce on-line measurements at a high frequency, in parallel with low-frequency laboratory measurements. Compared to laboratory practices, sensor data quality assessment and control practices are far less structured at most utilities. This leads to inaccurate sensor data with unknown uncertainty factors.

This chapter shows how to establish standard operating procedures (SOPs) to support sensor data quality assessment and control and subsequent maintenance actions by producing relevant sensor metadata. Furthermore, SOPs are provided for the most commonly used wastewater quality sensors, inspired by utility and academic best practices. This chapter builds on definitions provided in Chapter 3 and provides additional definitions specifically related to sensors maintenance. Chapter 6 complements the methods in this chapter, which are based on reference measurements, with data-analytical techniques.

5.1 HOW TO READ THIS CHAPTER

The goals of this chapter are:

(1) To illustrate how standard sensor validation procedures can be deployed to produce metadata that are useful to assess sensor's long-term performance and delivered data quality.
(2) To provide at least one standard operating procedure (SOP) for several common water quality sensors (dissolved oxygen, pH, ammonia nitrogen, and more).

(3) To provide a practical case of how automatic sensor calibration and metadata collection has been implemented in practice.

The scheme below shows the chapter structure and the dependencies between sections. Sections marked in blue or white are considered essential to understand the chapter. Grey sections may be skipped without jeopardizing a high-level understanding. We recommend revisiting the data quality definitions in Chapter 3 or inspect the list of definitions at the end of this book as needed.

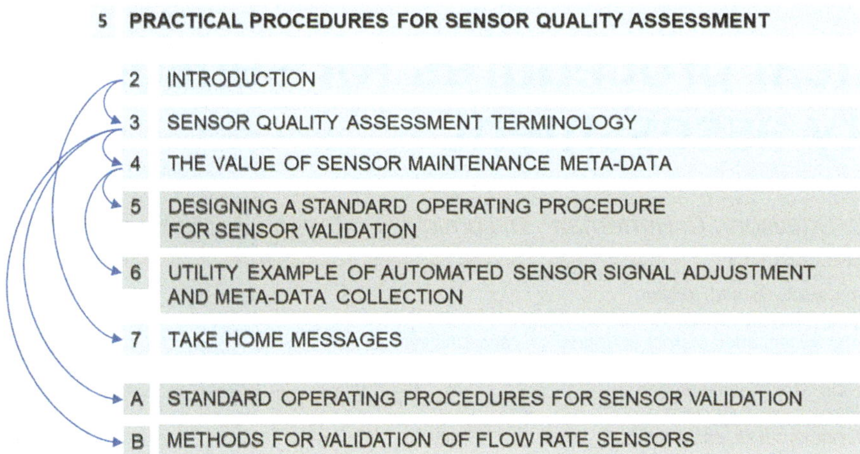

5.2 INTRODUCTION

As wastewater utilities are increasingly relying on sensor data for plant retrofitting, monitoring operations, plant optimization, and process automation, it becomes increasingly important to apply a systematic approach to sensor data quality assessment and quality control (QA/QC). This need becomes more pressing when sharing data within a large-scale organization and with external partners.

Sensors at wastewater treatment plants need to be regularly maintained. This can include cleaning, complete or partial replacement, calibration, and validation. Unfortunately, a systematic record-keeping of such actions is frequently lacking. As a result, it is sometimes hard or even impossible to properly judge the sensor state or the quality of the collected sensor data after prolonged periods of time, especially when personal memories fade or staff has moved on. When so, one necessarily relies on expert judgment, which may be subjective or difficult to come by.

In contrast, the quality of laboratory measurements is often controlled through use of standard methods or SOPs. Not only do many laboratories have SOPs in place, but the quality of SOPs themselves is often checked on a regular basis as well, for example, through yearly examination and revision. In this chapter, we provide a first step to correct this imbalance.

By providing templates and practical examples of SOPs for validation of online sensors, we hope that utilities can begin to apply standards for sensor-generated data that are on par with typical standards for laboratory-generated measurements and integrate these into their data QA/QC practices. These SOPs are designed to produce trusted reference measurements and online sensor signal under controlled conditions so that data quality assessment based on a comparative becomes feasible. In addition, the provided SOPs are considered an initial step towards sensor validation practices that are shared between utilities.

5.3 SENSOR QUALITY ASSESSMENT TERMINOLOGY

In this section, we define central concepts about sensor quality assessment. Practical aspects related to the definitions are emphasized in separate paragraphs (under the heading '*Utility*'). Finally, the key definitions are illustrated with two examples (E5.1 and E5.2). A few of the key terms have already been defined in Chapter 3 and are also listed in the Glossary of this book. This includes definition for bias, trueness, precision, variance, and accuracy. We recommend the reader to revisit these concepts as they will be used frequently in this chapter.

5.3.1 Definitions

Definition D5.1: A **sensor** is a device that communicates the value of a process variable (based on Asociación Española de Normalización, 2004).

By adopting this definition, we ignore certain elements of the data acquisition infrastructure that influence sensor data quality from this chapter. Examples of such factors are:

- The state of measurement environment (e.g., state of the flume for level-based flow rate measurement)
- The state of the other devices in the data pipeline, such as transmitters, cable integrity, IT infrastructure

The main reason to exclude these here is that quality assessment and control of these components does not require an approach that is unique to wastewater treatment. Standards from civil, electrical, and/or IT engineering should be adopted for validation of data acquisition infrastructure. Our focus is therefore on the management practices of the sensing element itself.

Definition D5.2: Sensor **quality assessment** (QA) consists of all actions performed to verify and document that a sensor is fully functional and produces data with the desired accuracy.

Sensor quality assessment may include, but is not limited to, calibration, metadata audit, signal verification, validation, and visual inspection.

Utility: Sensor quality assessment is in practice guided by (a) specific routines for different sensors, including the use of comparative measurements with laboratory tests, standard solutions, or additional sensors and (b) analysis of the results produced during such routines. As an example, analysing the trend in sensor trueness is a key activity in sensor quality assessment.

Definition D5.3: Sensor **maintenance** includes all actions required to maintain the sensor in a fully functional mode throughout its life cycle.

Common sensor maintenance actions include: calibration, cleaning, repair and spare part replacements, updating software, and recording the executed maintenance actions.

Utility: Records of sensor maintenance become sensor maintenance metadata. The executed and planned sensor maintenance may be recorded in dedicated maintenance management software, log books, or in a spreadsheet. Sensor maintenance metadata are key for sensor quality assessment. An important challenge today is the integration of sensor maintenance metadata with the sensor signals.

Definition D5.4: A sensor's **offset** is a constant deviation from the accepted reference value.

A bias caused by a non-zero offset results in the same bias regardless of the measured value in contrast to an error causing a change in the slope. The yellow line offset at 350 mg/L in Figure 5.1 illustrates this concept.

Definition D5.5: A sensor's **slope/sensitivity** is the amount of change of the measured variable (output) for a standardized change in the variable of interest (input).

If the input and output represent the same variable measured in the same scale (e.g., ammonia laboratory measurement and ammonia analyser signal), then the ideal slope is one. This is typical for direct sensors (see Definition D5.10). The increased sensitivity in the yellow line, as compared to the ideal sensitivity in Figure 5.1, illustrates this concept.

In many sensors, the measured variable is correlated with a variable of interest (see Definition D5.10). When so, the ideal slope will not be equal to one. As an example, a pH-sensor measures an electric potential (in mV) to correlate it with a pH value (log of proton activity). In this case, the ideal and theoretical sensitivity under standard conditions is equal to 59.15 mV per pH unit. The ideal slope is typically specified by the manufacturer or through a theoretical derivation.

Utility: The sensor slope may be an indication of the sensor's condition. For example, the slope for a galvanic Clark cell electrode (dissolved oxygen sensor) successively decreases due to normal wear-and-tear until the membrane needs replacement.

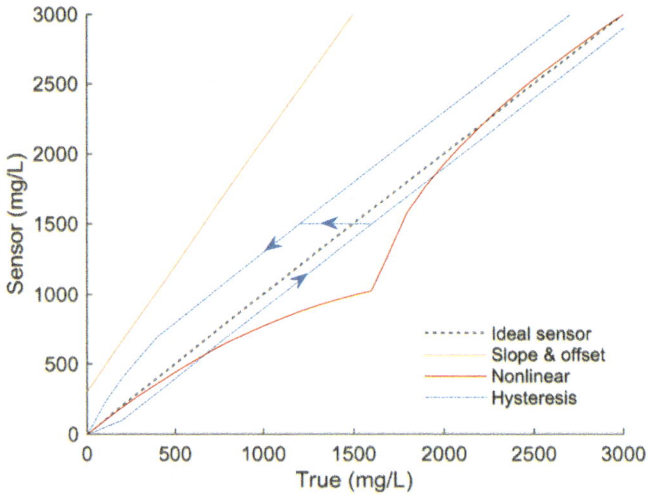

Figure 5.1 Example E5.1: Illustration of the sensor errors: slope and offset error, non-linearity, and hysteresis.

Definition D5.6: The **linearity** of a sensor indicates to what extent a change in the true component (input) is reflected as a linear change in the measurement (output) throughout its measurement range.

Highly accurate reference measurements in the full measurement range are needed to determine a sensor's linearity. For this reason, linearity is seldom assessed on site, but more often in the sensor suppliers' laboratories (factory calibration/validation). Substantial non-linearities can however be revealed during validation if multiple reference measurements are taken. The red line in Figure 5.1 exemplifies a non-linearity in a sensor.

Utility: Non-linear effects lead to errors in the measurement that can't be corrected for during offset or slope calibration. The presence of non-linear effects is crucial to understand the limitations on sensor accuracy of using a linear calibration curve or to mitigate those limits by establishing a non-linear calibration curve.

Definition D5.7: **Hysteresis** is the phenomenon where the sensor measurement value depends on the direction of change for the measured value.

The blue line in Figure 5.1 illustrates this concept. For an increasing true value, the measurement follows the lower blue dashed line. However, when the true concentration decreases, the sensor outputs a constant value until the upper blue dashed line is achieved and further follows this line for decreasing values.

Utility: As an example, airflow rate measurements can exhibit hysteresis with respect to the valve position measurement. This can be a consequence of backlash in the valve, which means the effective valve position is different depending on the direction of movement (hysteresis, closing vs. opening motion). Note that it is effective valve position that exhibits hysteresis in this case. In the absence of a direct valve position measurement, this becomes apparent as a hysteresis effect on the airflow rate.

Definition D5.8: Sensor **validation** or sensor **verification** refers to a systematic procedure for (a) collecting and recording measurement values under controlled conditions to quantify quality measures such as accuracy, bias, precision, offset, slope, and/or linearity and (b) to evaluate whether the quality measures satisfy predetermined specifications for the intended application. The simplest form of validation is to compare the sensor reading to a reference measurement.

The term validation is sometimes used interchangeably with verification and their definitions vary between domains and applications. ISO 9000 quality routines state that verification is the procedure of providing objective evidence that specified accuracy requirements have been fulfilled. Verification is achieved when the sensor is installed, cleaned, and calibrated as recommended by the manufacturer. Validation may be considered to additionally requiring local and end-user accuracy requirements. For the purposes of this study, **sensor validation** and **sensor verification** are considered synonymous.

SOPs for validation of the most common water quality sensors are provided in Appendix 5.A of this chapter. The most common validation exercise is to compare and record the deviation between a sensor reading and an accepted reference measurement.

Utility: The main benefit of sensor validation is to have access to quantified measures of sensor accuracy, in turn leading to a better understanding of the obtained sensor data quality in view of sensor maintenance and data use.

Definition D5.9: Signal **transmission verification** or **end-to-end testing** refers to a systematic procedure to verify that the sensor signal remains consistent during transfer from the sensor to the location for data storage such as a database.

As an example, Figure 5.2 shows the total daily volume (in m^3/day) for a wastewater treatment works over a 3-month period. At this wastewater treatment works there are two flow rate sensors with one at the front of the works downstream of the storm weir and the other at the final effluent of the treatment works. Ideally, both flow rate sensors should provide the same data points, apart from random measurement errors; however, as it can clearly be seen the flow-to-full treatment (FFT) (blue) is reading significantly less than the final effluent (pink). This suggests that the wastewater treatment works is gaining flow through the treatment process, which is not the case.

However, when we look at the ratio of the sensor signals, we can see that there is a common ratio which suggests a systemic error associated with the flow readings. By applying the ratio of the instantaneous flow readings, we can see that the resultant line (in red), which is the corrected FFT, follows the final effluent trend more closely (although not exactly). This is a very typical pattern seen in data quality errors where one of the flow meters is incorrectly scaled – which is to say the measurement range of the instrument does not match the measurement range assumed in the telemetry system. Transmission verification is used to detect and eliminate this kind of error.

Utility: Typical errors that can be detected during signal transmission verification include analog transmission noise and changed current (mA-scaling), programming errors, change of units, and unsynchronized time stamps.

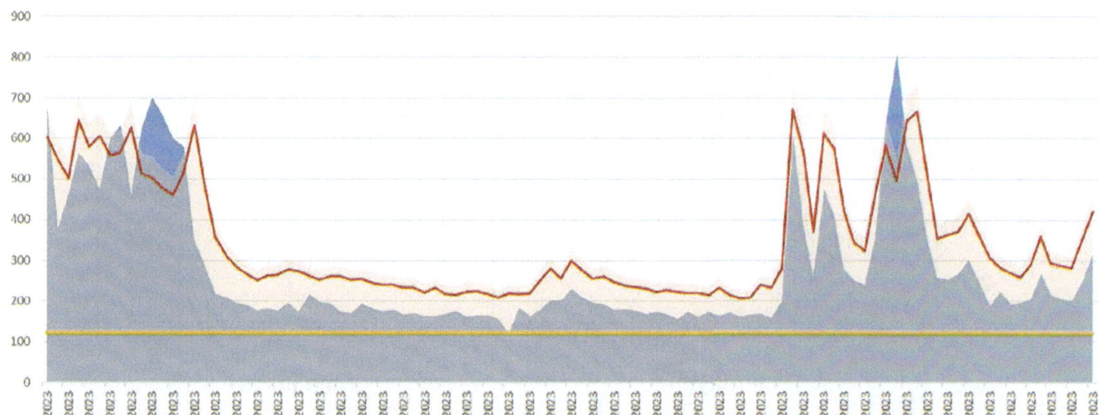

Figure 5.2 Flow rate measurements on a wastewater treatment plant: FFT is measured at the inlet and flow effluent (FE) is measured at the effluent. An error in transmission causes the FE to be systematically larger than the FFT where they should be the same. This kind of error can be detected and eliminated through transmission verification.

Definition D5.10: **Calibration** is the act of adjusting a sensor to achieve the desired accuracy by matching the sensor output to one or more accepted reference values.

Calibration is commonly executed following a sensor validation. The purpose of calibration is to reduce the sensor's bias. Calibration can be referred to as one-point, two-point, or multipoint calibration, depending on the number of used reference measurements. The sensor's offset and/or slope are adjusted during calibration depending on the root cause of the identified bias. Figure 5.3 illustrates calibration of a sensor's slope and offset.

The technical computations during calibration are often executed automatically via the sensor's transmitter unit, and only the reference measurement value(s) is (are) manually entered. However, the sensor-internal calibration curves, including offset and slope values, may be accessible as metadata, either by manually reading the transmitter or automatically via digital communication protocols (e.g., Modbus and Profibus). An alternative calibration method that does not require access to the internal memory of the sensor is explained through Example E5.2.

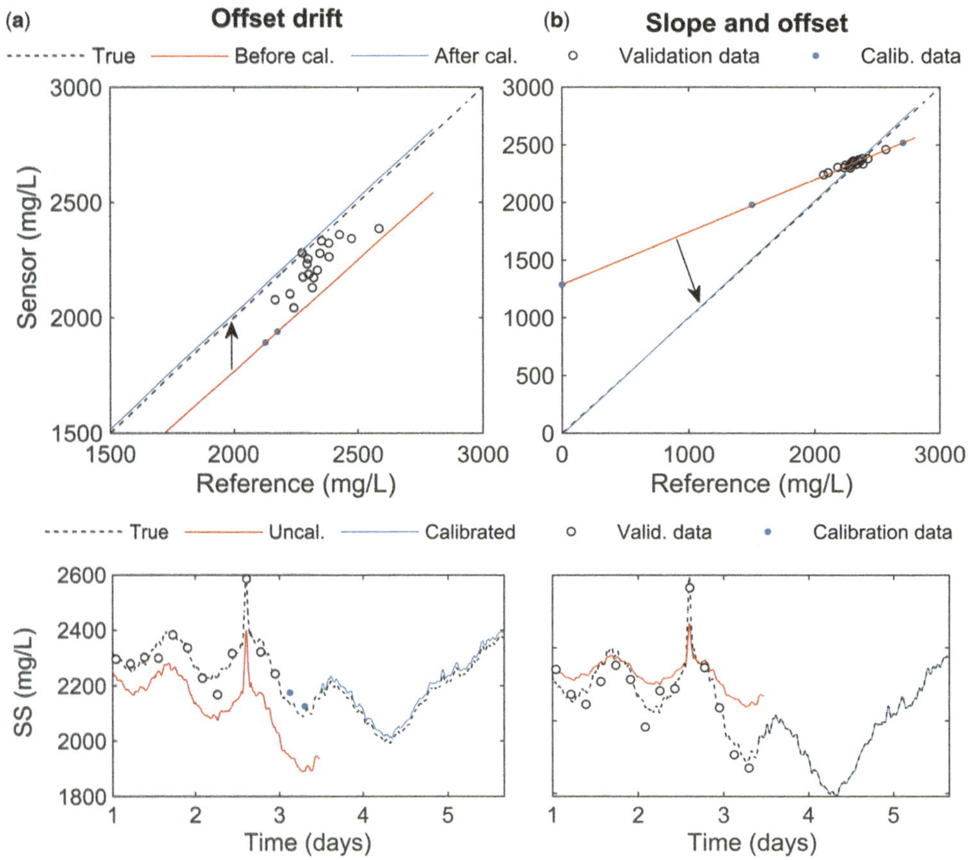

Figure 5.3 (a) Correction of a sensor offset drift with an offset calibration based on the mean value of two data points. (b) Slope and offset error corrected by a slope and offset calibration based on three data points (cfr. Chapter 2) spread out in the full measured range. Note that the true value is only shown for illustrative purposes in this simulated example and not in practice.

Utility: The main purpose of calibration is to ensure that the produced sensor signal matches the true value of the measured value as close as possible. Many sensors require calibration at regular time intervals to ensure that the effect of sensor ageing and changes in confounding factors, commonly observed as drift, can be compensated. Note that it is preferred to cover the whole measurement range as illustrated in Figure 5.3b, and not to concentrate the measurements in a narrow concentration range as in Figure 5.3a

Definition D5.11: A **direct measurement** is obtained when the sensor directly measures the component of interest.

An example of a sensor that uses a direct measurement is a conductivity sensor, which directly measures how well a current is transferred in the media. Most sensors, however, use indirect measurements.

An indirect measurement measures a correlated component, such as the light absorbance at a certain wavelength, as proxy for the desired water quality variable such as suspended solids. Similarly, ion-selective electrodes (ISEs) are based on the measurement of an electric potential which can be correlated with concentration of one or more ionic species through ISEs. This is commonly used for measurement of pH and concentrations of ammonia, nitrate, and potassium. The indirect measurement signal is then converted to the desired component within the sensor. This means that a change in a confounding factor may affect the variable that is measured while the variable of interest is not affected by this confounding factor, thus invalidating the assumed relationship between the two variables. Therefore it is important to understand the measurement technique and its limitations.

Utility: Most water quality sensors provide indirect measurements. Only conductivity and oxidation–reduction potential (redox) are considered direct measurements. Any sensor that requires some form of calibration prior to use is considered an indirect measurement. For example, pH measurements are based on the measurement of an electric potential and total suspended solids is based on the light intensity of scattered light. The calibration curve is used to correlate the measured variable (e.g., electric potential) with the variable of interest (pH). Naturally, calibration plays a key role in obtaining accurate readings.

5.3.2 Example E5.2: Calibrating offset and slope errors

Calibration is performed to reduce systematic deviations between the true value of the measured variable, which cannot be known perfectly, and the sensor signal. During calibration, the sensor's offset and/or its slope is adjusted. At least two reference measurements at different concentrations (sometimes referred to as two-point calibration) are needed for calibrating both the offset and slope (Figure 5.3b), whereas a minimum of one reference measurement is needed for a calibration of the offset or slope only (Figure 5.3a; see also the example in Section 5.4.1.2). Using more measurements than strictly required, preferably covering the range of values expected in the measured process, will improve the effectiveness of the calibration.

An offset calibration uses one reference measurement for adjusting the sensor signal. Note that one reference measurement can be based on the average of several measurements, to increase its precision. Offset calibration affects all subsequent sensor measurements by a change in the offset parameter of the calibration curve, thus leading to a momentary shift in the recorded values at the time of calibration.

The terms one-point and two-point calibrations are used frequently but can be confusing since they only indicate the number of reference measurements required for calibrations, and not how the actual calibration is executed. We recommend consulting the sensor instructions to fully understand the effects of calibration on a case-by-case basis. We also note that complex instruments may require more than two points for effective calibration. Examples include (a) address multicollinearity in multivariate spectrophotometers or (b) capturing non-linear effects in sensors that do not exhibit a linear response to the measured variable.

A calibration under unstable conditions, for example, when the sensor body temperature differs from the ambient temperature, can result in a calibration error, in turn leading to systematic measurement errors. Potential sources of disturbances should be identified and, where feasible, quantified. The validation procedures in Appendix 5.A either (a) eliminate such disturbances or (b) enable a corrective computation for different sensor types.

5.4 THE VALUE OF SENSOR MAINTENANCE METADATA

Trust in shared data requires a common understanding of the quality of large-scale sensor datasets. One useful step to achieve this is to document sensor maintenance tasks in a uniform and rigorous manner. This may include cleaning, calibration, validation, and part or whole-sensor replacement tasks. In this chapter, we suggest the use of SOPs for the specific task of sensor validation as this produces metadata that are particularly useful to track the state of online sensors and the quality of the data they produce. These SOPs can be integrated into existing frameworks and practices for data QA/QC.

In the short term, sensor validation practices geared towards the evaluation of any sensor quality measure, such as accuracy, bias, offset, precision, slope, and/or drift, can help establish sound cleaning and calibration practices, especially by determining an optimal frequency for this task. In the long term, well-documented sensor validation practices enable effective sharing of large-scale data sources with staff within the organization as well as with external project partners. The following sections detail some of the tangible benefits that can be expected from the collection of sensor metadata through well-documented sensor validation practices.

5.4.1 Optimizing calibration frequency

The simplest way to improve sensor management practices in most utilities is to systematically track how the calibration curve of a sensor changes over time. The primary purpose of calibration is counteracting drift. Consequently, when a two-point calibration procedure is executed on a regular basis, one can save the parameters of this curve (offset and slope) every time a calibration is executed. Modern sensor hardware can keep internal logs of changes in the calibration curve, which in many cases can be transferred securely to a utility-managed database by means of conventional protocols (e.g., Ethernet, Modbus).

5.4.1.1 Tracking drift of offset and slope in view of calibration frequency tuning

Importantly, the drift rate often depends on the medium that the sensor is exposed to. This means that the recorded changes in offsets and sensitivities are specific to the sensor and the location in a plant. In the case of severe changes in the wastewater composition, the drift rate may be time dependent as well. For this reason, the optimal frequency of calibration should be tuned specifically to the sensor and the medium it is exposed to. In the case of temporal changes of the sensor drift, an adaptive strategy may be useful by adjusting the time between calibrations from one calibration to the next. Clearly, optimal adjustment of the calibration frequency is only feasible if the calibration curves are recorded and stored in a systematic and reliable manner. Tuning this frequency with well-maintained calibration curves can reduce unnecessary sensor maintenance costs, assure that data quality specifications are met, and increase confidence in the produced data.

In Example E5.3, both the offset and the slope are tracked whenever the sensor is calibrated. Modified approaches may consist of:

(1) Evaluating whether a calibration is necessary by evaluating the offset and slope first and only updating the calibration curve when a sufficiently large change is noted.
(2) Replacing the monitoring of offset and slope with a procedure which tracks only one parameter of the calibration curve (offset or slope). This may be useful if (a) it is known that the sensor has no meaningful change in either its offset or slope during its lifetime, or (b) it is considered too cumbersome to obtain steady-state measurements for more than one reference value of the measured variables.

5.4.1.2 Example E5.3: Tracking sensor offset and slope for two pH sensors

We illustrate the systematic collection of sensor metadata for two pH sensors with pH-procedure-2 (version 1.0, see Appendix 5.A.2.2). For a typical ion-selective pH sensor used in the wastewater sector, the relationship between the measured variable (electric potential) and the variable of interest is described as a linear relation, named the Nernst equation. The offset is defined as the steady-state value of the electric potential provided by the sensor exposed to a neutral medium (i.e., pH 7). At standard atmospheric temperature and pressure (SATP), the theoretical value for the offset is 0 mV. Similarly, the slope is the change in electric potential for a unit change in the measured variable (pH), again assuming steady-state before and after such a change. At SATP, the theoretical value for the slope is 59.16 mV for most pH sensors.

Figure 5.4 shows measured values for the offset and slope for two commercial pH sensors installed in a nitrifying reactor for a period of 1 year. On a regular basis, pH-procedure-2 was used to estimate the offset and slope of the sensors. One can see that the offset increases to 60 mV or more after 1 year of operation. This would amount to a change of 1 pH unit if the sensor was never calibrated. In contrast, the slope is stable and remains within 2 mV/pH unit through the considered time window. This increases to a deviation of <0.05 pH units when the process pH is 6. The measured rate by which the offset changes (about 60 mV per year, or 1 pH unit assuming constant slope) can be used to determine an optimal frequency of calibration, trading accuracy off against effort. In this example, this means that a calibration every 30 days is more than sufficient to maintain an offset with a magnitude smaller than 6 mV ($\sim$0.1 pH unit).

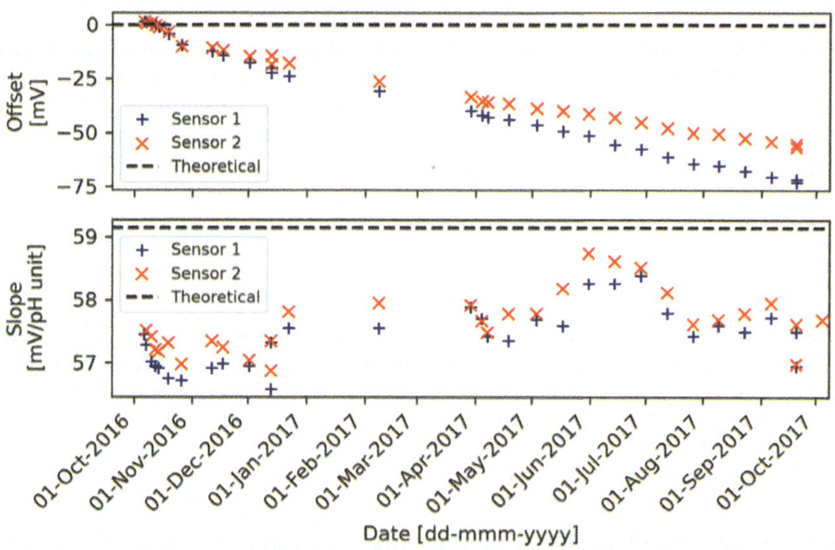

Figure 5.4 Offset and slope values obtained with pH-procedure-2 (version 1.0, see Appendix 5.A). The offset and slope values are recorded whenever the procedure is executed. In this case, the long-term trends in offset and slope suggest that accounting for drift of the offset may be sufficient during calibration. A one-point calibration may therefore be sufficient. The drift of the offset amounts to about 60 mV per year. This means a calibration every 30 days is more than sufficient to maintain the offset parameter of the calibration curve within 6 mV of its true value, thus maintaining the measured pH within 0.1 pH units of the true value.

5.4.2 Optimal cleaning frequency

Sensor cleaning is a basic but time-consuming maintenance action. Commonly, cleaning is executed as a fixed routine, which makes it important to optimize the cleaning interval.

Changes in sensor data quality due to fouling (e.g., offset and slope) can be quantified by recording the sensor value before and after cleaning – under constant measurement conditions. Systematic quantification of the cleaning effect enables a straightforward tool to optimize the cleaning interval. The systematic metadata collection simply requires adding the time for cleaning when the sensor values before and after are already stored in a database.

5.4.3 Intra-organizational data sharing

It is typical for the users of sensor data to be different from the staff responsible for sensor quality assessment and maintenance. When the set of sensors becomes large, a data user may find it challenging to fully trust the available data sources in the absence of additional metadata regarding routine assessment and maintenance tasks. Quantified effects of sensor validation, calibration, and maintenance action can assist in gauging the quality and fit-for-purpose of sensor data. Increased reliance on sensor data for plant operations and management warrants an increased attention to data quality. A systematic approach, based on the SOPs presented in the Appendix 5.A, is recommended to capture long-term trends in the state of sensing equipment and the quality of the data that they produce.

5.4.4 Inter-organizational data sharing and model identification

Sensor data collected by wastewater utilities are increasingly used by partners outside the fence. Data may be shared routinely with consultants, regulators, industry partners, and researchers. Metadata describing the procedures in use to assess and control sensor data quality can be instrumental to properly understand the shared data, evaluate their fit-for-purpose, and share a common language to identify and describe important events and trends in the sensor data.

As a simple example, consider the case where a data-driven model is built with data from two sensors, A and B, producing measurements, y_i^{A} and y_i^{B} $(i = 1,\ldots,N)$. A common approach is to weigh the measurements from the different sensors by their precision. Practically, the deviation between a model output (e.g., $\hat{y}_i^{\mathrm{A}}$) and the measured value (e.g., y_i^{A}) is divided by measurement error standard deviation for that sensor (σ_i^{A}) to produce standardized residuals. The standardized residuals are then equal to:

$$r_i^{\mathrm{A}} = \frac{\hat{y}_i^{\mathrm{A}} - y_i^{\mathrm{A}}}{\sigma_i^{\mathrm{A}}}, \quad r_i^{\mathrm{B}} = \frac{\hat{y}_i^{\mathrm{B}} - y_i^{\mathrm{B}}}{\sigma_i^{\mathrm{B}}}$$

These standardized residuals are then given equal weight during model training and testing. For example, the weighted sum of squared residual (WSSR) is a common measure to assess model quality and is computed as follows:

$$\mathrm{WSSR} = \sum_{i=1}^{N} \left(r_i^{\mathrm{A}}\right)^2 + \sum_{i=1}^{N} \left(r_i^{\mathrm{B}}\right)^2$$

The measurement error standard deviation typically varies from one sensor to the next and may vary in time as well. Unfortunately, it is difficult to obtain reliable estimates of the measurement error standard deviations directly from the measurements (y_i^{A}, y_i^{B}) under dynamic, uncontrolled conditions. Access to routinely evaluated estimates of the measurement error standard deviation facilitates the definition of sensible quality measures for any model, whether it is based on machine learning or first principles. A standardized approach to the regular and systematic evaluation of the sensor precision can thus be very valuable to evaluate model fit and forecast performance.

5.5 DESIGNING A STANDARD OPERATING PROCEDURE FOR SENSOR VALIDATION

Setting up a validation practice is easy. However, it can be difficult to assure that a sensor is validated in a consistent way (by different technicians) when validation procedures are lacking. In this section, we give some recommendations for what to include in a sensor validation procedure to enable a consistent approach to sensor validation and maintenance.

5.5.1 Recommended structure for a standard operating procedure for sensor validation

First, we emphasize three key components to successfully implement an SOP for sensor validation: purpose, repeatability, and metadata recording.

- **Purpose.** The purpose of the validation needs to be clear since this decides the level of detail needed in the validation. To define the purpose of the SOP, address the following questions:
  - What sensor data quality measures (e.g., accuracy, bias, or precision) are relevant for the application, and what is their acceptable range?
  - What actions are available (a) to avoid sensor data quality measures exceeding the acceptable value range and (b) to take corrective action when sensor data quality measures exceeding the acceptable range
  - What kind of actionable information does the SOP provide and how do the outcomes of the SOP influence sensor maintenance tasks?

- **Repeatability.** A repeatable practice is key to obtaining consistent and trustworthy validation results, which is the core element in an SOP. Much like cooking, the SOP recipe should be clear with a step-by-step guide for the complete procedure. This includes preparation, consumables, and the actual execution. Often, missed preparations or lack of consumables and/or tools lead to stress, unnecessary work time, and short-cuts with unreliable validation data as a result. With a clear SOP, the focus is shifted towards carefully executing the validation rather than trying to remember all equipment and how the validation should be executed.
- **Metadata recording.** Ideally, metadata are recorded as automatically as possible. For example, the periods of time that a sensor is placed in a standard solution can be recorded automatically through a push of a button in some modern sensors. This can serve two purposes. First, the data collection is conducted consistently, which minimizes human error and variability. In turn, this improves the repeatability of the validation results. Second, automatic data collection facilitates long-term data storage and reuse. In contrast, manual transfer of recorded metadata into a usable database often proves tedious and is vulnerable to a loss of information. Automatic data collection makes it so that the sensor validation metadata become part of the organizational memory, rather than inaccessible in isolated spreadsheets. When the metadata are easy to access, they allow straightforward tracking of the sensor life-time performance and quality assessment of historic data.

Ultimately, a standard validation procedure should contain some essential information, which is listed below. Information that puts the procedure into a practical context can be added as supplementary information to tailor it for site-specific conditions.

A template that satisfies the above and provides a generally applicable structure can be found in Appendix 5.A. We discuss the recommended elements of a sensor validation SOP next.

Essential elements for an SOP for sensor validation

We consider the following elements essential and suggest including them in this order in any SOP for sensor validation:

(1) State the purpose of the validation procedure.
(2) Determine how many measurement points and repetitions are needed.
(3) All hardware.

(4) All consumables.
(5) A detailed step-by-step description of the procedure.
(6) A method to compute and record validation measurements and sensor data quality metrics in the form of metadata.
(7) Define other metadata to record such as the lot number of the standard used or the person performing the validation.

Optional elements for an SOP for sensor validation
The following information may prove valuable to include in an SOP, depending on the complexity of the measurement principle, the measured medium, and site-specific conditions:

- Description of the measurement principle and confounding factors.
- History of the SOP.
- Purpose of the sensor measurement.
- Additional checks to perform prior to sensor validation, such as:
  - Verification of the integrity of the sensor mount, power supply, and signal cabling.
  - Verification of mixing conditions around the sensor.
  - Checking for and cleaning of debris.

5.5.2 General recommendations for reliable reference measurements

General success factors during sensor validation include representative water samples and unbiased measurements. Below, we list four common methods used to achieve unbiased comparisons between a sensor and its reference measurement.

In general, we recommend using a bucket subsampling procedure (procedure A). Based on the limited studies available today, this is the only approach that can guarantee an unbiased comparison between the sensor signal and the reference measurements. Any sensor data quality metadata produced with other methods (procedures B, C, and D) should be considered as less reliable. For this reason, all sensor validation procedures in Appendix 5.A make use of procedure A. These can be modified to adopt procedure B, C, or D depending on the applicable constraints.

(A) **The bucket subsampling procedure**
A bucket is used to sample the process at a representative location. The concentration may be changed by diluting or concentrating after sampling to cause the composition to change in a controlled manner. Any variations in the composition of the water sample relative to the original water sample are considered known. The bucket water sample is then mixed, where after the sensor is positioned in the bucket for an extended time until the measurement is stable. An automatic stirrer and air bubbling can be used to assure a homogenous mixture during the measurement and to avoid settling of suspended solids. A grab water sample can then be taken from the bucket and analysed in a laboratory as a reference of the sensor reading. This water sample preparation minimizes uncontrolled variations in the water composition so that any deviation between sensor data and reference data is considered caused by sensor measurement error only.

This method provides an unbiased validation since both the sensor signal and the water sample correspond to the same well-mixed medium. Also, the conditions in the bucket are constant, which makes it easy to obtain a stable mean value regardless of the sensor measurement noise magnitude.

The drawback of the bucket approach is that it can be difficult and time consuming to obtain a large grab water sample from the process.

(B) **The in-situ sensor comparison**
A reference sensor (e.g., a hand-held device) is positioned next to the stationary sensor for an extended time. The time should be sufficient to reduce any effects from potential smoothing and averaging filters in the hand-held sensor and the supervisory control and data acquisition (SCADA). The two measurements can be compared for mean values and differences during in-situ process variations.

A pre-requisite is that the reference sensor is accurate, that the timestamps in the data records for both sensors are synchronized with the same clock, and that the sensor signal dynamics due to sensor signal processing (e.g., signal filter) are negligible. An in-situ comparison is complicated by the fact that the measured variable may be changing during the validation procedure and that the dynamic response of the two sensors may be very different.

(C) **The 24-h-composite sampling approach**

The medium close to the sensor is sampled at a volume or flow-proportional rate. A laboratory analysis of the composite water sample can then be compared with the sensor signal for the same time period as the composite water sample was taken. The sensor signal needs to be adjusted akin to the water sample. That is, the sensor concentration measurement is adjusted to be flow proportional, before its mean value is compared with the laboratory result.

A new sampling device installation is seldom undertaken for sensor validation purposes. However, sometimes such sampling equipment is already in place for other reasons and can then be reused for obtaining a 24-h-composite reference measurement and sensor validation.

(D) **The quick grab samples**

The most common, but unfortunately also the least accurate reference measurement, is a grab water sample taken as close as possible to the sensor. The grab water sample result can easily be compared with the sensor reading at the time of sampling. The speedy result combined with the ease of leaving the sensor in-place explains its popularity. Unfortunately, differences between the grab water sample result and the sensor reading are influenced by many uncontrollable factors that prevent using this approach for an unbiased and precise evaluation of the sensor data quality. Large deviations between the two measurements may not be due to sensor inaccuracy only, thus leading to a pessimistic evaluation of the sensor data quality, in turn leading to excessive maintenance actions.

The main concern with this method is that it is difficult to assure that the grab water sample is representative of the medium. The sensor may be too deep to take a water sample right next to it. It may also be difficult to grab the water sample at the exact same location always. In addition, the dynamics of the measured process or online signal processing may cause delay and unwanted signal variations during validation, making the recorded value unrepresentative of the sensor measurements. Similarly, recording the sensor value during grab sampling in the SCADA (in contrast to the sensor display) can differ in time if the time for the grab water sample was recorded from an incorrect clock, such as the sensor display. Ideally, the recorded time matches that of a clock that is synchronized with the SCADA clock (the master timekeeper).

5.5.3 Templates and exemplary standard operating procedures for sensor validation

Appendix 5.A of this chapter includes 14 SOPs for five variables that are commonly measured in wastewater practice. The variables considered are:

(1) Dissolved oxygen (DO): three procedures
(2) pH: three procedures
(3) Total ammonia nitrogen (TAN): two procedures
(4) Total suspended solids (TSS): four procedures
(5) Ultraviolet–visible (UV/Vis) light absorbance: two procedures

These variables were selected by considering:

- Variables for which the sensor technology and the techniques for validation are mature
- Variables for which sensor calibration and validation is feasible on-site
- Variables for which sensor calibration or validation is regularly executed by on-site staff

For these reasons, certain variables were omitted, for example:

- Dissolved N_2O concentration: Techniques for sensor validation are limited and subject to change.
- Gas flow rates: Sensor validation is typically completed by the sensor manufacturer. Often, this can only be executed at dedicated off-site facilities (e.g., factory validation).

- Water flow rates and levels: We could not collect any step-wise procedure for sensor validation that is being used today at wastewater plants. In case of the electromagnetic flow rate measurement principle, this is typically assumed by the sensor manufacturer. Often, this can only be executed at dedicated off-site facilities (e.g., factory validation). For level measurement, correct placement of the level sensor, and removal of any debris is most important. For level-based flow rate measurement (flume), the physical integrity of the flume is critical also. The level sensor itself is often considered accurate and drift-free. This may explain the lack of existing sensor validation practices that could be collected for the purposes of this chapter. Appendix 5.B of this chapter provides notes on the validation this type of sensors. Flow rate sensors and level sensors can also be validated through flow balancing techniques, as explained in Chapter 6.
- Pressure and temperature: These sensors are typically considered accurate and drift-free and systematic measurement errors are mostly due to errors in installation. This explains why repeated validations are rarely executed. Appendix 5.B of this chapter provides notes on the validation this type of sensors.

5.6 UTILITY EXAMPLE OF AUTOMATED SENSOR SIGNAL ADJUSTMENT AND METADATA COLLECTION

Most typically, sensor calibration curves are stored locally in the sensor transmitter. Calibration occurs at the sensor, and only the internally processed signal is transferred to the centralized data acquisition system (e.g., SCADA, historian, or database). One alternative is to leave the calibration curve in the transmitter unchanged over a longer period and instead perform the calibration adjustment within the centralized SCADA environment. This approach has the benefit of retaining all metadata from the calibration within the existing database structure and using a more universal approach across all sensors, even when produced by different manufacturers.

This example shows how sensor signals are adjusted automatically after ingestion in the distributed controller system (DCS) at the Rya water resource recovery facility (WRRF) in Gothenburg, Sweden. The approach is named *auto-adjustment* and example views from the Rya SCADA are shown in Figure 5.5.

Auto-adjustment aims to minimize the effects of systematic and random errors to the process without resorting to intensive instrument calibration in the field. It consists of the following steps.

(a) Take a reference measurement (grab water sample) next to the sensor which is to be adjusted.
(b) After laboratory analysis of the grab water sample, store the metadata, that is, the reference measurement value in the laboratory information management system (LIMS).
(c) Automatically transfer the reference measurement from the LIMS to the DCS.
(d) Automatically validate the plausibility of the reference measurement from (a) with an algorithm in the DCS with predefined prerequisites.
(e) Automatically update how the sensor raw signal is adjusted in the DCS. This is achieved with a recursive least-squares algorithm with exponential decay. The adjustment of the raw sensor signal is performed by means of a breakpoint curve, which is a continuous but piece-wise linear curve (see left part in Figure 5.5). The number of segments (breakpoints) is user definable for each sensor making it possible to compensate for deficits in the sensor's linearity. This kind of curve is also known as a linear spline function (Ramsay & Silverman, 2005). The adjusted signal is stored in the DCS and is thus available for process control.

At the Rya WRRF, the auto-adjustment function reduces the need for manual calibration of the sensors, saving an estimated 10–40 h per sensor and year depending on the sensor type. Auto-adjustment of the sensor signal allows the sensor to remain in the process media during the whole

Figure 5.5 A SCADA screenshot showing the auto-adjustment faceplate for a suspended solid sensor. The sensor measures two media with different characteristics, direct precipitated water (data cluster near the origin and below 10 mg/L), and primary settled water (cluster at middle right, typically above 50 mg/L). The two different characteristics are handled with a non-linear calibration curve with a breakpoint at 12 mg/L.

procedure, hence maximizing the measurement up-time. This also minimizes effects of calibration with artificial media or media that may not be representative of the measured environment. By examining how the curves in the function change over time, phenomena such as fouling, sensor drift, and the sensor nearing the end of its useful life can be tracked and quantified. Thus, the auto-adjustment function helps to minimize process disturbances due to inaccurate sensor measurements which will generally improve process control.

It should be noted that the variability introduced by the reference measurement (grab water sample) is partly reduced by the recursive correction of several measurements,. However, this also delays correction of a sensor drift. A faster correction may be achieved with the bucket approach in combination with a more rapid calibration curve update.

5.7 TAKE HOME MESSAGES

In many wastewater utilities, only a limited set of data collected during sensor maintenance is recorded and inspected in a routine manner. As data are increasingly used across disciplines within the same plant, as well as shared with outside partners, confidence in the sensor data can be improved through several simple techniques:

- Match the level of rigour applied to sensor data quality assessment and control to the rigour applied to analytical laboratory measurements.
- Develop, deploy, and maintain standard operation procedures for sensor maintenance, validation, and calibration. This chapter provides templates and examples for the purpose of sensor validation.

- Ensure that measurement results obtained with reference materials (e.g., calibration media) or reference sensors (e.g., handheld sensors) are recorded and stored in a uniform manner and inspect these measurements to determine optimal frequencies for cleaning, calibration, and other maintenance tasks.

ACKNOWLEDGEMENTS

The standard operating procedures for sensor quality assessment are a compilation of existing procedures used in utilities and academia. We gratefully acknowledge the support from Ahlem Filali (INRAE, France), Angel Roblés (Universitat de València, Spain), Conor Heffernan and Fenghua Yang (Metropolitan Water Reclamation District of Greater Chicago, IL, USA), Kris De Gussem (Aquafin, Belgium), Linda Åmand (Käppala Association, Sweden), and Sara Nilsson and Gabriel Persson (Stockholm Vatten och Avfall, Sweden).

REFERENCES

Asociación Española de Normalización (2004). UNE EN 12255-12:2004 Wastewater treatment plants – Part 12: Control and automation. 18pp.

Mašić A., Santos A. T. L., Etter B., Udert K. M. and Villez K. (2015). Estimation of nitrite in source-separated nitrified urine with UV spectrophotometry. *Water Research*, **85**, 244–254, https://doi.org/10.1016/j.watres.2015.08.031

Ramsay and Silverman (2005). Functional Data Analysis. Springer, New York, NY, 429pp.

RealTechWater Inc. (2017). www.realtechwater.com/w_p/wp-content/uploads/2021/04/Spectrum.png. accessed 12th July, 2023

S::CAN GmbH (2020). Water Quality OnLine Waste Water. https://www.s-can.at/en/download-catalogs-brochures-and-flyer/, accessed 12th July 2023. 108pp.

Thomas O. and Burgess C. (eds). (2017). UV-Visible Spectrophotometry of Waters and Wastewater. Elsevier, Cambridge, United States, 526pp.

APPENDIX 5.A: STANDARD OPERATING PROCEDURES FOR WATER QUALITY SENSOR VALIDATION

This section provides sensor validation procedures for common water quality sensors used in WRRFs. These can be used in the current form or modified to achieve different goals to satisfy applicable constraints. We strongly recommend consulting manufacturer's recommendation before deploying any of these procedures.

5.A.1 Procedures for a dissolved oxygen sensor

This section provides three procedures for DO sensor validation. Figure 5.A.1 shows the data quality metrics that can be estimated with these procedures.

5.A.1.1 DO-Procedure-1: Bias quantification with a hand-held reference sensor

Version and dates

- Version: 1.0
- Last review and revision: 31st July 2023
- Next review and revision due by: 31st July 2025

Objective

The goal of this procedure is to cross-check the DO value of a DO sensor in its normal environment with a reference DO value produced by a hand-held reference DO sensor. The action is well suited for WRRFs with many DO sensors as the same bucket with wastewater can be reused for nearby sensors.

This procedure enables:

- A single evaluation of the measurement error at the current operating point
- Quantification of bias (see Chapter 3)

If repeated regularly, this procedure enables:

- Quantification of drift (i.e., drift of bias, see Chapter 3)

Hardware

- DO sensor to be validated
- DO sensor to be used as reference, that is, calibrated and validated against a standard measurement
- Bucket around 20 L/5 gallon
- Stirrer

Consumables

- 10 L/2.5 gallon of wastewater, collected at the sensor's location

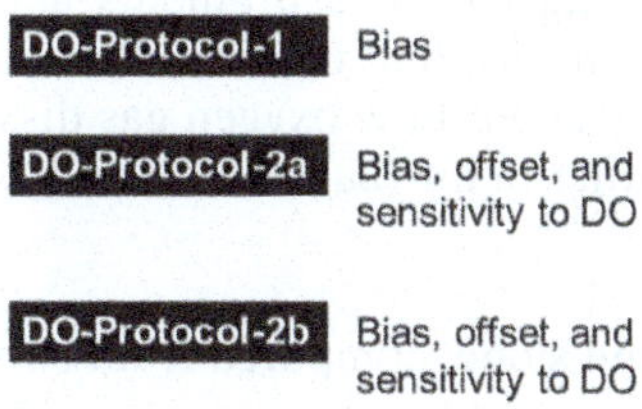

Figure 5.A.1 Overview of DO sensor validation procedures and the resulting sensor data quality metadata.

Experimental procedure

(1) Preparation (1–24 h prior to the validation)
   (a) Clean the DO sensor that is to be validated.
   (b) Validate the hand-held reference DO sensor with DO-protocol 2a or DO-protocol 2b.
(2) Validation
   (a) Collect a 10 L water sample with the bucket.
   (b) Add stirrer and continuously mix the water sample gently with the stirrer to obtain a homogenous mixture and avoid settling of suspended solids.
   (c) Visually inspect the DO sensor and assure it is clean and undamaged.
   (d) Place both the DO sensor and the reference hand-held DO sensor in the bucket.
   (e) Ensure that the temperature values of both sensors stabilize at the same value (normally 1–5 min).
   (f) Record the stabilized temperature values from both sensors.
   (g) Assure that the wastewater is continuously stirred.
   (h) Record stabilized DO values with both sensors.
   (i) Reposition the DO sensor back into its normal measurement position.
(3) Clean-up
   (a) Pour the water sample back (or continue with the next DO sensor).
   (b) Clean bucket and stirrer.
   (c) Return bucket and DO reference sensor to their normal location.

Data analytics and recording

(1) Store the values of temperature, DO for the sensor, and hand-held reference sensor in a database.
(2) The difference between the DO sensor and the DO reference sensor is an estimate of the absolute measurement error of the sensor under current operating conditions and can be used to determine whether validation with DO-Procedure-2a or DO-Procedure-2b or other actions are necessary.

Additional remarks

Repetition with multiple DO sensors using the same water sample is feasible and minimizes the time allocated for water sample preparation. To this end, repeat steps 1a and 2c–2i for every DO sensor.

5.A.1.2 DO-Procedure-2a: Two-point validation with saturated air

Version and dates

- Version: 1.0
- Last review and revision: 31st July 2023
- Next review and revision due by: 31st July 2025

Objective

This procedure is used for validating DO measurements at two reference conditions, zero DO and saturated DO. The reference conditions, that is, two different DO concentrations, are possible to obtain by using theoretical knowledge about how oxygen gas dissolve in air and water. For saturated DO, two approaches are available, either in air (2a) or water (2b).

This procedure enables:

- Quantification of bias, offset, and slope, computed from two DO-concentrations

If repeated regularly, this procedure enables:

- Quantification of drift (of bias, offset, and slope)

Hardware

- DO sensor to be validated
- Accurate temperature sensor (±0.5°C)
- Barometric pressure sensor (is only needed once)
- Conductivity sensor (may only be needed once if the influent is affected by sea water and large salinity variations)
- 20 L bucket

Zero DO in reduced water (validation at 0 mg DO/L)

- 3 L bucket containing tap water
- 10 g of sodium sulphite or equivalent strong reducing agent

DO in saturated air (validation at 9 mg DO/L and 20°C)

- 1 L beaker with a lid and similar diameter to DO sensor
- Straps for fixing sensor in the 1 L beaker

Consumables

- Paper towels (or equivalent)
- Clean tap water

Experimental procedure

(1) Preparation (1–24 h before validation)
  (a) Clean the DO sensor and resubmerge it in its original position.
  (b) Fill the 20 L bucket with 10–15 L tap water and locate at the site for the validation. Allow the tap water to achieve the same temperature as the ambient air.
  (c) Add 200 mL tap water to the bottom of a 1 L beaker and close the lid. Store the beaker at the same temperature as the ambient air (e.g., in the same room or outdoors). This will assure saturated air at the ambient air temperature. Shield the sensor from direct sunlight if executed outdoors.
  (d) Identify the height (metres above sea level) for the location where the validation is executed, alternatively measure the barometric pressure.
  (e) Measure the salinity or conductivity at the site where the sensor is located.

(2) Validation
  (a) Rinse the DO sensor (including its rod) with tap water.
  (b) Allow the sensor to stabilize at the current room (outdoor) temperature by submerging the clean sensor into the 20 L tap bucket with clean water at ambient air temperature.
  (c) Measure the temperature in the clean water bucket and compare with the temperature recorded by the DO sensor.
  (d) Record both temperature values.
  (e) Zero DO in water
    (i) Locate the clean sensor in a 3 L clean water bucket with a stirrer.
    (ii) Successively add sodium hydrosulphite (or the selected reducing agent) until the measured DO achieves a stable low value close to zero.
    (iii) Record the measured value at 0 mg/L DO.
    (iv) Record the sensor value and the current temperature indicted by the sensor and the external thermometer. Assure that the temperatures are equal.
    (v) Compare the recorded value with the theoretical value in the DO saturation table (Table 5.A.1) when correcting for barometric pressure and temperature.

Table 5.A.1 Correction factors for dissolved oxygen (DO) concentration at 100% saturation and 1 atm for different temperatures.

Saturation Dissolved Oxygen Concentrations						
	Conductivity (μS/cm)					
Temperature (°C)	**0**	**2000**	**4000**	**6000**	**8000**	**10,000**
10	11.28	11.20	11.12	11.04	10.95	10.87
11	11.02	10.94	10.87	10.79	10.70	10.62
12	10.77	10.69	10.62	10.54	10.47	10.39
13	10.53	10.46	10.38	10.31	10.24	10.16
14	10.29	10.22	10.15	10.07	10.00	9.93
15	10.07	10.00	9.93	9.86	9.79	9.72
16	9.86	9.79	9.72	9.65	9.58	9.52
17	9.65	9.58	9.51	9.46	9.39	9.32
18	9.45	9.38	9.33	9.26	9.19	9.13
19	9.26	9.20	9.14	9.07	9.01	8.95
20	9.08	9.02	8.96	8.90	8.83	8.77
21	8.90	8.84	8.78	8.72	8.66	8.61
22	8.73	8.67	8.62	8.56	8.50	8.44
23	8.56	8.51	8.45	8.39	8.34	8.28
24	8.40	8.35	8.29	8.24	8.18	8.12
25	8.24	8.19	8.13	8.08	8.03	7.98

(f) DO in saturated air
   (vi) Wipe the clean sensor with a dry paper towel and remove any water droplets on the sensor membrane/cap.
   (vii) Remove the lid from 1 L beaker and position the sensor by fixing it in a stable position with straps (or similar) in an upright position. Assure that the sensor is just above the water surface and has no contact with water.
   (viii) Allow the measurement signal to stabilize (10–30 min). Assure that the temperature is constant and equal for the sensor body as the air during the validation. Direct sunlight or strong cold wind should be avoided (worst case may require indoor validation).
   (ix) Record the sensor value and the current temperature indicted by the sensor and the air temperature by the external thermometer. Assure that the temperatures are equal.
(g) Reposition the DO sensor back into its normal measurement position.

(3) Clean-up
   (a) Pour the water sample back (or continue with the next DO sensor).
   (b) Clean hardware
   (c) Return hardware and consumables to their storage location.

Data analytics

(1) Define pressure correction factor from barometric measurement or the height of location.
(2) Validate the DO sensor's temperature correction by computing the difference between the recorded external temperature and the one recorded by the sensor.
(3) Verify that a correct salinity correction is applied for the measurements in the activated sludge tank.
(4) Compute the difference between the DO sensor value and the theoretical value during the reference conditions, adjusted for temperature, pressure, and salinity.

(5) Record all values.
(6) Validate the DO sensor by estimating its slope and offset using linear regression for zero and saturated DO values.
(7) Record the estimated slope and offset.

Additional remarks
Repetition with multiple DO sensors using the same water sample is feasible and minimizes the time allocated for water sample preparation. To this end, repeat steps 1a and 2c–2i for every DO sensor.

5.A.1.3 DO-Procedure-2b: Two-point validation with saturated water

Version and dates

- Version: 1.0
- Last review and revision: 31st July 2023
- Next review and revision due by: 31st July 2025

Objective
This procedure is used for validating DO measurements at two reference conditions, zero DO and saturated DO. The reference conditions, that is, two different DO concentrations, are possible to obtain by using theoretical knowledge about how oxygen gas dissolve in air and water. For saturated DO, two approaches are available, either in air (2a) or water (2b).
This procedure enables:

- Quantification of bias, offset, and slope, computed from two DO-concentrations

If repeated regularly, this procedure enables:

- Quantification of drift (of bias, offset, and slope)

Hardware

- DO sensor to be validated
- Accurate temperature sensor (±0.5°C)
- Barometric pressure sensor (is only needed once)
- Conductivity sensor (may only be needed once if the influent is unaffected by sea water and large salinity variations)
- Air compressor or access to compressed air
- Zero DO in reduced water (validation at 0, 0 mg DO/L)
    - 3 L bucket containing clean water
    - 10 g of sodium sulphite or equivalent strong oxygen reducing agent
- DO in saturated water (validation at 9 mg DO/L and 20°C)
    - Device allowing adding compressed air to 5 L bucket device, to allow aerating
    - Mixing rod or stir plate with a magnetic stir bar

Consumables

- Paper towels (or equivalent)
- Clean tap water

Experimental procedure

(1) Preparation (1–24 h before validation)
    (a) Clean the DO sensor and resubmerge it in its original position.
    (b) Fill 3 L and 5 L buckets with clean water and locate at the site for the validation. Allow the clean water to achieve the same temperature as the ambient air.

   (c) Identify the height (metres above sea level) for the location where the validation is executed, alternatively measure the barometric pressure.
   (d) Measure the conductivity at the site where the sensor is located.
(2) Validation
   (a) Rinse the DO sensor (including its rod) with clean water
   (b) Allow the DO sensor to stabilize at current room (outdoor) temperature by submerging the clean sensor in the 5 L clean water bucket with ambient air temperature.
   (c) Measure the temperature in the clean water bucket and compare with the temperature recorded by the DO sensor.
   (d) Record both temperature values.
   (e) Zero DO in water
      (i) Locate the clean sensor in the 3 L tap water bucket with a stirrer.
      (ii) Successively add sodium hydrosulphite (or the selected reducing agent) until the measured DO achieves a stable low value close to zero.
      (iii) Record the measured value at 0 mg/L DO.
      (iv) Record the sensor value and the current temperature indicted by the sensor and the external thermometer. Assure that the temperatures are equal.
      (v) Compare the recorded value with the theoretical value in the DO saturation table (Table 5.A.1) while correcting for barometric pressure and temperature.
   (f) DO in saturated water
      (i) Consult a DO solubility table (Table 5.A.1) and identify the 100% saturated DO concentration for the temperature of the 5 L bucket and barometric pressure.
      (ii) Rinse the DO sensor (including its rod) with tap water
      (iii) Allow the DO sensor to stabilize at current room (outdoor) temperature by submerging the clean sensor in the 5 L tap water bucket with ambient air temperature.
      (iv) Successively aerate the water in the 5 L bucket using compressed air until the sensor measurement stabilizes at or above the identified 100% DO saturation limit – stop aerating.
      (v) Allow the DO measurement to stabilize (3–10 min) while stirring.
      (vi) Record the DO at 100% saturation and the temperature.
   (g) Reposition the DO sensor back into its normal measurement position.
(3) Clean-up
   (a) Pour the water sample back (or continue with the next DO sensor).
   (b) Clean hardware
   (c) Return hardware and consumables to their storage location.

Data analytics

(1) Define pressure correction factor from barometric measurement or the height of location.
(2) Validate the DO sensor's temperature correction by computing the difference between the recorded external temperature and the one recorded by the sensor.
(3) Verify that a correct salinity correction is applied for the measurements in the activated sludge tank.
(4) Compute the difference between the DO sensor and the theoretical value during the reference conditions, adjusted for temperature, pressure, and salinity.
(5) Record all values, both measured and theoretical ones in a database.
(6) Validate the DO sensor by estimating its slope and offset using linear regression for zero and saturated DO values.
(7) Record the estimated slope and offset.
(8) Decide whether the sensor needs subsequent calibration.

Additional remarks
Repetition with multiple DO sensors using the same water sample is feasible and minimizes the time allocated for water sample preparation. To this end, repeat steps 1a and 2c–2i for every DO sensor.

5.A.1.4 Contextual information for dissolved oxygen sensor validation

The following paragraphs serve as examples of what could be included as contextual information in an SOP for DO sensor validation.

Purpose of DO measurement
The sensor is used to measure and control the DO in the activated sludge process. At the same time, the temperature can be recorded as this secondary measurement is included to compensate for the temperature influence on the DO concentration.

Available measurement principle
Two types of DO sensors are common: electrochemical and fluorescent. Both types measure the oxygen partial pressure, which is converted to DO concentration:

- *Electrochemical (polarographic):* Electrochemical DO sensors consist of an oxygen (gas) permeable membrane, an electrolyte, an anode, and a cathode. When oxygen molecules diffuse through the membrane, the anode is oxidized by the oxygen thereby creating a current that is proportional to the partial pressure of oxygen. An initial voltage needs to be applied to overcome the activation energy between the anode and cathode, that is, the polarization energy.
- *Optical (fluorescent):* Fluorescent DO sensors use fluorescence to measure the oxygen partial pressure. A fluorescence sensor consists of a fluorescent dye covering the sensor tip, a light source with specific wavelengths (commonly LED light), and a photodetector for recording the relaxed fluorescence-emitted light. The dye absorbs (is excited) by a light source and subsequently emits light at a longer wavelength (lower energy) as fluorescent light. When oxygen is present at the fluorescent dye, the remitted fluorescence light is delayed in proportion to the presence of oxygen molecules on the sensor.

Confounding factors
The DO concentration is measured by the partial pressure of oxygen gas in the solution. Hence, factors influencing the oxygen partial pressure will also influence the DO measurement. The following confounding factors may impact the DO measurement:

- *Angle of measurement*:
  - The angle of the sensor probe relative to the water surface is known to influence the measurement. Effects may include entrapment of air bubbles or solids.
  - *Mitigating action:* Unless specified otherwise by the manufacturer, we recommend to position the sensor with the membrane or optical window facing down while maintaining a 45° angle with the water surface.
- *Air bubbles*:
  - Air bubbles can give an erratic and increased oxygen reading.
  - *Mitigating action:* Adjust the sensor to a sharper contact angle with the water surface or re-locate the sensor.
- *Barometric pressure*:
  - The atmospheric pressure has a proportional impact on the partial oxygen pressure. Weather pressure changes are in the order of 950–1050 hPa, which has minor impact (<5%) on the reading. However, during sensor calibration, the barometric pressure can have a large impact and should be compensated for, especially if the WRRF is located higher than sea level.
  - *Mitigating action:* Correct for barometric pressure during calibration. For example, a plant at 1000 m above sea level corresponds to 899 hPa, relative to 1013 hPa at sea level, and hence needs an 11% correction.

- *Fouling*:
  - DO sensor's contamination prevents oxygen diffusion in the membrane and increases response time.
  - *Mitigating action:* Clean the sensor and revise best cleaning practice.
- *Salinity*:
  - Salinity decreases the oxygen solubility and hence the oxygen partial pressure. Sewers penetrated by sea water or combined sewer overflow may influence the salinity and in turn, the DO measurement.
  - *Mitigating action:* Adjust for common conductivity during calibration and monitor large conductivity changes in the influent.
- *Polarization current* (electrochemical sensor type):
  - An initial current is needed for polarization of the sensor (i.e., initiating the current driven by the reduced oxygen molecules). For this reason, an electrochemical DO sensor needs to be in operation with power for at least 30 min before it produces reliable data.
  - *Mitigating action:* Ensure that (a) signals are labelled as untrusted for 30 min after powering on or (b) signal data are only stored when the sensor has been powered on for at least 30 min.
- *Temperature*:
  - The solubility of oxygen in water is highly temperature dependent. For this reason, the DO measurement is automatically corrected by an internal temperature sensor, within the DO sensor. If the sensor body has a different temperature than the ambient water (or air), or if the temperature sensor is inaccurate, the DO measurement will in turn be incorrect (see Table 5.A.1).
  - *Mitigating action:* Assure sensor body temperature to stabilize before measurements. Validate the internal temperature sensor.
- *Direct sunlight* (fluorescent sensor type):
  - Direct sunlight can influence the measurement as well as the sensor body temperature.
  - *Mitigating action:* Shield the sensor from sunlight during validation.

Pressure correction formula

$$\text{Pressure correction factor} = p_{\text{Val}}/p_{\text{Atm}}$$

where p_{Val} is the barometric pressure at the site for validation and p_{Atm} is the atmospheric pressure at the sea level, that is, 1 atm.

DO solubility table for correction of temperature and conductivity

See Table 5.A.1.

5.A.2 Procedures for pH sensor validation

This section provides three procedures for pH sensor validation. Figure 5.A.2 shows the data quality metrics that can be estimated with these procedures.

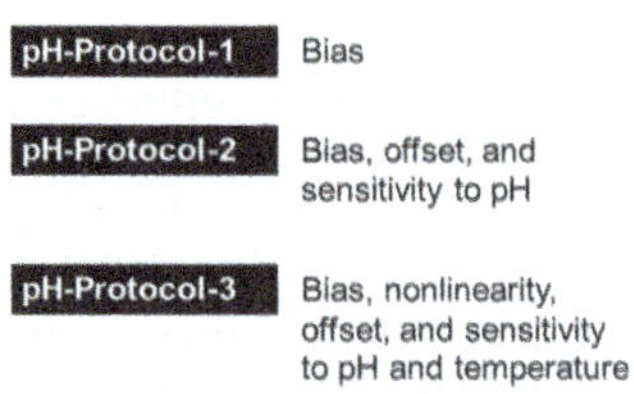

Figure 5.A.2 Overview of pH sensor validation procedures and the resulting sensor data quality metadata.

5.A.2.1 pH-Procedure-1: Procedure for one-point validation of pH sensors at current operating point

Version and dates

- Version: 1.0
- Last review and revision: 31st July 2023
- Next review and revision due by: 31st July 2025

Objective

This procedure aims to cross-check the pH value of a pH sensor in its normal environment with a reference pH value produced by a hand-held reference pH sensor.

This procedure enables:

- A single evaluation of the measurement error at the current operating point
- Quantification of bias (see Chapter 3)

If repeated regularly, this procedure enables:

- Quantification of drift (i.e., drift of bias, see Chapter 3)

This procedure does not enable:

- Quantification of non-linearity, offset, slopes (sensitivity to pH and/or temperature), sensor dynamics, or variance (see Chapter 3)

Hardware

- pH sensor to be validated (pH^{test}), connected to display, data logger, or transformer
- Reference pH sensor (pH^{ref}), cleaned and verified in a laboratory (see pH-procedure-2 or pH-procedure-3)
- Two buckets of 4 L (~1 gallon) or larger
- Non-abrasive cloth
- Stirrer

Consumables

- 4 L (~1 gallon) of wastewater, collected at the sensor's position
- 4 L (~1 gallon) of tap water

Experimental procedure

(1) Solution preparation:
  - (a) Collect water sample with bucket
  - (b) Add stirrer and continuously mix the water sample gently with the stirrer to obtain a homogenous mixture and avoid settling of suspended solids

(2) Clean sensor:
  - (a) Collect tap water with bucket
  - (b) Clean pH^{test} sensor with non-abrasive cloth and tap water

(3) Place pH^{test} and pH^{test} sensors in bucket

(4) Identify the unit of measurements with an X:
  - (a) Temperature: [__] °C [__] °F [__] K
  - (b) Electric potential: [__] mV

(5) Ensure the temperature values of both sensors are stable.

(6) Record values:
  - (a) Record temperature value with both sensors
    - (i) T^{test}: ____
    - (ii) T^{ref}: ____

(b) Record electric potential value with both sensors
   (i) P^{test}: ____
   (ii) P^{ref}: ____
(c) Record pH values with both sensors, given current calibration
   (i) pH^{test}: ____
   (ii) pH^{ref}: ____

(7) Pour the water sample back
(8) Place pH^{test} sensor back into its normal measurement position
(9) Clean buckets, stirrer
(10) Return bucket and pH^{ref} sensor to their normal location

Data analytics and recording

(1) Store the values T^{test}, T^{ref}, P^{test}, P^{ref}, pH^{test}, pH^{ref} in a spreadsheet or database.
(2) Check that T^{test} and T^{ref} are the same within an acceptable margin of error (e.g., 0.5 K). If not, repeat experimental procedure when all solutions have achieved stable temperature. If repeated attempts fail, validate the temperature sensors, and apply necessary steps to ensure acceptable measurement of the temperature sensors.
(3) The pH measurement difference ($pH^{test} - pH^{ref}$) is an estimate of the absolute measurement error of the sensor under current operating conditions and can be used to determine whether a more extensive validation is necessary (e.g., with pH-procedure-2 or pH-procedure-3).
(4) The potential measurement difference (P^{test}, P^{ref}) is valuable to study long-term effects on the wear-and-tear of the pH sensor, especially since this difference is unaffected by the sensor's calibration curve.

5.A.2.2 pH-Procedure-2: Procedure for two-point validation of pH sensors in buffered media

Version and dates

- Version: 1.0
- Last review and revision: 31st July 2023
- Next review and revision due by: 31st July 2025

Objective

This procedure aims to record the raw sensor output of ion-selective pH sensors under controlled conditions. In this case, controlled conditions mean that:

- Stable and buffered solutions are used to validate the sensor at specific pH values
- The temperatures of the solutions are stable and recorded with the same pH sensor

This procedure enables:

- Evaluation of the measurement error at two controlled pH values
- Quantification of bias (see Chapter 3), offset, and slope (sensitivity to pH)

If repeated regularly, this procedure enables:

- Quantification of drift (of bias, offset, and slope)

This procedure does not enable:

- Quantification of non-linearity, sensitivity to temperature, sensor dynamics, or variance (see Chapter 3)

Hardware

- pH sensor to be validated, connected to display, data logger, or transformer

- Three glass beakers of 200 mL
- Three magnet stirrers with magnet
- Buckets of 5 L (~1 gallon) or larger
- Non-abrasive cloth

Consumables

- 5 L (~1 gallon) of tap water
- Solutions with stable temperature, 100 mL each
  - Tap water [W]
  - Calibration solution at pH 4 [C4]
  - Calibration solution at pH 7 [C7]

Experimental procedure

(1) Solution preparation:
  (a) Label beakers with W, C4, and C7
  (b) Place each beaker on a magnetic stirrer
  (c) Add a magnet to each beaker
  (d) Pour each solution (W, C4, C7) into the beaker with matching label (W, C4, C7)
  (e) Start stirrer
(2) Remove sensor from its normal measurement position
(3) Clean sensor:
  (a) Collect tap water with bucket
  (b) Clean sensor with non-abrasive cloth and tap water
(4) Place sensor in beaker W
(5) Wait until sensor temperature stabilizes
(6) Ensure all solutions have achieved room temperature
(7) Identify the unit of measurements with an X:
  (a) Temperature: [__] °C [__] °F [__] K
  (b) Electric potential: [__] mV [__] __
(8) Record 0:
  (a) Record temperature value *T*0: ____
  (b) Record electric potential value P0: ____
(9) Step change 1:
  (a) Move sensor from beaker W to beaker C4
  (b) Wait 5 min
(10) Record 1:
  (a) Record temperature value T1: ____
  (b) Record electric potential value P1: ____
(11) Step change 2:
  (a) Move sensor from beaker C4 to beaker C7
  (b) Wait 5 min
(12) Record 2:
  (a) Record temperature value T2: ____
  (b) Record electric potential value P2: ____
(13) Step change 3:
  (a) Move sensor from beaker C7 to beaker C4
  (b) Wait 5 min
(14) Record 3:
  (a) Record temperature value T3: ____
  (b) Record electric potential value P3: ____

(15) Step change 4:
    (a) Move sensor from beaker C4 to beaker W
    (b) Wait 5 min
(16) Record 4:
    (a) Record temperature value T4: ____
    (b) Record electric potential value P4: ____
(17) Clean sensor with non-abrasive cloth and tap water
(18) Place sensor back into its normal measurement position
(19) Recycle/waste all solutions
(20) Clean glassware, stirrers, and magnets and return to storage

Data analytics and recording

(1) Store the values T0, T1, T2, T3, T4, P0, P1, P2, P3, P4 in a spreadsheet or database
(2) Check that T1, T2, T3 are all equal within an acceptable margin of error (e.g., 0.5 K). If not, repeat experimental procedure when all solutions have achieved stable temperatures.
(3) The value P2 is an estimate of the offset of the sensor
(4) Two estimates of the sensor's slope at standard temperature ($T_s = 298.15$ K) can be computed with the following steps:
    (a) Sort the values T1, T2, and T3 in increasing order and select T_a as the middle value (median temperature). Ensure T_a is expressed in Kelvin (K)
    (b) Compute the first estimate of the sensor's slope as

$$\mathrm{S1} = \frac{\mathrm{P1} - \mathrm{P2}}{3} \times \frac{T_s}{T_a}$$

    (c) Compute the second estimate of the sensor's slope as

$$\mathrm{S2} = \frac{\mathrm{P3} - \mathrm{P2}}{3} \times \frac{T_s}{T_a}$$

5.A.2.3 pH-Procedure-3: Procedure for multipoint validation of pH sensors in buffered media

Version and dates

- Version: 1.0
- Last review and revision: 31st July 2023
- Next review and revision due by: 31st July 2025

Objective

This procedure aims to record the raw sensor output of ion-selective pH sensors under controlled conditions. In this case, controlled conditions mean that:

- Stable and buffered solutions are used to validate the sensor at specific pH values
- The temperatures of the solutions are stable and recorded with the same pH sensor

This procedure enables:

- Evaluation of the measurement error at three controlled pH values
- Quantification of bias (see Chapter 3), non-linearity, offset, and slopes (sensitivity to pH and/or temperature)

If repeated regularly, this procedure enables:
Quantification of drift (of bias, non-linearity, offset, and slopes, see Chapter 3)

- This procedure does not enable:
- Quantification of sensor dynamics or variance (see Chapter 3)

Hardware

(1) pH sensor to be validated, connected to display, data logger, or transformer
(2) Temperature-controlled beakers of 200 mL: 7×
(3) Magnet stirrers with magnet: 7×

Consumables

(1) 4 L (~1 gallon) of tap water
(2) Seven solutions with stable temperature, 100 mL each
 (a) Tap water at standard temperature (T_s = 298.15 K) [W]
 (b) Calibration solution at pH 4, 7, 10 at standard temperature (T_s = 298.15 K) [C4s, C7s, C10s]
 (c) Calibration solution at pH 4, 7, 10 at application temperature (T_a = 288.15 K) [C4a, C7a, C10a]

Experimental procedure

(1) Solution preparation:
 (a) Label beakers with W, C4s, C7s, C10s, C4a, C7a, and C10a
 (b) Place each beaker on a magnetic stirrer
 (c) Add a magnet to each beaker
 (d) Pour each solution (W, C4s, C7s, C10s, C4a, C7a, and C10a) into the beaker with matching label (W, C4s, C7s, C10s, C4a, C7a, and C10a)
 (e) Start stirrer and ensure temperature control
(2) Remove sensor from its normal measurement position
(3) Clean sensor:
 (a) Collect tap water with bucket
 (b) Clean sensor with non-abrasive cloth and tap water
(4) Place sensor in beaker W and wait until sensor temperature becomes stable
(5) Ensure all solutions have achieved room temperature
(6) Identify the unit of measurements and record these as in Table 5.A.2.
(7) Record 0:
 (a) Record temperature value T0: ____
 (b) Record electric potential value P0: ____
(8) Repeat until every calibration solution has been measured:
 (a) Step change: Move sensor to the next solution in this sequence: W, C4s, C4a, C7a, C7s, C10s, C10a, W
 (b) Wait for 5 min

Table 5.A.2 Recorded values.

	Temperature	Potential
Measurement unit		
W		
C4s		
C4a		
C7a		
C7s		
C10s		
C10a		
W		

   (c) Record temperature value T in Table 5.1
   (d) Record electric potential value P in Table 5.1
(9) Clean sensor with non-abrasive cloth and tap water
(10) Place sensor back into its normal measurement position
(11) Recycle/waste all solutions
(12) Clean glassware, stirrers, and magnets and return to storage

Data analytics and recording

(1) Store the values of Table 5.A.2 in a spreadsheet or database
(2) Check that all temperatures for the media W, C4s, C7s, and C10s are equal within an acceptable margin of error (e.g., 0.5 K). Check that all temperatures for the media C4a, C7a, and C10a are equal within an acceptable margin of error. If not, repeat the experimental procedure when all solutions have achieved stable temperatures.
(3) Execute a linear regression with the temperature values and the known pH value. Use all data in Table 5.A.2 corresponding to the media C4s, C4a, C7a, C7s, C10s, and C10a. Do not use medium W. Inspect the offset and slope values (for temperature and pH) and compare with (a) previously obtained values (if available) and (b) theoretical values (Nernst equation). Determine whether the quality of the regression is acceptable by inspecting the residuals as a function of the measured potential, measured temperature, and known pH values. Use this inspection to evaluate whether the sensor's behaviour is sufficiently linear to be acceptable. If considered acceptable, use the computed regression coefficient to determine and record the sensor bias, offset, and slope.

5.A.3 Procedures for a total ammonia nitrogen sensor

This section provides two procedures for TAN sensor validation. Figure 5.A.3 shows the data quality metrics that can be estimated with these procedures.

5.A.3.1 TAN-Procedure-1: Procedure for one-point ammonia sensor validation at current operating point

Version and dates

- Version: 1.0
- Last review and revision: 31st July 2023
- Next review and revision due by: 31st July 2025

Objective

The goal of this procedure is to cross-check the TAN concentration value produced by an ammonium sensor in its normal environment with a reference concentration value produced through an analytical lab measurement. The procedure is designed for use with an ISE sensor. To modify the procedure for use with an ammonia analyser (colorimetric principle), see the additional remarks below.

This procedure enables:

- A single evaluation of the measurement error at the current operating point
- Quantification of bias (see Chapter 3)

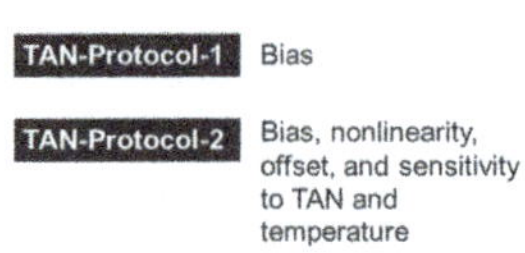

Figure 5.A.3 Overview of TAN sensor validation procedures and the resulting sensor data quality metadata.

If repeated regularly, this procedure enables:

- Quantification of drift (i.e., drift of bias, see Chapter 3)

Hardware

- Ammonia sensor to be validated
- Bucket around 20 L/5 gallon (or larger)
- Stirrer
- Vial for transport of a water sample to a laboratory setting

Consumables

- 10 L/2.5 gallon of wastewater, collected at the sensor's location
- 100 mg allylthiourea powder

Experimental procedure

(1) Preparation (1–24 h prior to the validation)
 (a) Clean the ammonia sensor that is to be validated
(2) Validation
 (a) Collect a 10 L water sample with the bucket.
 (b) Add stirrer and continuously mix the water sample gently with the stirrer to obtain a homogenous mixture and avoid settling of suspended solids.
 (c) Visually inspect the ammonia sensor and assure it is clean and undamaged.
 (d) Add allylthiourea to inhibit the oxidation of ammonia (e.g., 10 mg N/L is typically well above the inhibition level).
 (e) Place the ammonia sensor in the bucket.
 (f) Assure that the wastewater is continuously stirred.
 (g) Ensure that the pH and temperature value of the sensor are at a steady value (normally 5–10 min).
 (h) Record the stabilized pH value from the ammonia sensor.
 (i) Record the stabilized temperature value from the ammonia sensor.
 (j) Record the stabilized TAN concentration value from the ammonia sensor.
 (k) Take a 20 mL water sample from the bucket for laboratory analysis.
 (l) Reposition the ammonia sensor back into its normal measurement position.
(3) Clean-up
 (a) Pour the water sample back.
 (b) Clean bucket and stirrer.
 (c) Return bucket and ammonia sensor to their normal location.
 (d) Transfer the 20 mL water sample to the laboratory for analysis with a standard measurement or accepted analytical method for reference measurement.

Data analytics and recording

(4) Store the values for pH, temperature, and TAN concentration for the sensor in a database.
(5) Store the ammonia nitrogen concentration value obtained through analytical measurement in the laboratory.
(6) The difference between the ammonia sensor value and the analytical measurement is an estimate of the absolute measurement error of the sensor under current operating conditions and can be used to determine whether further actions are necessary.

Additional remarks

(7) Repetition with multiple ammonia sensors using the same water sample is feasible and minimizes the time allocated for water sample preparation. To this end, repeat steps 1a and 2c–2l for every ammonia sensor.

(8) Collection of ammonia sensor values through an automated data acquisition system rather than manual recording will facilitate quantification of sensor precision (see Chapter 3).

(9) If the ammonia sensor is equipped with additional electrodes (e.g., K^+) to compensate for confounding effects, then add a step after step 2j to also measure the stabilized value obtained with these electrodes. It may of interest to then modify step 3d as well so that the confounding ion concentration is also measured.

(10) To modify this procedure for use with an ammonia analyser (colorimetric measurement principle), change steps 1a through 2k by accounting for the following aspects:
   (a) Adjust cleaning procedure according to analyser manual.
   (b) Rather than changing the position of the sensor, change the position of the sampling line so the analyser collects water from the bucket during validation.
   (c) Adjust wait times to account for the fact that the analyser measures the concentration with delay (due to time needed for colorimetric reaction) and at a lower frequency (e.g., 10-min interval).

5.A.3.2 TAN-Procedure-2: Procedure for multipoint ammonia sensor validation at current operating point

Version and dates

- Version: 1.0
- Last review and revision: 31st July 2023
- Next review and revision due by: 31st July 2025

Objective

The goal with this procedure is to cross-check the TAN concentration value produced by an ammonium sensor in its normal environment with a reference concentration value produced through an analytical lab measurement. The procedure is designed for use with an ISE sensor. To modify the procedure for use with an ammonia analyser (colorimetric principle), see the additional remarks below.

This procedure enables:

- Evaluation of the measurement error at five distinct ammonia concentration values
- Quantification of bias (see Chapter 3), non-linearity, offset, and slope (sensitivity to ammonia)

If repeated regularly, this procedure enables:

- Quantification of drift (of bias, non-linearity, offset, and sensitivity to ammonia)

This procedure does not enable:

- Quantification of sensor dynamics or variance (see Chapter 3)

Hardware

- Ammonia sensor to be validated
- Beaker of 20 L/5 gallon (or larger)
- Beaker of 2 L/0.5 gallon (or larger)
- Stirrer
- 20 mL vial for transport of a water sample to a laboratory setting

Consumables

- 10 L/2.5 gallon of wastewater, collected at the sensor's location
- 1 L of ammonia stock solution with a concentration of 440 mg N/L
- 100 mg allylthiourea powder

Experimental procedure

- Preparation (1–24 h prior to the validation)
  (a) Clean the ammonia sensor that is to be validated.
- Validation
  (a) Collect a 10 L water sample with the 20 L beaker. Decant as necessary to obtain a precise volume.
  (b) Add stirrer and continuously mix the water sample gently with the stirrer to obtain a homogenous mixture and avoid settling of suspended solids.
  (c) Visually inspect the ammonia sensor and assure it is clean and undamaged.
  (d) Add allylthiourea to inhibit the oxidation of ammonia (e.g., 10 mg N/L is typically well above the inhibition level).
  (e) Place the ammonia sensor in the bucket.
  (f) Assure that the wastewater is continuously stirred.
  (g) Ensure that the pH and temperature value of the sensor are at a steady value (normally 5–10 min).
  (h) Take a 10 mL water sample from the bucket for laboratory analysis.
  (i) Record the stabilized pH value from the ammonia sensor.
  (j) Record the stabilized temperature value from the ammonia sensor.
  (k) Record the stabilized TAN concentration value from the ammonia sensor.
  (l) If five measurements have been obtained, go to step 2n. Otherwise go to step 2m.
  (m) Step-wise addition of stock solution (repeated four times):
    (i) Add 200 mL of the ammonia stock solution.
    (ii) Ensure that the pH and temperature value of the sensor are at a steady value (normally 5–10 min).
    (iii) Go back to step 2i.
  (n) Reposition the ammonia sensor back into its normal measurement position.
  (o) Pour the water sample back.
- Clean-up
  (a) Clean bucket and stirrer.
  (b) Return bucket and ammonia sensor to their normal location.
  (c) Transfer the 20 mL water sample to the laboratory for analysis with a standard measurement or accepted analytical method for reference measurement.

Data analytics and recording

- Store the values for pH, temperature, and TAN concentration for the sensor in a database.
- Store the ammonia nitrogen concentration value obtained through analytical measurement in the laboratory.
- Compute the expected concentrations corresponding to each ammonia nitrogen concentration value obtained by the sensor:

$$\begin{array}{l}\text{expected}\\ \text{concentration}\end{array} = \frac{\text{analytical concentration} \times \text{sample volume} + \text{stock concentration} \times \text{stock volume}}{\text{stock solution volume} + \text{sample volume}}$$

- Execute a linear regression with the expected ammonia concentrations and the sensor values. Determine whether the quality of the regression is acceptable by inspecting the residuals as a function of the expected concentrations. Use this inspection to evaluate whether the sensor's behaviour is sufficiently linear to be acceptable. If considered acceptable, use the computed regression coefficients to determine and record the sensor bias, offset, and slope.

Additional remarks

- Repetition with multiple ammonia sensors using the same water sample is feasible and minimizes the time allocated for water sample preparation. To this end, repeat steps 1a and 2c–2k for every ammonia sensor.
- Collection of ammonia sensor values through an automated data acquisition system rather than manual recording will facilitate quantification of sensor precision (see Chapter 3).
- If the ammonia sensor is equipped with additional electrodes (e.g., K^+) to compensate for confounding effects, then add a step after step 2j to also measure the stabilized value obtained with these electrodes. After step 2m, step-wise addition of stock solution of the confounding ions (e.g., K^+) can be included in the procedure to also measure the offset and slope of these compensating electrodes. Care should be taken to not introduce new confounding factors. For example, high concatenations of Cl^- will confound the ammonia electrode signal so KCl solutions should be avoided for this purpose.
- To modify this procedure for use with an ammonia analyser (colorimetric measurement principle), change steps 1a through 2k by accounting for the following aspects:
    (a) Adjust cleaning procedure according to analyser manual.
    (b) Rather than changing the position of the sensor, change the position of the sampling line so the analyser collects water from the bucket during validation.
    (c) Adjust wait times to account for the fact that the analyser measures the concentration with delay (due to time needed for colorimetric reaction) and at a lower frequency (e.g., 10-min interval).

5.A.3.3 Contextual information for ammonia sensor validation

The following paragraphs serve as examples of what could be included as contextual information in an SOP for ammonia sensor validation.

Purpose of ammonia measurement

The sensor is most frequently used to measure and control the ammonia concentration in the activated sludge process. A pH and temperature measurement is included to compensate for their influence on the electrode potential. Ideally, variations of these compensating measurements should either be controlled or measured during ammonia sensor validation.

Available measurement principles

Two types of ammonia sensors are common: ion-selective and colorimetric:

- *Ion-selective electrode:* ISE is used to measure an electric potential between two electrodes. The reference electrode and measuring electrode are both placed in media with a known composition and are both separated from the measuring environment with membranes that enable electrons to pass but minimize ion transfer. The measuring electrode is designed to enable molecular ammonia to pass and change the pH of the measuring electrode. This results in an electric potential that can be measured. In equilibrium, the electric potential is proportional to the logarithm of the molecular ammonia concentration in the measured environment. Thus, for measurement of TAN, the pH and temperature must be measured as well.
- *Colorimetric:* Colorimetric method is based on a reaction of ammonia with a reagent, in turn producing a colourful chemical. The resulting change in the light absorbance of the medium is proportional to the original TAN concentration.

Measurement range

Adequate selection of an ammonia ISE or ammonia analyser requires consideration of the confounding factors (see below) and the measurement range. An ammonia ISE is considered reliable from 1 to 1000 mg N/L. This suffices for monitoring and automation of conventional biological nitrification processes but may not suffice for effluent water quality monitoring.

Confounding factors (ammonia ion-selective electrode)

The measured electrode potential is influenced by the ammonia ion (NH_4^+) concentration. The ammonia ion (NH_4^+) concentration changes as a function of the TAN concentration, the pH, and temperature. Thus, to measure the TAN concentration, the electrode potential must be compensated for these effects. Many commercial ion-selective sensors for ammonia include additional electrodes and apply automatic compensation.

In addition to pH and temperature, the electrode potential is also influenced by changes in the concentration of ions that are similar in size and charge to the ammonia ion (NH_4^+). The most frequent confounding is caused by variations in the potassium (K^+) concentration. Most typically, this results from application of de-icing salt to the road system during the winter season. Some ammonia sensors will include electrodes that are sensitive to such ions. This then enables compensation for their effect of the measured electrode potential. Contact your sensor vendor and the local city administration to evaluate the need for such a compensation scheme or check [1] and [2]. When applicable and considered relevant, the above sensor validation procedures can be modified to capture the measurement error for these compensating electrodes as well.

Confounding factors (ammonia analyser)

The colorimetric measurement principle is far more specific to ammonia and exhibits little to no confounding factors when used within the measurement range provided by the manufacturer.

5.A.4 Procedures for a total suspended solids sensor

This section provides four procedures for TSS sensor validation. Figure 5.A.4 shows the data quality metrics that can be estimated with these procedures.

5.A.4.1 TSS-Procedure-1: Sensor integrity check

Version and dates

- Version: 1.0
- Last review and revision: 31st July 2023
- Next review and revision due by: 31st July 2025

Objective

The goal of this procedure is to check the deviation from factory calibration with a physical reference plate with known optical properties. Since the sludge properties impact the measured suspended solids, this check should only be regarded as a sensor integrity check prior to a complete validation.

This procedure enables:

- Quantification of bias during standardized non-process conditions

If repeated regularly, this procedure enables:

- Quantification of drift due to the sensor electronics

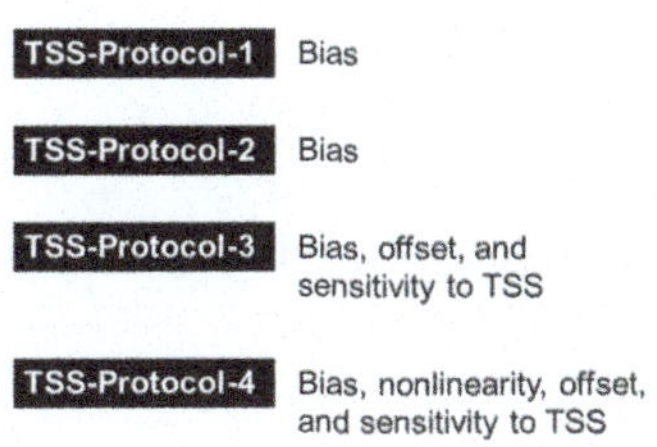

Figure 5.A.4 Overview of TSS sensor validation procedures and the resulting sensor data quality metadata.

Hardware

- TSS sensor to be validated
- A physical TSS reference disc applicable for the TSS sensor to be validated. The reference disc is a solid piece of a material such as plastic with fixed optical properties. Since the disc needs to fit the sensor measurement window, it is commonly tailored by the sensor manufacturer and not a generic component.

Consumables

- Bucket with hot water (for cleaning)
- Sponge or brush (for cleaning)
- Diluted acid of 1–3% HCl (for cleaning)
- Detergent (for cleaning)
- Paper towel

Experimental procedure

(1) Clean the sensor with water, a detergent, or diluted acid until any visible residues have been removed.
(2) Dry the sensor, and in particular, the measurement glass windows with paper.
(3) Mount the physical TSS reference disc to the sensor.
(4) Allow the sensor reading to stabilize at the current ambient temperature.
(5) Record the raw TSS signal. Note: the standard TSS value should not be used since this changes during calibration.
(6) Compare the raw signal with the previous reading (or factory value) to assess the drift.

5.A.4.2 TSS-Procedure-2: Bias quantification

Version and dates

- Version: 1.0
- Last review and revision: 31st July 2023
- Next review and revision due by: 31st July 2025

Objective

The goal of this procedure is to cross-check the TSS value of a TSS sensor in its normal environment and current state with a reference TSS value produced by a hand-held reference TSS sensor. The action is well suited for WRRFs with many TSS sensors as the same bucket with wastewater can be reused for nearby sensors.

This procedure enables:

- Quantification of bias computed from two TSS sensors

If repeated regularly, this procedure enables:

- Quantification of bias drift

Hardware

- TSS sensor to be validated
- A hand-held TSS sensor to act as reference, recently (within 1 week) calibrated and validated with TSS-procedure-3 or TSS-procedure-4
- Bucket around 20 L/5 gallon
- Stirrer
- A piece of dark cloth (when the TSS sensor is located outdoors)

Consumables

- 10 L/2.5 gallon of wastewater collected at the TSS sensor's location

Experimental procedure

(1) Preparation (up to 1 week prior to the validation)
  (a) Validate and/or calibrate the hand-held reference TSS sensor using TSS-procedure-3 or TSS-procedure-4.
(2) Validation
  (a) Visually inspect the TSS sensor and assure that the glass windows measuring absorbance are clean and undamaged.
  (b) Clean the TSS sensor that is to be validated or omit cleaning if the current (potentially fouled sensor) measurement is to be validated.
  (c) Collect a 10 L water sample with the bucket.
  (d) Continuously mix the water sample gently with the stirrer to obtain a homogenous mixture and avoid settling of suspended solids.
  (e) Place the hand-held TSS sensor in the bucket and gently swirl in a circular motion until the reading stabilizes.
  (f) Cover the bucket with the dark cloth when validating in the outdoors.
  (g) Record the value for the reference TSS sensor.
  (h) Continuously mix the water sample gently with the stirrer to obtain a homogenous mixture and avoid settling of suspended solids.
  (i) Place the TSS test sensor in the bucket.
  (j) Allow the TSS reading to stabilize while assuring that the wastewater is continuously stirred. Note that this may require an extended time (several minutes) if a long time constant for the low-pass filter is applied in the sensor's digital communication box.
  (k) Cover the bucket with the dark cloth when validating in the outdoors.
  (l) Record the stabilized TSS value.
  (m) Reposition the TSS sensor back into its normal measurement position.
  (n) Pour the water sample back (or continue with the next TSS sensor and thereby avoiding repetition of steps 2b–2f).
  (o) Clean bucket and stirrer.
  (p) Return the bucket and hand-held reference TSS sensor to their normal location.

Data analytics and recording

(1) The difference between the mean value from the reference TSS sensor (steps 2e–2g) and the TSS sensor (step 2l) is an estimate of the absolute measurement error of the sensor under current operating conditions (either clean or as is).

5.A.4.3 TSS-Procedure-3: Two-point validation at zero and a selected concentration

Version and dates

- Version: 1.0
- Last review and revision: 31st July 2023
- Next review and revision due by: 31st July 2025

Objective

The goal with this procedure is to validate the TSS value of a TSS sensor at zero and the commonly measured TSS concentration (about 2000–4000 mg/L in activated sludge basin and 5000–12,000 mg/L in the return sludge). The procedure can be used to validate both permanently located and hand-held TSS sensors.

This procedure enables:

- Quantification of bias, offset, and slope, computed from two TSS concentrations

If repeated regularly, this procedure enables:

- Quantification of drift (of bias, offset, and slope)

Hardware

- TSS sensor to be validated
- 2 buckets around 20 L/5 gallon
- 1 bucket around 10 L/5 gallon
- Stirrer
- Cleaning equipment: brush and sponge
- Safety gear: safety goggles and protective gloves (during acid cleaning)
- A piece of dark cloth (if the TSS sensor is located outdoors)

Consumables

- 5 L/1 gallon of wastewater, collected at the TSS sensor's location
- 5 L/1 gallon of tap or distilled water
- Paper towel
- Diluted acid of 1–3% HCl (for cleaning)
- Water sample bottles for laboratory measurements of suspended solids. Use 250 mL bottles unless otherwise requested by the laboratory.

Experimental procedure

(1) Preparation (up to 24 h prior to the validation)
  (a) Fill the 10 L bucket with tap or distilled water and allow to de-aerate indoors. Air bubbles from tap water may impact the reading and the filled bucket should be allowed to de-aerate for at least 4 h, but desirable for 24 h.
  (b) Fill one of the 20 L buckets with 5 L hot water (to be used for cleaning).
(2) Validation
  (a) Clean and record the sensor value at zero TSS.
    (i) Clean the TSS sensor that is to be validated with clean water.
    (ii) Visually inspect the TSS sensor and assure that the glass windows measuring absorbance are clean and undamaged.
    (iii) Position the cleaned TSS sensor in the 10 L bucket with de-aerated clean water and allow the measurement to stabilize.
    (iv) If the measurement is above zero, clean the sensor again with prepared acid solution to remove calcium and other precipitations.
    (v) Rinse the sensor with clean water.
    (vi) Repeat step ii and assure that the sensor measures zero, otherwise repeat steps (iii) and (iv) until reading is between 0 and 10 mg/L.
    (vii) Record the zero clean water value.
  (b) Sensor value at current operating point.
    (i) Collect a 10 L water sample with the 20 L bucket at the sensor location (or any desired concentration).
    (ii) Continuously mix the water sample gently with the stirrer to obtain a homogenous mixture and avoid settling of suspended solids.
    (iii) Place the TSS sensor in the bucket and gently swirl in a circular motion until the reading stabilizes.

(iv) Cover the bucket with the dark cloth when validating in the outdoors.
(v) Record the value.

(c) TSS reference measurement.
(i) Continuously mix the water sample gently with the stirrer to obtain a homogenous mixture and avoid settling of suspended solids.
(ii) Take a water sample (using the 250 mL water sample bottle) and send to a lab for analysis.

(d) Reposition the TSS sensor back into its normal measurement position.
(e) Pour the water sample back.
(f) Clean bucket and stirrer.

Data analytics and recording

(1) Record the sensor measurement in tap water and at current operating point.
(2) Obtain and record the TSS reference measurement value (from a lab).
(3) Compute the offset and slope deviation by least-squares regression of the recorded values.
(4) Store the offset and slope values together with the recorded values obtained in steps 2a–2c as validation metadata.

5.A.4.4 TSS-Procedure-4: Multipoint validation of total suspended solids sensors

Version and dates

- Version: 1.0
- Last review and revision: 31st July 2023
- Next review and revision due by: 31st July 2025

Objective

This procedure aims to record the sensor output of optical TSS sensors at three different TSS concentrations by successively diluting a concentrated base water sample.

This procedure enables:

- Quantification of bias, non-linearity, offset, and slope (sensitivity to TSS)

If repeated regularly, this procedure enables:

- Quantification of drift (of bias, offset, slope, and non-linearity)

Hardware

- TSS sensor to be validated
- 1 L beaker
- 4 buckets around 20 L/5 gallon
- 1 bucket around 10 L/5 gallon
- Stirrer
- Cleaning equipment: brush and sponge
- Safety gear: safety goggles and protective gloves (during acid cleaning)
- A piece of dark cloth or plastic (when the TSS sensor is validated outdoors)

Consumables

- 30 L (two buckets with 15 L each) water sample at a concentration in the middle of the sensor's measurement range.
- 3 water sample bottles for laboratory measurements of suspended solids. Use 250 mL bottles unless otherwise requested by the laboratory.

Experimental procedure

(1) Preparation
  (a) Prepare a zero TSS solution (up to 24 h prior to the validation). Fill the 10 L bucket with tap or distilled water and allow to de-aerate indoors. Air bubbles from tap water may impact the reading and the filled bucket should be allowed to de-aerate for at least 4 h, but desirable for 24 h.
  (b) Fill a 20 L bucket with at least 15 L tap water (to be used for diluting the water samples).
  (c) Prepare the TSS base solution (up to 1 h prior to the validation).
    (i) Fill two 20 L buckets with 15 L wastewater each with a concentration in the middle of the desired validation range.
    (ii) Allow the sludge to settle for 30–60 min.
  (d) Prepare tap water for cleaning. Fill the remaining 20 L bucket with 5 L hot water.
  (e) Prepare the concentrated (150%) base water sample.
    (i) Gently decanter about 30% (5 L) of the low TSS water in both buckets in step 1c.
    (ii) Dilute the concentrated water sample into one of the buckets to a total volume of about 16 L (at least 15 L + a water sample for laboratory analysis). The concentration should now be ~50% higher than in the original water sample.
    (iii) Pour the remaining sludge into the second bucket.
    (iv) Continuously mix the water sample gently with the stirrer to obtain a homogenous mixture and avoid settling of suspended solids, then take a water sample for laboratory analysis marked as *TSS 150%*.
    (v) Continuously mix the water sample gently with the stirrer to obtain a homogenous mixture and avoid settling of suspended solids, then remove excess wastewater until 15 L remains in the bucket.

(2) Validation
  (a) Clean and record the sensor value at zero TSS.
    (i) Clean the TSS sensor that is to be validated with clean water.
    (ii) Visually inspect the TSS sensor and assure that the glass windows measuring absorbance are clean and undamaged.
    (iii) Position the cleaned TSS sensor in the 10 L bucket with de-aerated clean water and allow the measurement to stabilize.
    (iv) If the measurement is above zero, clean the sensor again with prepared acid solution to remove calcium and other precipitations.
    (v) Rinse the sensor with clean water.
    (vi) Repeat step ii and assure that the sensor measures zero, otherwise repeat steps (iii) and (iv) until reading is between 0 and 10 mg/L.
    (vii) Record the zero clean water value.
  (b) Measure the concentrated base solution.
    (i) Position the TSS sensor in the bucket centre containing the 150% TSS concentration. The sensor should be at 10 cm away from any wall or bottom and allow the measurement to stabilize before recording.
    (ii) Continuously mix the water sample gently with the stirrer to obtain a homogenous mixture and avoid settling of suspended solids.
    (iii) Record the sensor value at 150% TSS of the original TSS concentration.
  (c) Prepare diluted solutions at 1/3 and 2/3 of the concentrated base water sample solution.
    (i) Dilute 5 L of the concentrated base water sample with 10 L tap water (or effluent). This solution will be in between zero and the initial TSS concentration.
    (ii) Take a water sample for laboratory analysis marked as *TSS 50%*.

   - (iii) Dilute the remaining 10 L base water sample with 5 L tap water. This will be the same concentration as the initial (or slightly lower due to any decanted suspended solids).
   - (iv) Take a water sample for laboratory analysis marked as *TSS 100%*.
- (d) Measure TSS in two concentrations.
   - (i) Position the TSS sensor in one of the buckets and gently swirl in a circular motion until the reading stabilizes. Assure that the sensor is placed at least 10 cm away from the wall and the bottom of the bucket to avoid light scattering effects. Cover the bucket with the dark cloth when validating in the outdoors.
   - (ii) Record the sensor value.
   - (iii) Repeat for the second bucket.
- (e) Reposition the TSS sensor back into its normal measurement position.
- (f) Pour the water sample back.
- (g) Clean bucket and stirrer.

Data analytics and recording

(1) Record the sensor measurement in tap water and at the three concentrations.
(2) Obtain and record the TSS reference measurement value (from lab).
(3) Compute the offset and slope deviation by least-squares regression of the recorded sensor and lab values.
(4) Store the offset and slope values together with the recorded values as validation metadata.

It is also possible to check the lab measurements when the dilution in buckets has been executed carefully. Then, the lab measurements should follow the dilution of 50, 100, and 150% of the original water sample.

5.A.4.5 Contextual information about total suspended solids sensor validation

The following paragraphs serve as examples of what could be included as contextual information in an SOP for TSS sensor validation.

Available measurement types

The TSS is obtained from an optical and infra-red (IR) measurement in the non-visible wavelength (860–890 nm). Two types exist, where the first one measures the light scattering at a certain angle (90° for low solids and 135° for higher solids), and the second type measures the absorption through a small column distance (180°).

Confounding factors

Daylight can impact IR readings and the sensor should therefore be shielded from direct sunlight during measurements outside the process tank.

The emitted sensor light can be reflected on a close wall or nearby sensor (specifically in case of the light scattering measurement type). For this reason, most manufacturers recommend a 10 cm minimum distance to any wall or bottom.

Air bubbles can reflect light and impact the reading. De-aerated water for zero calibration is therefore advised.

Many TSS sensors contain a temperature sensor to correct for large temperature changes. Thus, the sensor needs to adapt to any temperature change when repositioning the sensor, for example, when the sensor is validated in air before the sensor temperature has reached the ambient temperature.

The correlation between the optical measurement (scattered or absorbed light) and TSS may be non-linear. Since this non-linearity can vary throughout the year it is not possible to correct for this within the sensor's measurement computations. Instead, multipoint calibration is applied to adapt to the local sludge properties. How this potential non-linear relationship is handled during calibration within the sensor differs between manufacturers, and it is therefore especially important to understand the calibration procedure in detail for the sensor at hand. It should be noted that the provided multipoint validation procedure is sufficient to detect an inaccurate calibration.

General recommendations for TSS reference sampling

It is challenging to measure the TSS in wastewater for two reasons. First, the sludge is heterogenous and varies greatly with space and time in an activated sludge basin. Taking a representative water sample is difficult if the goal is to achieve a representation of the sensor's ambient condition. Therefore, although it is possible to drop a handheld sensor in next to the permanent sensor, **we recommend to always take a subsample in a bucket** to assure that both the reference and validated sensors measure the same TSS concentration. Second, laboratory analysis of TSS can be biased and with substantial random errors for high TSS concentrations (above 3000 mg/L). Therefore, well-developed laboratory practices (APHA and ISO) and replicate analyses (2–3 analyses for each water sample) are instrumental to achieve a high accuracy during TSS sensor validation. It is likely that the laboratory practice is not applicable (accredited) for high TSS concentrations, for example, samples in the return active sludge. If so, the original water sample should be diluted to an accepted concentration. Most often this is executed with distilled or tap water, although this may dilute light-absorbing components and result in a bias. Effluent water is generally not advised as it may contain suspended solids. However, the dissolved light-absorbing components are likely to be present in similar concentrations so that the dilution of light-absorbing components can be reduced by doing so.

The concentration range in the activated sludge basin varies during day and season depending on the influent flow. It can therefore be difficult to obtain both low and high TSS concentrations when the full measurement range needs to be validated. Two simple tricks can be used to obtain a spread in concentration range. First, if the water sample concentration is too high, it can be diluted with effluent water (not tap water). Second, if the water sample concentration needs to be heightened, a water sample from the return sludge (RAS) can be diluted to the desired concentration. Also when the RAS is too diluted (e.g., during high-flow situations), a large water sample (20 L) can be taken and allowed to settle, and the clean water phase removed by gently pouring or siphoning it out. This can be repeated to achieve the desired concentration.

TSS sensor manuals commonly suggest keeping the sensor tip submerged at least 5 cm deep while keeping the tip more than 10 cm from the sides and bottom of the validation bucket. This usually requires the use of a 20 L bucket to be filled with at least 10 L of wastewater.

5.A.5 Procedures for an ultraviolet-visible light absorbance sensor for the purpose of concentration measurements

This section provides two procedures for UV/Vis light absorbance sensor validation. Figure 5.A.5 shows the data quality metrics that can be estimated with these procedures. This concerns the use of light absorbance sensor for the estimation of concentrations of wastewater contaminants, most typically by collecting a multivariate light absorbance spectrum. These procedures are not designed for use in UV/AOP (advanced oxidation processes) treatment systems but could be modified for this purpose.

This procedure aims to provide the guidelines for quality assurance and quality control of measurements.

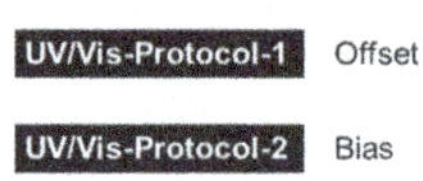

Figure 5.A.5 Overview of UV/Vis light absorbance sensor validation procedures and the resulting sensor data quality metadata.

5.A.5.1 UV/Vis-Procedure-1: Sensor integrity check

Version and dates

- Version: 1.0
- Last review and revision: 31st July 2023
- Next review and revision due by: 31st July 2025

Objective

The goal with this procedure is to check the blank spectrum. A blank spectrum is collected after each of three cleaning steps to quantify the effect of cleaning and to quantify changes in the sensor by comparison with a history of blank spectra.

This procedure enables:

- Quantification of offset in standardized non-process conditions

If repeated regularly, this procedure enables:

- Quantification of drift of offset due to fouling, scaling, or permanent damage to the optical components

Hardware

- UV/Vis sensor to be validated
- Bucket around 10 L (~2.5 gallon) or larger
- Non-abrasive cloth or brush
- Stirrer

Consumables

- 10 L (~2.5 gallon) distilled water
- Detergent (for cleaning)
- Diluted acid of 1–3% HCl (for cleaning)
- Soft tissue (for drying)

Experimental procedure

(1) Preparation
  (a) Deactivate any automatic cleaning devices on the sensor (e.g., air sparging, wiper).
  (b) Take the sensor out of its measurement place.
  (c) Remove any debris from the optical path of UV/Vis sensor and clean the sensor with water and cloth, then dry with soft tissue.
  (d) Clean the compressed air channels (if present) with tap water and with the help of Q-tips or interdental brushes.
(2) Validation an incremental cleaning
  (a) Visually inspect the UV/Vis sensor and assure it is clean.
  (b) Place the UV/Vis sensor in the stirred bucket with distilled water.
  (c) Assure that the distilled water is continuously stirred.
  (d) Record the UV/Vis spectrum (spectrum 1).
  (e) Take the UV/Vis sensor out of the bucket.
  (f) Clean the UV/Vis sensor with water, detergent, and cloth, and dry using soft tissue.
  (g) Rinse with distilled water.
  (h) Place the UV/Vis sensor in the stirred bucket with distilled water.
  (i) Assure that the distilled water is continuously stirred.
  (j) Record the UV/Vis spectrum (spectrum 2).
  (k) Clean the UV/Vis sensor with diluted acid and dry using soft tissue.
  (l) Rinse with distilled water.

(m) Place the UV/Vis sensor in the stirred bucket with distilled water.
(n) Assure that the distilled water is continuously stirred.
(o) Record the UV/Vis spectrum (spectrum 3).
(p) Reposition the UV/Vis sensor back into its normal measurement position.

(3) Clean-up
(a) Empty the bucket and clean the stirrer.
(b) Return bucket and UV/Vis sensor to their normal location.
(c) Reactivate automatic cleaning devices as needed (e.g., air sparging, wiper).

Data analytics and recording

(1) Store the three spectra obtained successively after each cleaning step (spectra 1–3: water, detergent, acid).
(2) Spectrum 3 should correspond to a blank spectrum and is ideally flat and close to zero over the whole range of wavelengths. Compare this blank spectrum with a history of blank spectra (visually or numerically) to see if additional cleaning or maintenance may be warranted or whether permanent damage has been incurred.
(3) The difference between spectra 1 and 2 is an indicator of how much material can be removed with light cleaning. Qualitatively, this corresponds to the removal of fouling. Ideally, this difference remains rather small. If not, increase the frequency of cleaning until the difference is acceptable.
(4) The difference between spectra 2 and 3 is an indicator of how much material can only be removed with chemicals, acid in this case. Qualitatively, this corresponds to the removal of scaling. Ideally, this difference remains rather small. If not, increase the frequency of chemical cleaning until the difference is acceptable.

Additional remarks

- Repetition with multiple UV/Vis sensors using the same water sample is feasible and minimizes the time allocated for water sample preparation. To this end, repeat steps 1a–d and 2c–k for every ammonia sensor.
- Collection of all sensor values through an automated data acquisition system rather than manual recording will facilitate quantification of sensor precision (see Chapter 3).
- Some manufacturers provide a transparent piece of solid reference material that can be placed in the measurement window (see below). This is also known as a multifunctional slide. Using this material avoids the need to use distilled water as reference medium, thus enabling a faster and more reliable result.

5.A.5.2 UV/Vis-Procedure-2: Procedure for one-point sensor output validation at current operating point

Version and dates

- Version: 1.0
- Last review and revision: 31st July 2023
- Next review and revision due by: 31st July 2025

Objective

The goal with this procedure is to cross-check the sensor output value (e.g., chemical oxygen demand (COD), nitrate concentration) in its normal environment with a reference concentration value produced through an analytical lab measurement.

This procedure enables:

- A single evaluation of the measurement error at the current operating point
- Quantification of bias (see Chapter 3)

If repeated regularly, this procedure enables:

- Quantification of drift (i.e., drift of bias, see Chapter 3)

Hardware

- UV/Vis sensor to be validated
- Bucket around 10 L/2.5 gallon (or larger)
- Stirrer
- Vial for transport of a water sample to a laboratory setting

Consumables

- 10 L/2.5 gallon of wastewater, collected at the sensor's location
- 100 mg allylthiourea powder
- Sodium chlorate salt

Experimental procedure

(1) Preparation (1–24 h prior to the validation)
  (a) Deactivate any automatic cleaning devices on the sensor (e.g., air sparging, wiper).
  (b) Take the sensor out of its measurement place.
  (c) Remove any debris from the optical path of UV/Vis sensor and clean the sensor with water and cloth, then dry with soft tissue.
  (d) Clean the compressed air channels (if present) with tap water and with the help of Q-tips or interdental brushes.
(2) Validation
  (a) Collect a 10 L water sample with the bucket.
  (b) Add stirrer and continuously mix the water sample gently with the stirrer to obtain a homogenous mixture and avoid settling of suspended solids.
  (c) Visually inspect the UV/Vis sensor and assure it is clean and undamaged.
  (d) Add allylthiourea to inhibit the oxidation of ammonia (e.g., 10 mg N/L is typically well above the inhibition level).
  (e) Add chlorate to inhibit the oxidation of nitrite (e.g., concentration of 20 mM).
  (f) Place the UV/Vis sensor in the bucket. Place the sensor so that the measuring unit is at least 2 in deep while keeping the sensor tip more than 10 cm (~4 in) from the sides and bottom of the validation bucket.
  (g) Assure that the wastewater is continuously stirred.
  (h) Record the UV/Vis spectrum.
  (i) Record the UV/Vis sensor output for the variable of interest (e.g., COD concentration).
  (j) Take a 20 mL water sample from the bucket for laboratory analysis.
  (k) Reposition the UV/Vis sensor back into its normal measurement position.
(3) Clean-up
  (a) Pour the water sample back.
  (b) Clean bucket and stirrer and return to storage.
  (c) Transfer the 20 mL water sample to the laboratory for analysis with a standard measurement or accepted analytical method for reference measurement.

Data analytics and recording

(1) Store the values for the spectrum and the sensor output in a database.
(2) Store the concentration value obtained through analytical measurement in the laboratory.
(3) The difference between the sensor output value and the analytical measurement is an estimate of the absolute measurement error of the sensor under current operating conditions and can be used to determine whether further actions are necessary.

Additional remarks

- Repetition with multiple UV/Vis sensors using the same water sample is feasible and minimizes the time allocated for water sample preparation. To this end, repeat steps 1a–d and 2c–k for every ammonia sensor.
- Collection of all sensor values through an automated data acquisition system rather than manual recording will facilitate quantification of sensor precision (see Chapter 3).

5.A.5.3 Contextual information about ultraviolet-visible light absorbance measurements

UV/Vis spectrophotometer sensors are based on the measurement of the amount of UV/Vis light absorbed by a water sample. In commercial sensors, a single data point (for definition see Chapter 2) includes either (a) the light absorbance at a single wavelength (e.g., 254 nm) or (b) the light absorbance at multiple wavelengths in the region of 200–750 nm depending on the sensor provider. Figure 5.A.6 shows a typical absorbance value as a function of wavelength.

Under ideal measurement conditions, the amount of light absorbance due to one light-absorbing component is proportional to its concentration. This is known as the Beer–Lambert law. In practice, a water sample will contain multiple light-absorbing components. While this complicates sensor calibration, the theoretical relationship between the component concentrations and the resulting light absorbance remains linear (Thomas & Burgess, 2017).

The simultaneous measurement of absorbed light at multiple wavelengths provides the capability to estimate the concentration of multiple components at once with a single data point. Depending on the exact water composition, one may be able to estimate the concentrations of the following components in water:

- Biochemical oxygen demand (BOD)
- Benzene, toluene, xylene (BTX)
- Chemical oxygen demand (COD)
- Dissolved organic carbon (DOC)
- Nitrate (NO_3)
- Nitrite (NO_2)
- Total organic carbon (TOC)
- Total suspended solids (TSS)

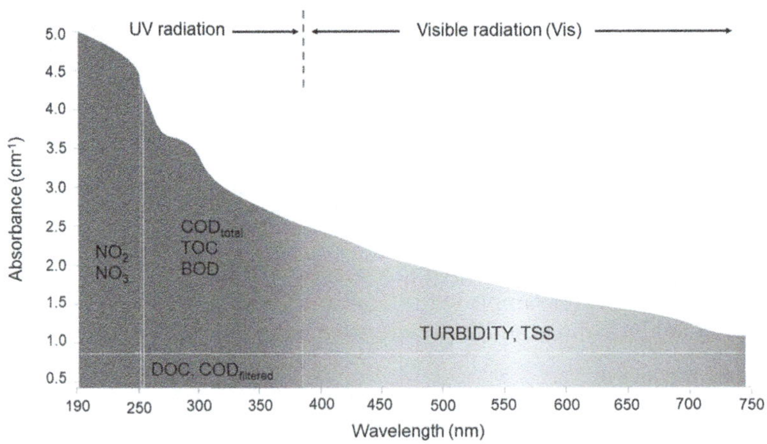

Figure 5.A.6 UV/Vis spectrum and water quality related parameters. *Source*: Adapted from RealTechWater Inc. (2017).

The exact relationship between the absorbance spectrum and the concentrations of interest may vary from one wastewater stream to another and may also vary in time. One reason for this is that lumped water quality variables (e.g., BOD, COD, DOC, and TOC) correspond to a multitude of light-absorbance components whose individual spectrum is different and whose relative concentrations are subject to change. Also, suspended solids do not only absorb light but also scatter it. As a result, the relationships between the light absorbance measurement and the concentrations of interest may become non-linear.

Most sensor manufacturers provide models within the spectrophotometric equipment to relate the absorbance measurements to the component concentrations of interest. Depending on the required accuracy (bias and/or precision, see Chapter 3), a local correction of the model may be needed to account for site-specific conditions. In the case of a unique wastewater composition (e.g., urine, Mašić *et al.*, 2015), a specialized model for this wastewater can be constructed.

5.A.5.4 General recommendations for validation of ultraviolet-visible light absorbance sensors

Optical light path. In general, every spectrophotometer sensor consists of an emitter (light source), the measuring window (with water sample), and a receiver (light sensor). This is illustrated in Figure 5.A.7. Typically, a light beam is sent through the measuring window with water present as well as through a reference optical path that does not contain water. The relative difference between the intensity of the light beam is used to quantify the light absorbance at one or more light wavelengths.

The operational status of the measuring window is critical for a good performance of the sensor. The presence of dirtiness and air bubbles will jeopardize the accuracy and correctness of the measurements. Maintaining this measuring window in a good status is the basis for reliable measurements and the core of the maintenance activities for this type of sensors.

Correlation-based measurements. UV/Vis sensors provide correlation-based estimates of concentrations of interest. It is important that the correlations that the sensor relies upon are not inadvertently destroyed during water sample preparation. This is especially important for variables that represent concentrations of multiple components (e.g., BOD, COD, DOC, and TSS).

It is challenging to measure lumped concentrations (e.g., BOD, COD, DOC, and TSS) in wastewater for two reasons. First, these concentrations represent multiple species whose individual concentrations can change while the lumped concentration remains the same. In practice, the lumped species are heterogenous and vary greatly with space and time in an activated sludge basin. Taking a representative water sample is difficult if the goal is to achieve a representation of the ambient conditions that the sensor is typically exposed to. For this reason, we recommend to always validate the sensor by placing it in a well-mixed bucket so that absorbance measurements and the reference measurements represent the same water sample composition.

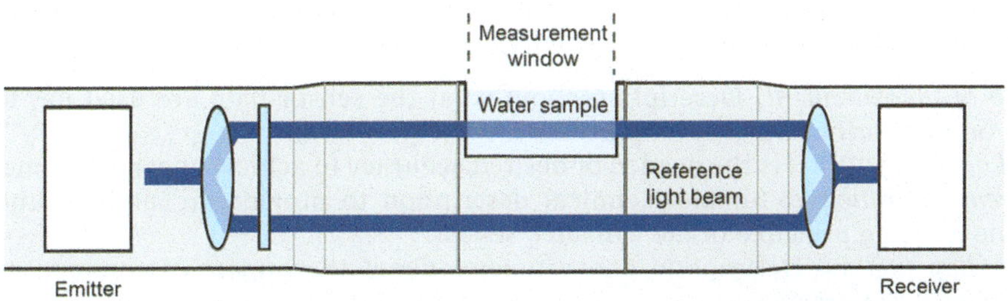

Figure 5.A.7 Typical structure of an online UV/Vis sensor. An emitter produces white light that is split between two beams. One beam (top) passes through a measurement window containing the water sample and is collected at the receiver. The other beam (bottom) arrives at the receiver directly and is named the reference light beam. The difference of light intensity between the two beams is measured for a series of wavelengths, which are arranged into a light absorbance spectrum. *Source*: Adapted from S::CAN GmbH (2020).

5.A.6 Procedure template

This blank procedure can be used to guide tailored and sensor-specific procedures that are not covered in this book. It includes the elements that are considered essential to a procedure that is useful for sensor quality assessment and systematic registration of the metadata it produces.

5.A.6.1 Template procedure for sensor validation

Version and dates

At the top of the SOP, add the following elements:

- Version number of the SOP
- Last time the SOP was reviewed and updated
- Next time the SOP should be reviewed

Objective

Describe the aims of the procedure. Include the sensor data quality metrics that can be quantified through the procedure (e.g., bias, non-linearity, offset, drift, slope, precision)

Hardware

List hardware (e.g., sensor to be validated, reference sensor, glassware, mixing equipment) needed for the validation. Include storage location if relevant.

Consumables

List chemicals and disposable tools required for the validation. Include storage location if relevant. Add preparation instructions if needed in advance.

Experimental procedure

Describe the validation in a step-wise manner. Be explicit and detailed. Newly hired staff can be a great resource for checking that the validation procedure is sufficiently detailed and well-structured.

(1) Solution preparation.
(2) ...

Data analytics

Describe how the recorded validation data should be processed to finalize the validation. Indicate how the sensor data quality metrics should be computed. Indicate where and how the resulting measurement values and sensor data quality metrics should be stored.

5.A.6.2 Contextual information

In addition to the actual validation procedure, some background information can be added to complete the understanding of the sensor's purpose and practical use. The list below provides examples of valuable contextual information in a sensor validation routine document:

- *Purpose of measurement*: Description about what the sensor data are used for, for example, control or monitoring of a distinct process.
- *Needed accuracy:* What is the needed or desired accuracy to achieve the measurement purpose.
- *Measurement principle*: Simple technical description to provide an intuitive understanding about the working principle of the validated sensor.
- *Confounding factors*: Description and estimates about the impact of common confounding factors relevant in practice.
- *Validation* procedure *metadata*: Information describing the validation procedure, for example, procedure version, original developer, contacts for assistance (sensor company/engineer at site/local champion), expected time duration for execution.

- *Sensor location*: A photo or illustration is beneficial to instruct new staff about the location of the sensor and associated equipment.
- *Metadata storage*: Instruction about what metadata produced through the validation procedure that should be stored, and in which location.
- *Common errors and troubleshooting*: Pitfalls and known issues with specific sensors, locations, and so on can be highlighted here. Solutions for dealing with common pitfalls: This part can summarize instrument technician experience and be the location for continuously documenting local sensor experiences.
- *Support and service*: Links and product numbers for spare parts/consumables.Information about how to retrieve added support from the sensor asset and maintenance system.

APPENDIX 5.B: VALIDATION OF WATER FLOW RATE, LEVEL, PRESSURE, AND TEMPERATURE SENSORS

5.B.1 Introduction

In the field, general validation techniques include:

Primary verification: Basically the electronic verification that an instrument is working with standard tolerances. Nowadays, this is normally built within the instrument and is outputted from the instrument on a periodic basis using highway addressable remote transducer (HART) protocols. Intelligent sensors will use this to automatically warn of instrument problem or failure using standard NAMUR protocols. Primary verification is normally simple to do but has limited use apart from ensuring correct operation of the instrument and so does not consider site factors such as fouling of the measurement point.

Secondary verification: It is the comparison of the instrument output reading to a traceable source. For example, for a closed-pipe electromagnetic flow meter comparing the flow reading with a clamp-on time-of-flight flow meter reading that has been calibrated externally and is traceable back to either national and/or international standards from a recognized testing facility. Secondary verification is usually more complex to carry out but checks the accuracy of the instrument in-situ. This evaluates both the correct operation of the instrument and any problems with the measurement point.

End-to-end testing: This method is also sometimes sensor to screen testing and it checks that the output of the instrument is reading correctly from the sensor end of the process via the controller output to the screen end of the process. This is especially important when using analog or pulsed outputs through a corporate telemetry system. This is because the scaling on the site of the data transmission has to match with the scaling on telemetry system end of the process to prevent errors due to differences in instrument scaling.

Calibration: Calibration is the verification and adjustment of an instrument against a known standard. This is often carried out in a laboratory environment. For example, electromagnetic flow meters are removed from their measurement location and mounted on a flow calibration rig. The meter is compared against a known standard and factors programmed within the instrument that changed to ensure correct operation. Calibration will not check the actual measurement installation in-situ. Quality measurement can be calibrated in-situ depending upon the functionality of the instrument and is undertaken by using a traceable calibration solution and protocols within the instrument to correct. This is often undertaken using standard laboratory protocols such as a simple standard addition method.

5.B.2 Validation of water flow rate sensors

The method for flow measurement validation will depend on the applied measurement principle. There is a large variety of measurement techniques, each requiring a different method of validation and thus a variety of skill sets. In this section, we will cover a selected set of common measurement techniques used within the water and wastewater industry.

5.B.2.1 Electromagnetic flow rate sensors

Electromagnetic flow rate sensors are meters that measure flow within a closed-pipe. The vast majority of this type of flow rate sensors is under a pipe-full condition and will only measure conductive fluids. In essence, the technique measures the velocity of the fluid within a fixed pipe.

There are technologies that will measure in part-full pipes but in essence these use an area–velocity type technique by measuring the level within the pipe. In this case both the velocity and the level need to be checked.

There are potential interferences with this measurement technique as it presumes a stable flow profile and so anything that takes the flow profile away from a standard velocity profile will result in deviation from the true figure. This is why an electromagnetic flow meter normally requires at least five upstream pipe diameters and two downstream pipe diameters of straight pipe. Where there is an influence such as valve or a severe pipe bend, these normally require 15 upstream pipe diameters and 5 downstream diameters.

As long as an electromagnetic flow meter is installed correctly then there is relatively little that will affect the sensor itself and, in this case, a primary verification of the instrument electronics should suffice as a validation technique along with a periodic end-to-end test. However, a risk-based approach should be examined in terms of the need for carrying out secondary verification. This is due to the risk of measurement point fouling.

Figure 5.B.1 shows the fouling immediately upstream of an electromagnetic flow meter where the bore of the pipe had been reduced from 300 to 125 mm in places within a 3-year period. If this degree of fouling was uniform (which it wasn't) the flow meter would record 5.76 times higher than the true flow meter reading. Basically, the flow meter assumed that it had a surface area of 0.283 m^2 (the surface area of a 300 mm electromagnetic flow meter), but in reality this was reduced to 0.05 m^2 which would increase the velocity by 5.76 times.

Where there is a risk then secondary verification of the measurement installation will pick up from that point. This is an extreme case which could be picked up relatively simply but in reality, the changes in flow pattern could be a lot more subtle and difficult to detect without using a secondary verification technique.

Time-of-flight flow measurement is often used as a secondary verification technique as it is more sensitive to fouling and signal failure will normally occur where fouling starts to become an issue.

Despite reduced risks from fouling, it is still prudent to secondary verification on a periodic basis as there are still risks of sensor drift that can affect measurement accuracy. This is especially important on larger flow meters where any uncertainty is likely to cause poor assumptions to be made.

Figure 5.B.1 Fouling of pipework immediately upstream of an electromagnetic flow meter.

5.B.2.2 Open-channel flow rate sensors

Open-channel flow is basically a level-based measurement technique and so the level part of the measurement can be verified via a simple level measurement technique either using a staffing gauge or for a crude check a metal rule. However, the effectiveness of the measurement technique also relies on the correct operation of the primary measurement device which is normally a flume or a weir. For these types of structures there are some basic techniques that can be employed to check for correct operation. These include:

- Is the primary measurement structure showing any signs of deformation? Is the weir plate bent?
- Is there any obstruction in or near the flume or weir? Is the flow pattern being disturbed at all?
- Is the structure discharging freely?

Figure 5.B.2 shows a classic example of both situations.

In this figure, the weir plate at the bottom shows significant distortion which due to the outward angle of the weir plate would contribute to an under-reading of the flow rate where this the only problem with the measurement chamber. However, it is not the only problem and the flow downstream of the weir plate cannot escape sufficiently and has flooded back meaning that the flow does not discharge freely. This would cause the flow meter to over-read.

In addition to these problems there is also a return pipe that will cause surging of the weir plate as well as foam within the measurement chamber which can adversely affect the ultrasonic flow meter that is being used as well as multiple different instruments within the path of the ultrasonic head and the measurement chamber which will cause vortices and incorrect measurement and poor echo confidence as the ultrasonic echo bounces off all of the instruments within the chamber.

In contrast, Figure 5.B.3 shows a perfect free-discharging profile as evidence by the spear-shaped pattern at the exit of the flume.

What is important with any primary measurement device is that it is free of obstruction (including anything that may be fouling the flume), is not showing damage of any kind, and is not being flooded for any reason.

In addition to this, the secondary measurement device that is measuring the flow by measuring the level should be secure, level, and free of any vibration or movement. If an ultrasonic level meter is being used, then it should be protected from sunlight as this can cause solar gain by increasing the temperature which affects the speed of sound.

Figure 5.B.2 Flooded, distorted, and cluttered flow measurement point.

Figure 5.B.3 Free-discharging flume.

These are simple and very quick checks that any operator can undertake in a couple of minutes using simple observational techniques.

When it comes to checking the accuracy of the actual level of device be it a flume or a weir for the primary device and ultrasonic, radar, or some other level measurement technique for the secondary device, the simplest method is to simply use a steel rule and compare it to the reading that the level monitor is receiving. This is not as accurate as using a surveyor's level which would be the preferred option, but it will work. By comparing the depth of water measured by the secondary device and a steel rule the accuracy can very quickly be checked. The alternative is to use a verification plate which needs to be pre-calibrated to the correct height and then use a standard plate to check the accuracy of measurement (if using the commonplace non-contact technique).

For non-contact techniques, which make up the majority of measurements that are undertaken, what all of these methods can inform the operator is how far the instrument's empty distance has drifted. If there is a discrepancy, then the empty distance (i.e., the total distance from the face of the level sensor to the bottom of the measurement point) needs to be recalibrated.

An alternative to this is to use a secondary verification technique with either a flow loop using a certified electromagnetic flow meter or a certified area–velocity device.

5.B.2.3 Area–velocity flow rate sensors

Area–velocity flow measurement differs from most flow measurement techniques insofar as there are actually two unknowns in the universal flow equation being measured. The first of these is the height of water representing the only variable in the area of the part of the equation and the velocity of the water. The height of the water is easily checked (as in the level technique described above) but the in-situ velocity measurement is more difficult to measure without using some sort of technique. It could be checked using dye or salt techniques, but this is often not available to operators within the field. The only alternative would be some sort of alternative area–velocity measurement technique or velocity technique such as a velocity gun.

The lack of a suitable validation technique for area–velocity flow measurement remains a problem that is yet to be addressed by the water industry.

5.B.2.4 Time-of-flight flow rate sensors

Time-of-flight flow measurement has long been a technique used to verify that flow measurement is working within a closed pipe. An electromagnetic flow meter basically works by measuring the velocity of the flow within the closed pipe. Practically, there are only a few techniques that can be employed to check the accuracy of the meter.

Figure 5.B.4 Time-of-flight flow measurement on a 900 mm pipe being verified by an insertion velocity meter (foreground).

An example of a technique that can be utilized is shown in Figure 5.B.4. This technique uses an insertion velocity meter that is used to verify the velocity in the pipework. The disadvantage of this technique is that an insertion point does have to hot-tapped into the pipeline and this can be relatively costly. Equally, if the pipeline is under pressure, then a pressurized insertion tool is required to insert the meter into the pipe safely.

5.B.2.5 Other techniques for flow rate sensor validation

There are many other techniques available for the measurement of flow; however, the principle of validation of the accuracy of the flow measurement will rely upon the basic principles of validation insofar as the following are routinely needed:

- A primary verification technique to ensure that the sensor is working correctly.
- A secondary technique to traceably verify that the sensor in the measurement system is working correctly.
- Some observational techniques to ensure that the physical structures are sound.
- An end-to-end test to ensure that the telemetry scaling is correct.

5.B.3 Validation of level sensors

The methods for level sensor validation are very similar to those for open-channel flow as both technologies require verification of the height of the water, but it depends very much upon the level technology that is being used. If it is non-contact level measurement, then the validation is based upon the measurement of the air implying the level of the water by deduction from the empty distance.

In this case there is both some physical and some measurement protocols to take into account.

The physical protocols are based upon the soundness of the installation. In Figure 5.B.5 we see an application of level measurement on a storm tank. In this case, the ultrasonic sensor that is being used has been placed on a hinged bracket. With wear on the hinge there is a noticeable dip in the measurement which obviously affects the measurement accuracy. This sort of installation is susceptible to this level of uncertainty and this either needs to be rectified or considered in the measurement uncertainty.

Figure 5.B.5 Inaccuracy in measurement caused by worn bracketry.

In an ideal world the bracketry should be braced in at least two places at two different points, for example, over a channel. In the case of Figure 5.B.5, a three-point bracing system is advisable or a locking system to ensure stability of the level measurement. Specifically, for ultrasonic level measurement it should be checked that the level of fluid cannot enter the empty distance zone of measurement by installing the level measurement at least 350 mm above the coping level. In addition to this if a fine level of accuracy is required then shielding of an ultrasonic sensor from solar gain should be considered. Good practice can be seen in Figure 5.B.6 on a flow measurement application showing both indirect (via ultrasonic and radar level measurements) and also by pressure measurement.

Where pressure measurement is used for level measurement and a direct level is measured then a direct validation of the level is needed. This can use a direct measurement of the level or an indirect method such as validation of the empty distance, but it can also use a pressure measurement technique

Figure 5.B.6 Level measurement for flow including radar (black sensor), ultrasonic (blue sensor with sunshield), and pressure (metal bar on right-hand side) all in front of a flume acting as the primary measurement device.

utilizing a traceable pressure source and a manifold system that allows the pressure measurement to be taken offline from the online measurement point.

In short, the protocols for checking the measurement of level are:

- Weekly check on the cleanliness of the measurement device ensuring that the sensor face is clear of any obstruction and debris.
- Quarterly check on the physical security of the bracing and bracketry and an end-to-end check to ensure that what is being recorded locally is what is being received in the telemetry system.
- An annual check on:
  - Primary verification that the instrument is functioning correctly.
  - Secondary verification of the actual level being measured, ideally using a traceable measurement technique.
  - If an analog communication method is being used then this should be checked on an annual basis using a calibrated analog reference instrument. This should include an end-to-end test as part of the validation.

5.B.4 Validation of pressure sensors

Pressure measurement whether it is used for water treatment and distribution or wastewater treatment and collection is utilized in many different forms. Each form of measurement will have checks that are unique to the measurement application. However, there are some common checks that also apply to all pressure measurement applications. What has to be said is that safety, especially where pressure is involved, is paramount. Even low-pressure systems can be dangerous. Operators should ensure that the monitoring point is isolated from any pressure system that can cause a health and safety issue.

Once it is safe to do so there are some basic physical checks, that apply to any pressure measurement, that should be undertaken; these include examining:

- The cap and housing's condition
- The diaphragm's condition – check for rips or tears
- The water tightness of the cable gland – it should not leak
- Condensation in the housing – there should not be any
- Check the connections to the electric module
- Check the seal – it should not be pressing on the diaphragm

In Figure 5.B.7 we see a pressure sensor with a ceramic membrane that has the advantage of being resistant to the majority of problems that less-resistant membranes can suffer from.

For a pressure sensor, it is helpful to know about the theory of operation.

All pressure sensors are essentially differential pressure measurements because the readings are relative to a reference pressure. The three primary measurements in the wastewater industry are

- Absolute pressure – pressure is referenced to a perfect vacuum.
- Gauge pressure – pressure is referenced to atmospheric pressure.
- Differential pressure – pressure is referenced to secondary process pressure.

Figure 5.B.7 Pressure sensor featuring a ceramic membrane (*source*: courtesy of Vega Control Systems).

Pressure sensors using diaphragms have replaced traditional bellow-sensing elements for many applications in the wastewater industry. The diaphragms are flat or concentrically corrugated metal or ceramic discs that deflect in response to pressure changes. As the diaphragm deflects, the deflection is measured using various methods such as capacitive, inductive, piezoelectric, piezoresistive, or optical. For example, the deflection changes the associated electrical property (e.g., the distance between the plates in a capacitor, piezoelectric response, loop reluctance, or optical transmittance), which is converted to a pressure reading by the instrument.

Diaphragm-seal systems consist of a pressure sensor, a diaphragm seal, and a fill fluid. They are either direct-mounted or connected via capillary tubing. The diaphragm seal provides a physical barrier between the process and the sensor and also transmits pressure through the fill fluid in the capillary tubing to the sensing element.

When a pressure transmitter is installed in a system with rapid pressure perturbations, surges, or fluid hammers, a snubber should also be installed to protect the sensing element from damage. A snubber works like a shock absorber.

It operates by driving a piston up against an orifice during rapid pressure fluctuations, choking the flow and thus dampening the pressure perturbations reaching the sensing element. By changing the piston diameter, the snubber can be sized to accommodate various viscosities and magnitudes of pressure surges.

Porous-disc-style pressure snubbers are not recommended for most wastewater applications because they clog easily; however, they work effectively in clean water applications.

Pressure-sensing elements are typically made of stainless steel, Hastelloy, or ceramic. When the process being measured is chemically incompatible with typically available materials, then a diaphragm seal should be considered. A diaphragm-seal system provides a chemical barrier while still transferring pressure to the sensing element via a fill fluid. Typical fill fluids include glycerine, silicone, and halocarbon.

The sensing element should be accessible for maintenance, using diaphragm seals with capillary tubing as required to mount the sensor at a serviceable height. Isolation valves should be provided to facilitate servicing the sensor. If the sensor is mounted higher or lower than the process connection, then an offset will need to be applied to the measurement to compensate for the elevation difference.

Flanged sensors should be used for processes that are prone to fouling. For applications involving solid-bearing liquids, flushing provisions and/or diaphragm isolation may be needed. Diaphragms for solid applications are available with nearly every conceivable process connection; the connections should be at least 25 mm (1 in) in diameter. Diaphragm flanges are available with integral flushing connections or diaphragm extensions to extend the sensing diaphragm into the process.

The ability to check on pressure sensors should be thought of in the design of the installation of the pressure sensor and where it is necessary then either an isolation technique or a manifold should be installed so that the pressure monitor can be taken offline, and a calibration loop can be set-up so that a traceable external pressure monitor can be used to calibrate the pressure monitor itself. This follows the principles of physical, primary, and secondary verification techniques.

5.B.5 Validation of temperature sensors

Temperature sensors are rarely validated or calibrated especially where it is part of another measurement, for example in flow rate sensors or water quality sensors where there is some element of temperature compensation. Where temperature measurement should include some element of temperature validation this includes temperature measurement on the final effluent of wastewater treatment works where there is a regulatory implication. There is also a need for temperature validation where it is part of a critical process, such as anaerobic digestion, and temperature measurement is needed as part of a quality control process.

It is helpful to understand some of the principles of temperature measurement.

There are a number of different ways of measuring temperature in an online measurement application. However, of these three methods are the most commonly used within wastewater treatment systems. These are:

- Thermocouples
- Resistance temperature detectors (RTDs)
- Thermistors

There are other methods that are not commonly used in wastewater treatment systems including using IR as a non-contact method, but these tend to be limited to hand-held units that can be used to check the accuracy of temperature measurement devices. So let's look at the principles of these three techniques.

5.B.5.1 Thermocouples

Thermocouples operate on the Seebeck effect, that is, the flow of current caused by heating a metallic junction composed of two dissimilar metals. The open-circuit voltage is a non-linear function of the junction's temperature and the wires' compositions.

All junctions of dissimilar metals produce the Seebeck effect, although some combinations produce more voltage per degree than others. As such, junctions other than those at the point of interest must be prevented or compensated for.

For example, thermocouple extension wires may be made of the same metal as the thermocouple or they may be maintained at a known temperature so unwanted voltage can be taken into account. All thermocouple circuits have at least two junctions: one at the point of measurement and the other to obtain two copper leads to connect to a voltmeter, which measures voltage caused by the junction of interest.

Two compensation methods are used for secondary junctions: softwire and hardwire. In softwire compensation, the thermocouple wires are terminated at an isothermal block, which keeps the two terminations at the same temperature.

This temperature is measured by an RTD, thermistor, or integrated circuit. The sensor's resistance is measured, converted to the reference temperature, and then converted to a reference junction voltage. This voltage is subtracted from that measured by the voltmeter to obtain the thermocouple voltage, which is converted to the measured temperature. Multiple thermocouples at the same reference temperature can be added to the measuring unit to reduce the cost of conversion per thermocouple.

In hardwire compensation, a voltage is generated to drive the reference junction voltage to equal 0°C (32°F); this is also called *electronic ice point reference*.

5.B.5.2 Resistance temperature detectors

RTDs operate on the principle that the resistivity of a metal changes with its temperature. By measuring the resistance at a point of interest, a temperature can be determined. This is the most common type of temperature measurement used where a separate temperature probe is used to thermally compensate a measurement device. Quite often a PT100 probe is being used which is basically a type of RTD (see Figure 5.B.8).

RTDs consist of a thin wire or film supported by a substrate with matching thermal-expansion characteristics. The thin metal element is necessary to obtain adequately high resistance (typically, about 100 Ω at 0°C [32°F]). It is supported so the RTD will be rugged enough for process measurement.

To measure the resistance change at the point of interest, lead resistance from the instrument wires must first be eliminated. One method for doing this involves using a constant current source and measuring the voltage change across the element.

This technique requires a four-wire lead or a three-wire lead and a Wheatstone bridge, which will introduce some error unless the bridge is balanced for the temperature being measured. A self-heating

Figure 5.B.8 Flow measurement with a separate RTD temperature sensor for temperature compensation, shielded from solar gain for additional accuracy.

error is also introduced by passing a current through the resistive element. The heat dissipated by the element will raise its temperature.

5.B.5.3 Thermistors

Thermistors are semiconductors whose resistivity changes in response to temperature changes. Most thermistors' resistances decrease as temperature increases.

Their temperature coefficients are higher than those for RTDs; a typical thermistor element has a resistance of 5000 Ω at 25°C (77°F). Therefore, lead resistance errors are not appreciable, and three-wire leads are typically used. Self-heating errors can be significant because of the thermistor's small mass.

Integrated circuit sensors are semiconductors whose resistivity increases as the temperature increases; otherwise, they are similar to thermistors. These sensors have not been widely used for process measurement but are widely used where temperature compensation is internal to a sensor head (e.g., temperature compensation inside an ultrasonic transducer).

5.B.5.4 Installation

All temperature sensors inserted to process streams need protection. Thermocouples, RTDs, and thermistors typically are inserted to a thermowell mounted in contact with the process fluid. A thermowell is a thin-walled metal tube capped at the process end so it can be removed without disturbing the process. Sensors are spring-loaded to achieve a positive contact with the thermowell's wall to minimize sensing lag time. Conductive paste is often coated on the temperature sensor in the thermowell to assist in heat transfer from the thermowell to the sensor. Care should be taken to ensure that the conductive paste is appropriate for the measured temperature range.

In some applications (e.g., those involving high pressure), penetrating the process vessel or piping is not desirable. As such, the temperature probe is mounted externally without a thermowell; however, higher measurement errors and lack of response need to be considered, especially if the temperature must be controlled.

In all applications, the probe should be protected from corrosive or erosive environments because physical changes in the sensor will cause it to lose calibration.

Thermocouple sensors can also lose calibration if wires become annealed at high temperatures. Annealing happens when a large temperature change suddenly occurs (e.g., removing the unit from a hot process to a cold ambient temperature). It can also occur because of vibration or if wires have been installed under tension.

5.B.5.5 Validation and calibration of temperature sensors

Realistically temperature is not something that is usually calibrated but it is something that should be verified to make sure that the temperature that your process is being controlled to is actually recording accurately or is something that can be checked if the accuracy of an instrument is being called into doubt. This is especially the case for digestion processes where the temperature of the digester is absolutely critical or for regulatory purposes where the temperature could be used for legal purposes. It is not necessary where temperature is being used for temperature compensation purposes.

So how do you verify the temperature is correct? If you are in a laboratory, you would normally compare it to a reference thermometer or put your glass bulb thermometer in a beaker with salty-iced water. Practically, you can't do this in the field so you either use a reference thermal measurement method such as using an IR thermometer or a calibrated RTD or you use a calibrated traceable temperature source such as a dry-block calibrator. The calibrator can then be sent off for calibration itself each year to ensure that it is operating correctly. These are expensive methods but where they are used to check on temperature measurement that are process critical then the cost of making sure that you are operating correctly is cheaper than developing a problem with your temperature measurement. Caution does have to be taken when verifying digester temperature measurement that the calibrators used hold the appropriate hazardous area certifications.

doi: 10.2166/9781789061154_0163

Chapter 6

Analytical methods for online data quality assessment

D. Aguado[1], J. Alferes[2], F. Arteaga[3], L. Belia[4], J. B. Copp[5], L. Corominas[6,7], F. Corona[8], A. Ferrer[9], H. Haimi[10,11], P. Kazemi[12], Q. H. Le[13], I. Miletic[14], M. Mulas[15], A. Robles[16], M. V. Ruano[16], S. Russo[17], O. Samuelsson[18], J. P. Steyer[19], K. Villez[20], E. I. P. Volcke[13], M. J. Wade[21,22] and J. Zambrano[23]

[1]CALAGUA – Unidad Mixta UV-UPV, Institut Universitari d'Investigació d'Enginyeria de l'Aigua i Medi Ambient – IIAMA, Universitat Politècnica de València, València, Spain
[2]VITO – WaterClimateHub, Ostend, Belgium
[3]Universidad Católica de Valéncia, València, Spain
[4]Primodal, Inc., Hamilton, ON, Canada
[5]Primodal, Inc., Hamilton, ON, Canada
[6]Catalan Institute for Water Research (ICRA-CERCA), Girona, Spain
[7]University of Girona, Girona, Spain
[8]School of Chemical Engineering, Aalto University, Espoo, Finland
[9]Multivariate Statistical Engineering Research Group, Universitat Politècnica de València, València, Spain
[10]FCG Finnish Consulting Group Ltd, Helsinki, Finland
[11]School of Engineering, Aalto University, Espoo, Finland
[12]Departament d'Enginyeria Química, Universitat Rovira i Virgili, Tarragona, Spain
[13]BioCo Research Group, Department of Green Chemistry and Technology, Ghent University, Gent, Belgium
[14]inCTRL Solutions Inc., Dundas, ON, Canada
[15]Department of Teleinformatics Engineering, Federal University of Ceará, Fortaleza, Brazil
[16]CALAGUA – Unidad Mixta UV-UPV, Departament d'Enginyeria Química, Universitat de València, Burjassot, Spain
[17]EcoVision Lab, Photogrammetry and Remote Sensing, ETH Zurich, Switzerland
[18]IVL Swedish Environmental Research Institute, Stockholm, Sweden
[19]INRAE, LBE, Univ Montpellier, Narbonne, France
[20]Oak Ridge National Laboratory, Oak Ridge, TN, USA
[21]UK Health Security Agency, Data Analytics & Surveillance, Canary Wharf, London, UK
[22]School of Engineering, Newcastle University, Cassie Building, Newcastle, UK
[23]The MathWorks AB, Knarrarnäsgatan 7, Kista 164 40, SE, Sweden

Keywords: Data-driven methods, data processing, data visualization, quality assessment, redundancy, mechanistic methods.

This chapter provides a comprehensive overview of the main steps for algorithmic sensor signal quality assessment, which can enhance the decision-making process for water resource recovery facility (WRRF) operation and optimization. It introduces the concept of redundancy as the basis for data quality assessment. It also explains the typical data processing pipeline, which consists of preliminary analysis, data pre-processing, and specific algorithmic approaches. Each of these processes is presented and discussed in three separate sections. Importantly, this chapter introduces the main approaches for data quality assessment, provides guidelines for selecting the most suitable one and the key performance indicators to evaluate them and explains how to collect metadata through such an algorithmic approach.

6.1 HOW TO READ THIS CHAPTER

The main purpose of this chapter is to provide an introduction to the most popular tools for algorithmic sensor data quality assessment and their selection and tuning within the context of wastewater treatment. The scheme below indicates dependencies within this chapter and can be used to navigate this chapter efficiently. The element shaded in grey is considered particularly helpful to select the specific data processing pipeline to be applied.

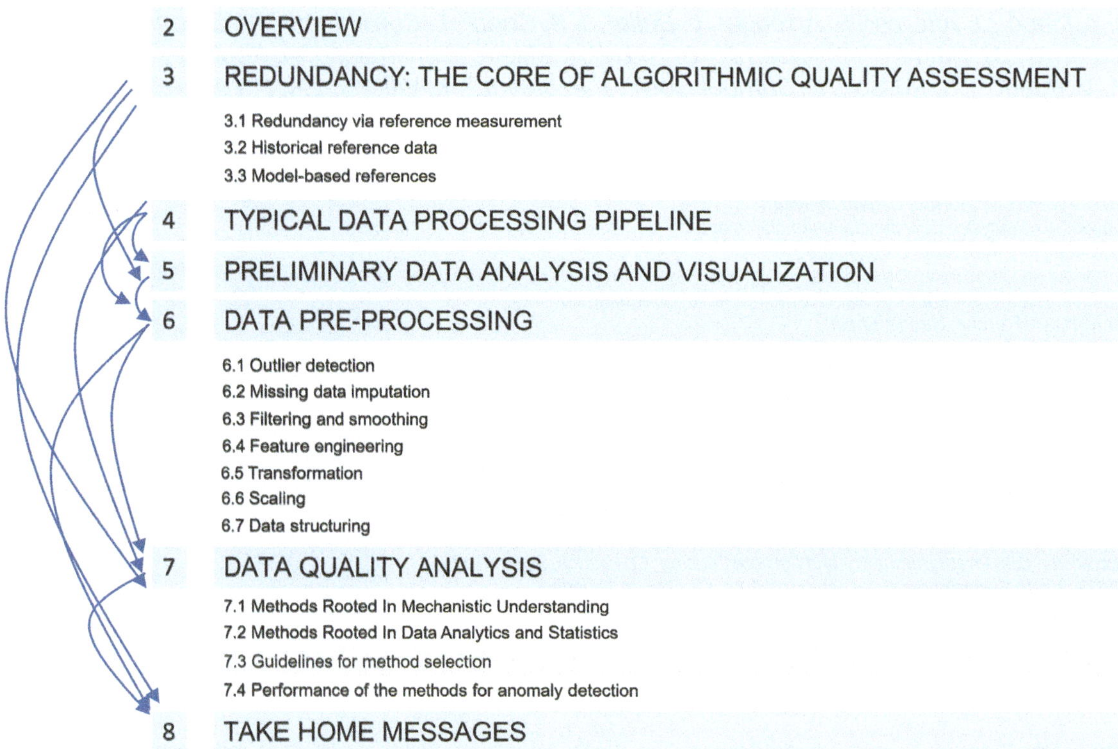

6.2 OVERVIEW

Water resource recovery facilities are increasingly realizing the benefits of high-frequency water quality data for plant operation and optimization. As a result, these facilities are installing more and more online instruments. Yet data quality is often overlooked during that decision-making process and often the most basic data quality questions go unanswered.

- Are data acquisition procedures checked (is the correct data being retrieved)?
- Is lab data quality checked?
- Is signal data quality checked?

- Are data quality procedures (algorithmic or manual) applicable, in place, being used?
- Is data quality a known issue?
- When data are used/requested, is the data quality made available to the user?

Furthermore, as the generated datasets have increased in size and complexity, it has become difficult to identify problems in a timely manner either manually or using simple rules that might have been sufficient in the past. To realize the benefits from the investments required for instrument installation and maintenance, it is necessary to avoid the collection of data of sub-par quality. However, basic fault detection approaches may fail to differentiate between real faults and typical process variations due to a limited ability to adapt to changes in normal data patterns. In addition, any delays in the detection of faults can be problematic if the data are being used for process decisions (i.e., in automatic control algorithms or manual operator actions).

Signal data quality assessment is increasingly recognized as an essential part to achieve intelligent WRRF operation and optimization. However, proper metadata collection is paramount to successfully apply data analytical techniques, and to truly improve the decision-making process for smart operation. This chapter will be devoted to the description of the principles that form the basis for basic and advanced data quality assessment techniques. Practical criteria for technique selection will be condensed in a set of general guidelines for the selection of methods for detection of anomalous events and sensor faults. The specific aims of this chapter are to:

- Introduce the concept of redundancy and its types as the core of algorithmic quality assessment.
- Define the typical data processing pipeline for quality assessment and emphasize its importance.
- Highlight the importance of data visualization for data quality assessment.
- Describe the main techniques of data pre-processing to obtain data that are suitable for subsequent use of data analytical techniques.
- Introduce the most popular data analytical techniques for algorithm data quality assessment, including both mechanistic and empirical methods.
- Provide useful guidelines for data analytical technique selection and its metadata requirements, which input data are needed, and how to interpret the results of each technique.
- Describe the key performance indicators to assess data analytical techniques.

6.3 REDUNDANCY: THE CORE OF ALGORITHMIC QUALITY ASSESSMENT

The main task in sensor quality assessment is to compare the sensor measurement with a reference value that represents an accurate estimate of the true value. The term *accepted reference value* is used in International Organization for Standardization (ISO, 1994) to emphasize that it is not always possible to obtain a perfect reference, although its accuracy should at least be quantified and appropriate for the given purpose. Many types of reference values exist that provide redundant information, about sensor measurements and, hence, can be used for validation purposes. This section outlines three common redundant information sources that can be used as references values, with varying degrees of accuracy: (1) redundant measurements, (2) historical reference data, and (3) model-based references. Seven examples are provided in Sections 6.3.1–6.3.3 to indicate how reference values and redundancies can be applied for validating WRRF data. The seven examples (a)–(g) are also illustrated in Figure 6.1.

Comparisons between redundant measurements are closely related to residual generation, which is the central step in fault detection and diagnosis. Hence, general overviews in that area are also applicable for sensor validation based on redundant measurements (see e.g., Basseville, 1993, 2001; Gustafsson, 2000; Pimentel *et al.*, 2014).

6.3.1 Redundancy via reference measurement

The most common strategy to assess signal data (e.g., sensor measurement) validity is to compare it with a duplicate measurement. This reference measurement can be obtained on the basis of the same measurement principle or an entirely different one.

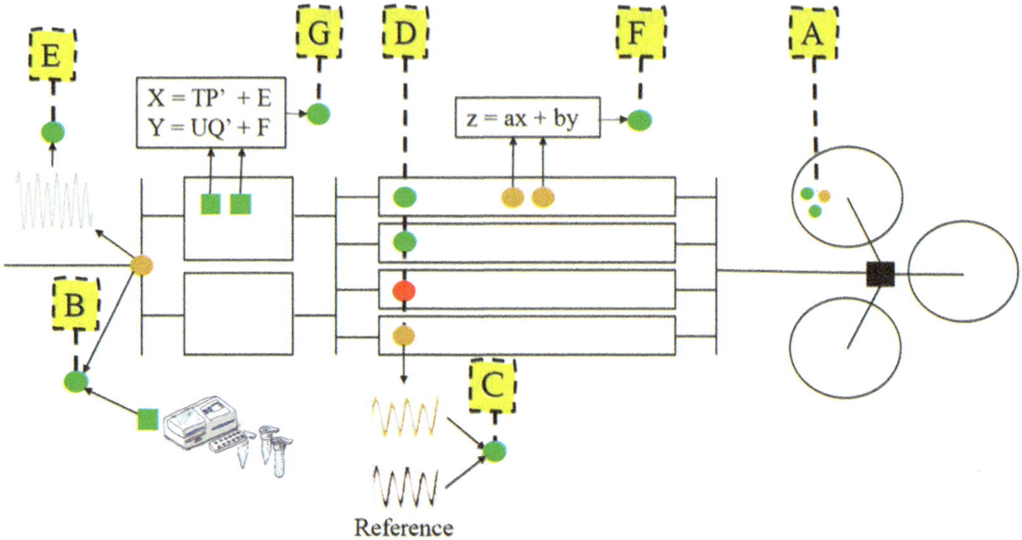

Figure 6.1 Scheme showing the different types of redundancy associated with wastewater resource recovery facilities. The diagram shows a simplified wastewater resource recovery facility consisting of two primary sedimentation tanks, four aeration lanes, and three secondary clarifiers. (A) Hardware redundancy; sensors providing identical measurements in space and time. (B) Measurement redundancy; an online sensor measurement is compared to laboratory data for the same measurement. (C) Typical reference data; a measurement is compared with historical reference data to assess adherence to process mean and variance. (D) Spatial redundancy; identical sensors in several parallel aeration lanes. (E) Qualitative/pattern redundancy; historic influent trend data are used to determine noise profile of influent sensor. (F) Theoretical reference models; a theoretical model describing the relationship between measurement variables is used to check both sensor and process conformity to expected trends. (G) Data-driven models; a representation of a PLS soft-sensor. Circles = measurement of interest, squares = indirect measurements for reference or inference, measurement/sensor is green = within expected limits, orange = cause for concern, red = error/failure.

(A) Hardware redundancy

The most straightforward way to obtain a reference measurement is to duplicate an existing sensor with an additional sensor of the same type, thereby achieving hardware redundancy (Figure 6.1A). This may be used for critical data signals or for cheap and ubiquitous sensors, such as pressure and temperature. Note that the hardware redundancy is not necessarily successful for monitoring sensor drift or fouling when this occurs simultaneously in both sensors and is subsequently unobserved. Further details on aspects related to redundant sensor measurements can be found in Haimi *et al.* (2015); Jung (2015); Nyberg and Frisk (2020); Ohmura *et al.* (2019); and Villez *et al.* (2016).

(B) Measurement redundancy

Most typically, sensor validation is based on comparison of a standard sensor measurement with an alternative, but appropriate, type of measurement for the same component, commonly performed off-line (Figure 6.1B). This procedure was exemplified in Chapter 5 (Section 5.5.2). In addition to the common grab water sample comparison, weekly flow-proportional lab water samples can also be used to assess the 7-day average for online sensors. Note that in such a situation, the online sensor values should be transformed into a flow-proportional integral value to match the expected laboratory water sample for a correct comparison. The benefit is a smaller sampling error, than for a grab water sample.

Some components can be measured with different techniques, each one with their own benefits and drawbacks. For example, ammonium can be measured using a spectrophotometric analyser or gas- and ion-sensitive probes, in addition to laboratory analysis. Sensor validation protocols related to the different measurement techniques were provided in Chapter 5 (Section 5.5) together with references.

6.3.2 Historical reference data

This section outlines four types of reference measurements containing redundant information when collected and analysed over time. We denote this group historic reference data.

(C) Typical reference data

Typical diurnal, weekly, and seasonal patterns are present in most influent data (Figure 6.1C). Knowledge of these patterns can be used to determine a process mean that describes a historical reference dataset. An essential step is to estimate the variance around the mean, which indicates its usefulness and precision as a reference value.

The main challenge with historical reference values is that the wastewater influent and processes will change over time. The historic reference datasets will become less representative of the current. This can be partially addressed by correlating process-related data with changes in the influent. Essential for any historical reference is the availability of sufficient metadata describing the historic influent and process conditions. Also, sensor disturbances and erroneous readings in the historic data should be identified and removed to reduce the noise in the typical reference data.

Additional examples of how historic data can be used as benchmarking references, commonly in the form of performance indicators, are provided by Balmér and Hellström (2012); Longo *et al.* (2016); and Samuelsson *et al.* (2017).

(D) Spatial redundancy

WRRFs are typically designed with several parallel and sequential treatment units. This feature can be used to compare equivalent measurements to identify deviations (Figure 6.1D), which we here denote spatial redundancy. The challenge is to assess the expected difference between, for example, parallel treatment lines, which in practice may depend on the load and plant hydraulics. Further, sequential measurement can contain information about the preceding process unit. Spatial redundancy has been shown to be useful for validating flow measurements (Villez *et al.*, 2016) and to detect deviating air flow rates (Carlsson & Zambrano, 2016).

(E) Qualitative redundancy

Qualitative redundancy is a data-driven type of redundancy, which uses historic process information embedded in sensor signal patterns as reference for the expected signal behaviour. For example, underlying patterns that reflect the influent dynamics, sensor noise, or some specific process behaviour can be observed in many measurements (Figure 6.1E). The pattern does not need to have physical meaning to be useful, but simply describe how the signal typically behaves. Univariate signal filtering (Alferes & Vanrolleghem, 2016), sedimentation profile tracking (Zambrano *et al.*, 2019), and automatic process monitoring and control (Thürlimann *et al.*, 2015; Wade *et al.*, 2005) are examples that illustrate the wide applicability of exploiting historic patterns as redundant information.

(F) Theoretical reference models

The literature provides reference values that are either based on theoretical computations or a compilation of extensive datasets that indicate a reference and typical or expected range (Figure 6.1F). Theoretical reference values are most common for process rates, which describe the expected relationship between several measurements. For example, a specific respiration or denitrification rate that strongly deviates from the theoretical assumptions can indicate bias in the underlying measurements, rather than a process disturbance. Standard WRRF-focused textbooks often provide reference values, see for example, Henze *et al.* (2002); Loosdrecht *et al.* (2016); and Metcalf and Eddy (2004).

6.3.3 Model-based redundancy

(G) Data-driven reference model predictions

In addition to the previous types of reference measurement, models that are explicitly derived to produce reference values for validation purposes can also be utilized as a source of redundant information. Here there is no distinction between mechanistic (models derived from first principles) or black-box models (purely data-driven) as either model type can produce a reference measurement through its prediction, which can be used for sensor validation (Figure 6.1G). A benefit with data-driven references is that the comparison can be conducted automatically and in real time once a model has been derived. The construction of such a model is the focus for the subsequent sections in this chapter, which includes many practical examples.

6.4 TYPICAL DATA PROCESSING PIPELINE

As previously evidenced, there is a need for a structured and, where possible, automated data processing approach. We name the series of data processing steps a data pipeline. Quite often, a multitude of pipelines exist, each designed for a specific aim. The most appropriate data pipeline for any given data source should be dictated by the purpose of the data.

A pipeline can be built selecting the suitable processing steps from those included in Figure 6.2, that is, (1) initial review, (2) data pre-processing, and (3) data quality assessment divided into methods rooted in mechanistic understanding and methods rooted in data analytics and statistics. Specifically, initial review comprises a series of approaches to gain insight into an initial understanding of the raw data such as data visualization (Section 6.5), data grouping, and descriptive statistics. The second step comprises different pre-processing techniques that can be applied (outlier detection, transformation, data imputation, scaling, filtering, and smoothing, feature generation and data structuring) depending on the raw data characteristics and the final data quality assessment technique to be used. For instance, imputation of absent or missing data will only be necessary if the necessary data values are not available, or a transformation will be applied to a variable if it exhibits a marked skewed distribution. The final step of the pipeline consists of one or more data quality assessment techniques. These techniques are generally grouped in mechanistic approaches: sanity checks and mass balances (Section 6.7.1); and empirical approaches: supervised and unsupervised approaches (Section 6.7.2).

The execution of these three steps can require human intervention (e.g., historical data visualization) or be completely automatically in the case of real-time data use (e.g., outlier detection). Moreover,

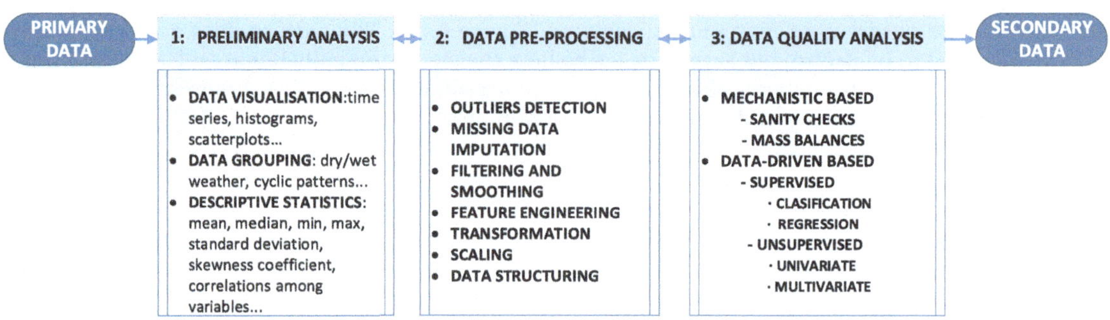

Figure 6.2 Steps that can be included in the development of a pipeline for data quality assessment.

these steps are not necessarily executed linearly; different feedbacks are possible within the pipeline. For instance, outliers can be detected in the last step by a data quality assessment approach, then this information can be retrofitted to the pre-processing step, for instance to fill in the gaps with reasonable values by a data imputation technique. In addition, not all the steps collected and shown in Figure 6.2 are always required. Therefore, there is no single valid pipeline for all cases, rather the pipeline is case-specific, as it will depend on what our data quality assessment is performed for: e.g., automatic control, regulatory oversight, simulation, and so on. The purpose also dictates whether the data pipeline permits human interaction (e.g., simulation) or not (e.g., online control). Figure 6.3 shows a data pipeline that could be used for automatic control purposes. The processed signal data of the process can be used as input of the control algorithm, either directly (e.g., the oxygen concentration) or indirectly via the online estimation of another variable or parameter using for example a data-driven model (e.g., artificial neural networks (ANNs)).

To our knowledge, there are few facilities that have implemented real-time automated raw data processing such as the typical pipeline shown in Figure 6.2. Rather, sensors are installed, maintained, and the data – good and bad – are recorded in a data management system of some kind. In most cases, a data processing step is not included at the time of collection. Instead, facilities typically rely on simple rule-based logic or data-use criteria to compensate for data variability and unknown errors. It must be recognized that all signals will have variability and error components (i.e., a signal is never perfectly correct) and the goal of any signal processing is to minimize the impact of those issues when using the data. We recommend that data of poor quality is not omitted but label led or marked instead (see also in Chapter 4).

Although facilities within the water sector are not at this level of data quality assessment, the embracement of some of the steps previously indicated will push forward the transition towards smart process operation and optimization. In the following paragraphs, some examples of data processing currently deployed at facilities are briefly described.

Figure 6.4 shows a simple example of an airflow signal data along with its 5-minute moving average (top), the calculated residuals for the same period (bottom left), and the distribution of those residuals (bottom right). This example shows how very simple analysis techniques (data visualization and descriptive statistic together with a filtering pre-processing step) can provide meaningful information about a signal's variability. In practice, this analysis would be executed with a larger dataset over a longer period of time. A smaller data set is used here for didactic purposes.

Figure 6.5 shows the response of a DO control sequence in a WRRF where a filtering technique has been applied to the control signal to smooth a control signal. In this example, the monitored DO (circular dots) has a rapid response and associated noise. Within the control structure, a 2-min moving average is calculated (solid line) and hardcoded thresholds are defined at 1.75 (upper horizontal line) and 1.9 mg/L (lower horizontal line). The valve response to the fluctuating moving average shows how the valve opens or closes as the moving average breaches the two thresholds (square dots).

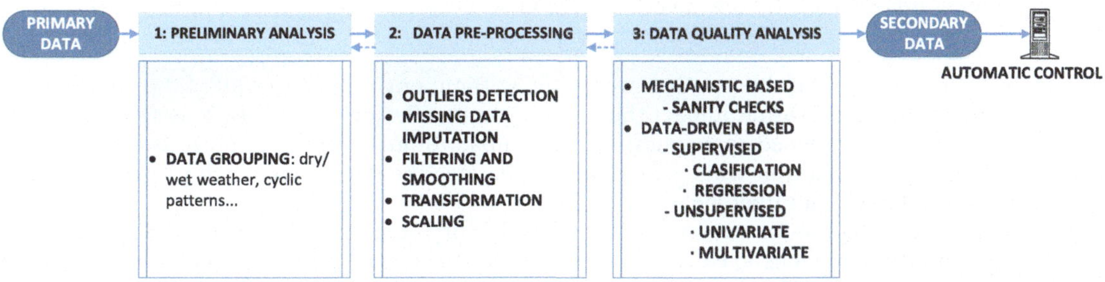

Figure 6.3 Steps that can be included in the development of a pipeline for data quality assessment for automatic control purposes.

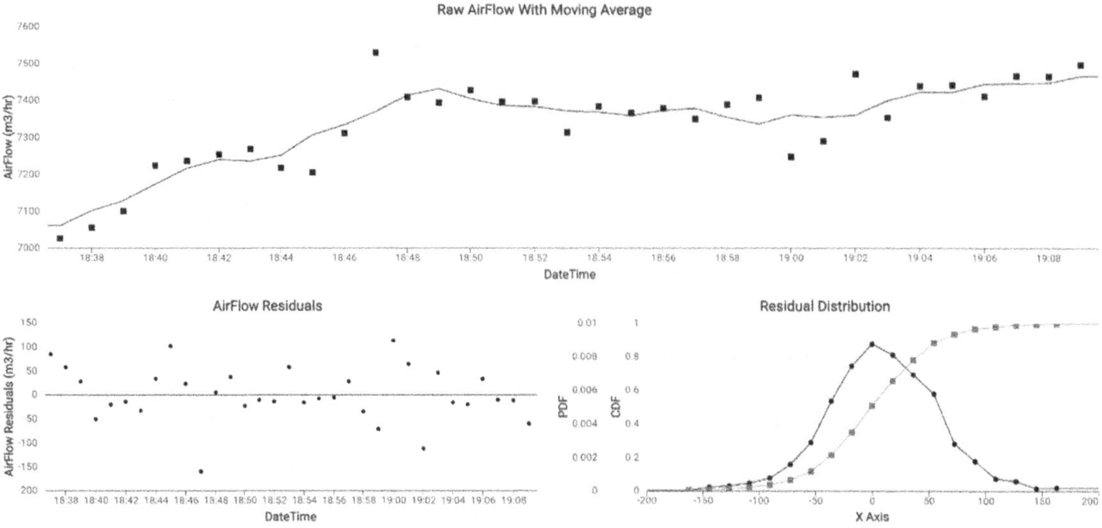

Figure 6.4 Simple residual analysis of an airflow signal data from a wastewater resource recovery facility. Raw signal (top, squares with 5-min moving average (top, line). Calculated residuals – moving average minus raw signal (bottom left, dots) along with the distribution frequencies of those residuals (bottom right, PDF [black dots] and CDF [grey squares]).

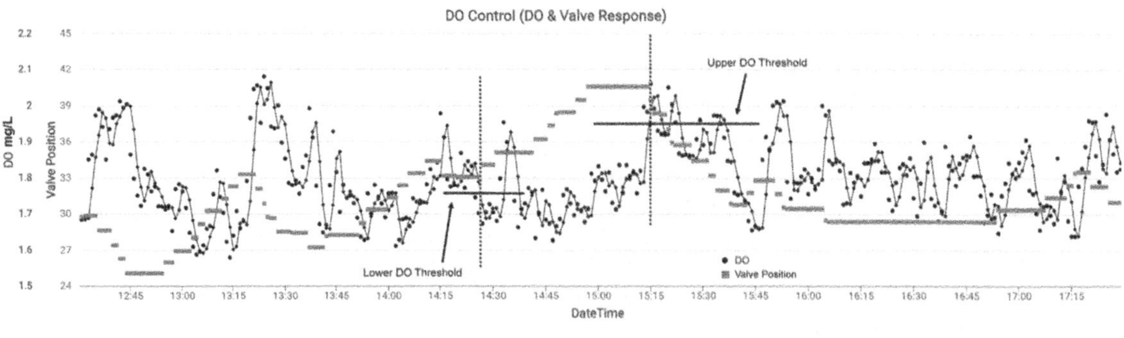

Figure 6.5 Illustrative example of a DO controller in a wastewater resource recovery facility. Monitored DO (black dots), 2-min moving average DO (solid line), valve opening (square dots).

Figure 6.5 illustrates that if the raw data were used without any processing, all of the natural error and variations in the signal would impact the resulting controller behaviour. By averaging the input signal, the variability is lessened somewhat, and the impact of an errant data point is avoided. This illustrates a customized solution where the data use dictates the degree to which a raw signal (and ultimately the response) is modified before use.

Filtering a signal is a common way to smooth out variability, remove gross errors, and reduce noise, but other use-based approaches exist as well to improve the signal's fitness-for-purpose. Figure 6.6 illustrates two other examples from the same dataset. First, it shows the use of a dead band whereby the DO is allowed to vary between 1.75 and 1.9 without any valve control action. Figure 6.6 also

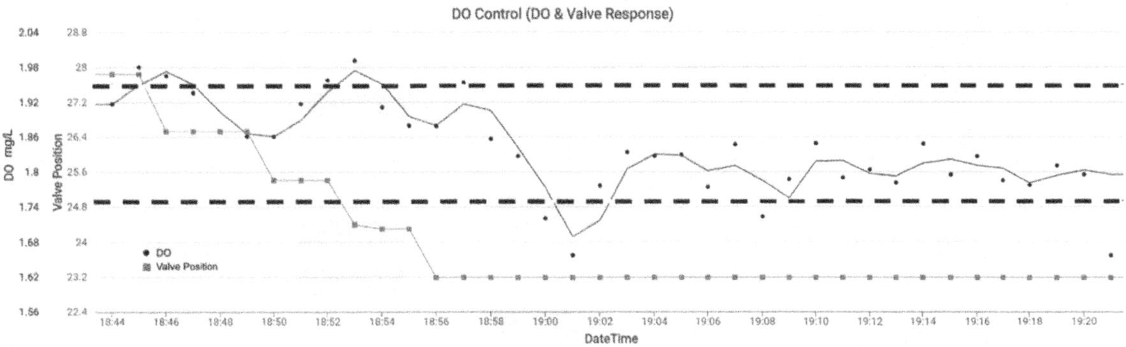

Figure 6.6 Illustrative example of a DO control performance with a dead band and time limited control behaviour (DO moving average (solid line), monitored DO (dots), control valve opening (squares with dotted line).

shows the use of time-related dampening whereby the control action is initiated only after a defined period time. The valve behaviour in the figure (square dotted line) shows a step-wise response, where the control action is initiated only after a minimum 3-min hold. That is, the valve responds, waits a minimum of 3 min, and then, if necessary, responds again. Although these constraints are often used to limit the wear and tear on actuators and neither of these approaches, strictly speaking, involves data processing, the lack of control action in these two cases is also a way to acknowledge that variations in the data are known to exist and acceptable because either: (i) the process would not be significantly affected by variations in this range; or, (ii) the signal variation is within a reasonable level of expected error/noise and more information (i.e., further data) is required to ensure that the change in the monitored variable corresponds to a meaningful change in the process.

Another example of dampening a control action occurs when the control structure requires that a threshold is breached for a defined amount of time. One of the reasons to introduce delays in control is to protect the actuators from continuously turning on and off. In this case, if the signal breaches the threshold for less than the defined period, then it is assumed the breach was due to error/noise or alternatively, the breach was unlikely to affect the process in that timeframe. Thus, the idea is that a breach is known to have occurred only if the threshold is exceeded for a defined period. Again, this is not signal processing per se, but illustrates a common approach to limit the impact of the variability, errors, and/or noise in a signal.

In the previous example, rolling 2-minute averages were used to control the movement of the aeration valves due to the rapid system dynamics involved in changing DO conditions, but other processes in the treatment plant have much longer time constants and require different data processing. Take for example a solids calculation included in a sludge retention time (SRT) controller (Figure 6.7) at a four-lane activated sludge (AS) plant. In this case, 6-hour averages for influent flow rates ($Q_{\text{Ave\_INF}}$), mixed liquor total suspended solid (TSS) concentration ($Q_{\text{Ave\_TSS}}$), wasted activated sludge flow rate ($Q_{\text{Ave\_WAS}}$), and returned activated sludge flow rate ($Q_{\text{Ave\_RAS}}$) are used to estimate a clarifier underflow concentration ($\text{TSS}_{\text{Clarif\_Under}}$) and ultimately the wastage flow rate ($Q_{\text{Clarif\_Under}}$) to maintain an SRT set point. Averaging the signals over 6 h is more than sufficient here as the time constant for this process is in the order of days or weeks with mixed liquor solid concentrations naturally varying throughout the day.

Any data processing (averaging, filtering, etc.) is application-specific and should be optimized to filter out unwanted variability while capturing the applicable system dynamics. The use cases here have illustrated that the degree with which a signal is processed is typically a function of the signal characteristics and the time constant associated with its use.

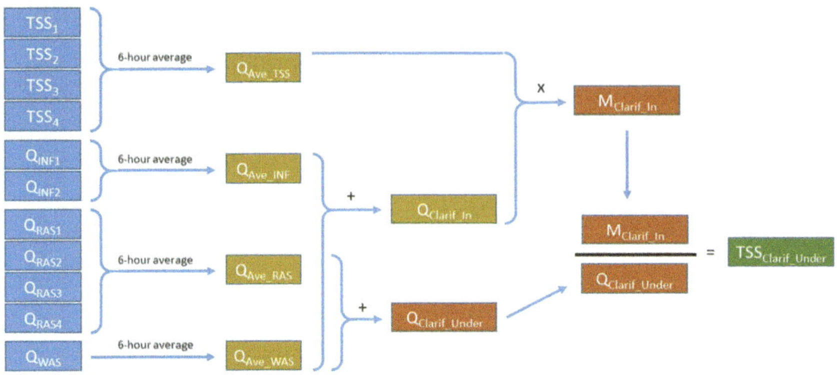

Figure 6.7 Example of a solids balance included in an SRT controller with moving average as a filtering pre-processing step of the raw sensor signals depicted in blue.

6.5 PRELIMINARY DATA ANALYSIS AND VISUALIZATION

There are numerous ways that historical and real-time data can be displayed and/or depicted. The choice will be application-specific and should be chosen to convey the required information in a concise and appropriate manner. Typical ways that have been used historically include:

- Statistical tables (which might include number of points, means, medians, standard deviations, percentiles, etc.)
- Bar charts
- Pie charts
- Histograms
- Box plots
- Heatmaps, etc.

Regardless of the application, the method of data visualization should be chosen to suit that application. Designers and operators may require different information from the same dataset, so the method used for visualizing the data to these groups may very well be different. The key to a successful choice is that the consumer of that data clearly understands the information being displayed and can extract the information they want efficiently. Time series, scatter plots, and statistical tables are widely used, but other methods are also available and may be more appropriate. The best choice will be tailored to the application and the audience for the data.

Checking the data quality is often initiated by plotting the signal data as a function of corresponding laboratory measurements (see also section 6.3). This can be easily visualized by a scatterplot. The agreement between the signal data and the laboratory data will be greater when more data conform to a 45° straight line that passes through the origin. The assessment can be quantified by a linear regression, checking the slope (the closer to 1 indicates better agreement), the ordinate at the origin (the closer to 0 is better), and the determination coefficient (R^2) of the fit (the closer to 1 is better). Figure 6.8 shows a scatterplot of the signal data from a nitrate sensor against the laboratory data. This figure clearly shows that the nitrate sensor was in very good agreement with the laboratory data as the resulting regression is nearly identical to the 45° line (green line). Simple plots similar to this are used in practice and can convey a lot of information very efficiently.

When historical data are being reviewed and analysed, other forms of visualization may be more suitable. For example, Figure 6.9 shows a frequency distribution plot depicting the distribution of some

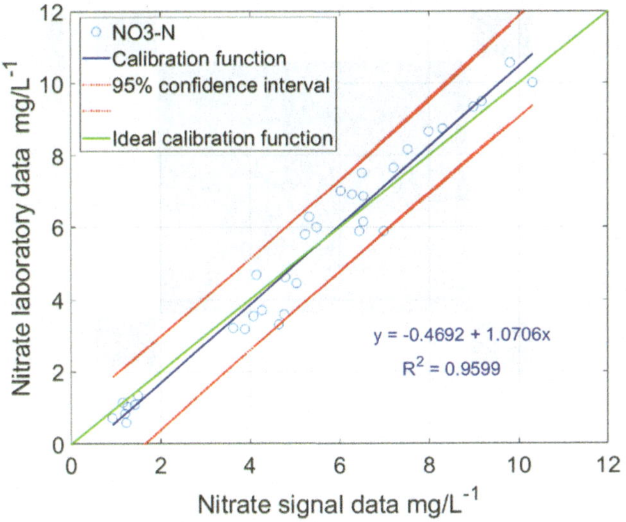

Figure 6.8 Scatterplot to assess the signal data quality from a nitrate sensor by redundancy via reference measurement from laboratory data.

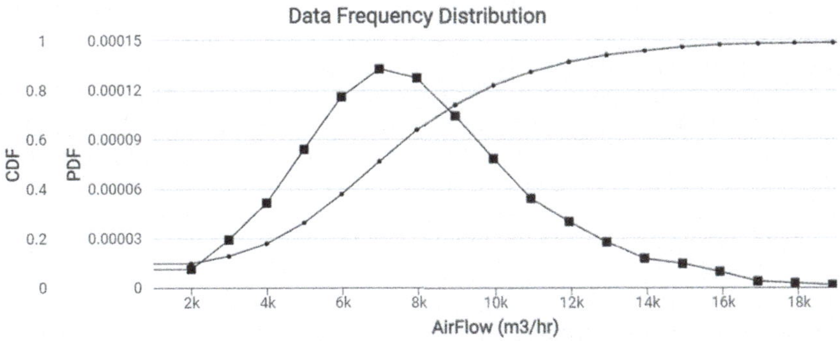

Figure 6.9 Distribution plot of raw airflow data to assess historical data variance (PDF [black squares] and CDF [black circles]).

raw airflow data. This might be used, for example, to verify that the data are normally distributed or could be used to quickly assess if the data are skewed.

Another visualization technique is the heatmap. This technique allows the visualization of the values of a data table (primary or secondary data, see Chapter 2) as different colours in two dimensions. The colour variation within the map facilitates the interpretation of the data, giving some visual clues on the data: how it is clustered or how it varies over the map. It provides a quick and effective way to convey the information contained in the data to the human eye, which facilitates detection of patterns of interest. As an example, Figure 6.10 shows a heatmap of the correlations among the signals recorded at the influent of a WRRF. It can be rapidly seen a negative correlation between the flow and the wastewater pollutants, and a highly positive correlation among the pollutants.

Today, data visualization at most treatment facilities is quite rudimentary due to the need to display signal data clearly, reliably, and succinctly to enable an accurate and fast response by human

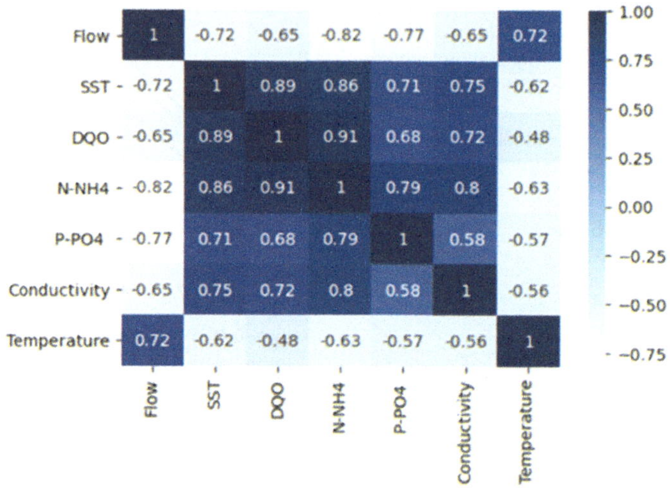

Figure 6.10 Heatmap of the correlations among the signal data from the influent of a wastewater resource recovery facility.

operators during troubleshooting. For the most part this is achieved with supervisory control and data acquisition (SCADA) screens that display the most recent value for key process variables as text often combined with figures that display the signal value as a function of time.

SCADA screens tend to depict the facility in a multi-layered graphical user interface, where a user can drill down into a particular area of the facility and retrieve specific information. For example, one might want to check the air flow to a particular basin (Figure 6.11). This might be achieved from the root screen by clicking on the AS object, then on the lane in question, then on the aeration object. At each step in this drilling process, different variables are displayed that relate to that query level. For example, it may be only necessary to click on the AS object to view the current DO values. Each SCADA application is plant-specific, but the basic functionality tends to follow this architecture.

The screens themselves vary from plant to plant and can take on various forms and configurations with an array of dials, gauges, and output fields (Figure 6.12).

Data trending options are also common. Take for example an effluent quality variable (e.g., TSS). It might be of interest to view the current value (and this might be possible using the SCADA system described above), but the current value provides no context with respect to the past or expected immediate future for that variable. So, depending on the use case, a plot showing a measured variable as a function of time might provide that context (Figure 6.13). Is the current behaviour normal or abnormal? If abnormal, can the trend be used to pre-emptively alert staff to an impending problem and proactively signal for a mitigating action?

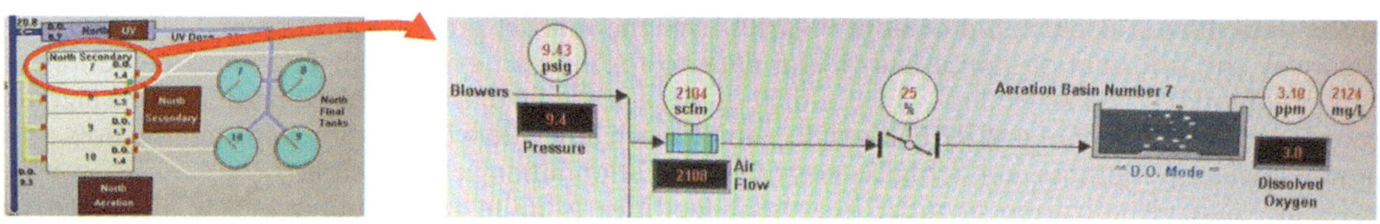

Figure 6.11 SCADA screenshot example.

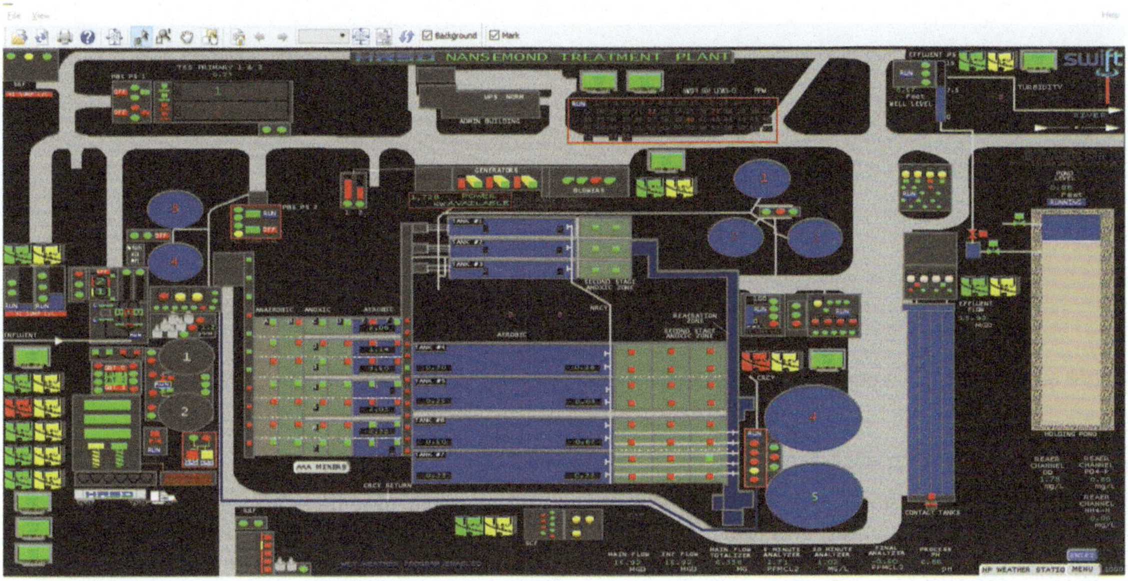

Figure 6.12 SCADA screen examples.

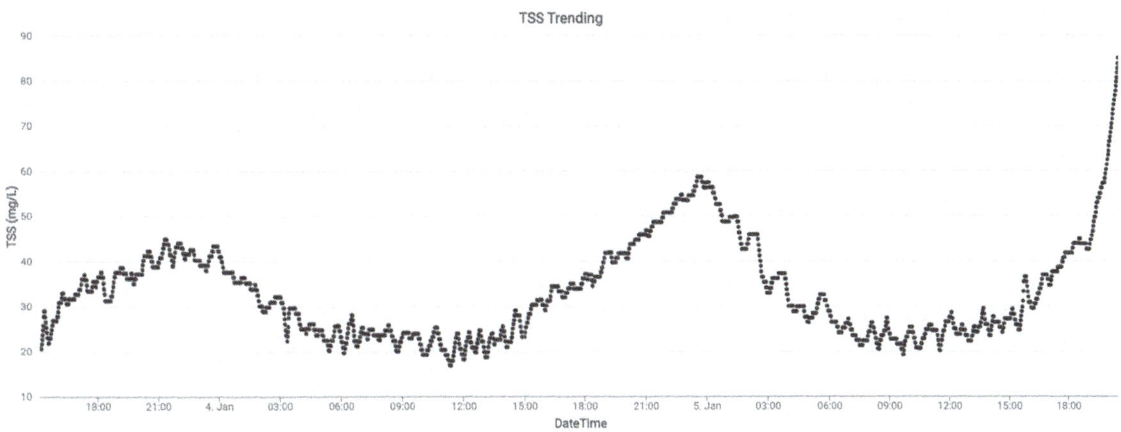

Figure 6.13 Time plot evolution of the effluent TSS of a wastewater resource recovery facility.

Where applicable, other charting options have been successfully used. Figure 6.14 shows a control chart example where upper and lower limits are defined, and the breaching of those limits might trigger an alarm or warning (see Section 6.7.2 for further details on this type of chart).

The display of current values provides a snapshot of the process state whereas time-series plots can effectively convey the most recent process dynamics and be used as an early warning if the process state is trending in an unfavourable direction.

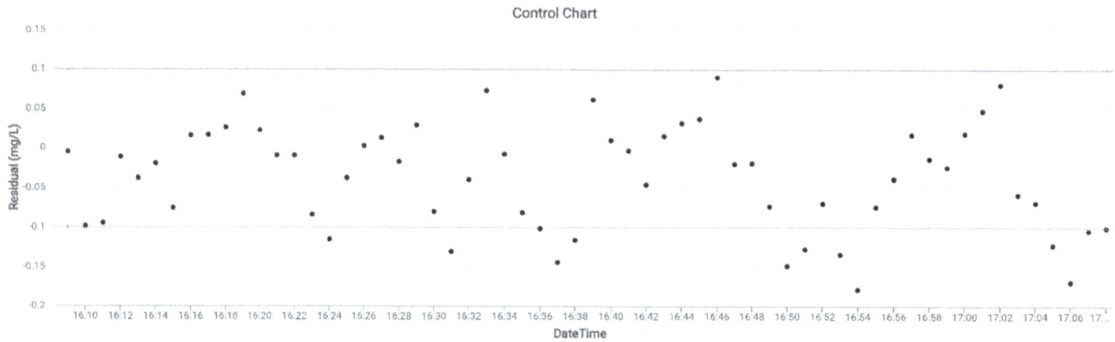

Figure 6.14 Example of a control chart applied to the residual of a signal data.

6.6 DATA PRE-PROCESSING

Signal data from the sensors deployed in WRRFs are often noisy, may have absent or missing values, outliers, out-of-range values, and may have a distribution with heavy tails. Thus, the information that the data convey about the underlying processes is hindered and not in an ideal form to be extracted through analytical methods. Through the pre-processing stage, the gathered data are converted or reshaped into a more suitable form for its further analysis making it easy for the detection of the underlying patterns. It is an essential stage for the later application of data quality analysis techniques (e.g., mass balances, machine learning, etc.). In fact, this stage can make the difference between developing a meaningful model or a useless model.

The main motivations to carry out the data pre-processing are:

- To enhance interpretability of the data and the information extracted from them.
- To apply data quality assessment methods, which may not accept the data in their original form. This may hold for unstructured data and datasets with absent or missing data entries.
- To achieve better performance of the algorithms. For example, the gradient descent algorithm tends to converge faster when all the variables have a similar scale.
- To satisfy important model constraints or assumptions.

Data pre-processing includes many techniques serving one or more of the above purposes. Some methods are applied commonly (e.g., centring and scaling) whereas others are used only in specific situations (e.g., data structuring). In this section, we focus on the most popular techniques.

6.6.1 Outlier detection

6.6.1.1 Purpose

The purpose of this technique is to detect anomalous observations in comparison with the rest of observations in the dataset. These non-typical observations have an unusually large or small value relative to the other values in the dataset. Usually, these atypical observations are investigated to figure out what caused such an unexpected value because they can be due to a real phenomenon that took place (e.g., related to the treatment plant dynamics), but they can also simply be consequence of an error or a sensor malfunctioning and thus they should be excluded from the dataset prior to further processing.

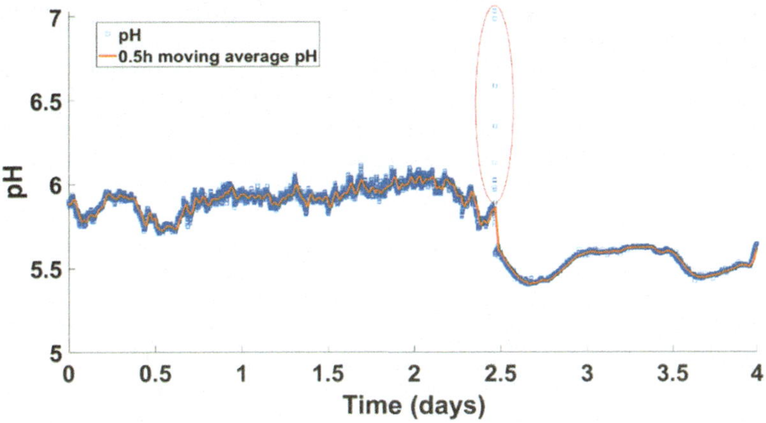

Figure 6.15 pH signal data from the AS basin. Outliers are detected after filtering the pH raw data.

6.6.1.2 Exemplary methods

Outliers can be detected in different ways:

- Through visual inspection of the data displayed in a graph, together with a technical interpretation of the process regarding the occurrence of the non-typical value.
- Through statistical methods, such as (a) by defining lower and upper limits for their detection based on the interquartile range, or statistical process control charts, such as the Shewhart chart and the exponentially weighted moving average (EWMA) control chart or (b) with multivariate methods, including multivariate statistical process control methods.

We strongly advise to store information about identified outliers by means of annotation of the data (see Chapter 4). This should include at least the method used to detect this outlier (e.g., *EWMA*, *visual inspection*, stored as method). We strongly advise against removal of outliers from centralized databases. Figure 6.15 shows an example of outlier detection after applying a filtering pre-processing step by moving averaging the pH signal data.

6.6.2 Data imputation

6.6.2.1 Purpose

Data imputation is the estimation of absent or missing data (see Chapter 2, Section 2.4.1.3) with a value based on other available information. This available information could be the preceding and subsequent values within the same signal. These could also be the data points in other signals recorded simultaneously, thus allowing us to exploit cross-correlation among the variables.

Data may be absent or missing for different causes: values outside the detection range, all the observations are not measured due to economic reasons, instrumentation failures (Grung & Manne, 1998), communication failures, periods of maintenance, samples not measured in required times, data dropped due to severe measurement errors, and sensors with different sampling periods (Arteaga & Ferrer, 2002). In online batch process monitoring and control, at any given time, future process behaviour can be treated as absent data (Nomikos & MacGregor, 1995).

Although it would be possible to analyse only the complete observations, that is, those observations of the dataset without absent or missing data, this simple approach can lead to a significant loss of information. When the amount of missing data is very small relative to the size of the dataset (<5%), discarding a few observations that contain absent or missing values may be a reasonable strategy which do not bias the analysis and only leads to a little loss of information. However, the greater the amount of absent or missing data, the greater the loss of information, and this is what makes the imputation methods valuable. Moreover, these methods are especially worthy when the absent or missing data appear in variables that are used as input in control algorithms.

6.6.2.2 Exemplary methods

Different methods have been proposed in the literature to impute absent or missing data. Their selection depends on several factors, such as the length of the break (i.e., the period with several sequential data absent or missing), the missing data pattern, the mechanism that led to absent or missing data, or even the aim of the analyst. Mulder *et al.* (2018) suggested five data imputation methods that can be used for typical data from WWTPs: (1) interpolation, (2) cross-correlation with other available variables, (3) fill in with the corresponding values from the average daily pattern of the variable, (4) repeat the last value (or the value from the preceding day), and (5) fill in with data from a mathematical model (e.g., from an influent generator to complete the gaps in influent variables, or from a calibrated activated sludge model (ASM) to fill in concentrations in the biological reactors or in the effluent).

For illustrative purposes, in Figure 6.16 the influent flow rate is used to showcase the possibility of filling in the data by two data imputation methods. First, by cross-correlation with the collected variable itself (for 2 weeks) using a principal component analysis (PCA) model (1), thus taking into account the autocorrelation pattern (day 17 from 11 : 00, dashed orange line). Second, by using the average daily pattern (3), calculated with the 2 weeks of data (from day 20, dashed green line).

The absent or missing data pattern describes what values are known and what values are absent or missing in the data matrix $\boldsymbol{X}$, along with the mechanism, that is, the possible relationship between the absence/missingness and the unknown value of the variable or other values within the same data point.

The most problematic case occurs when certain set of variables are never jointly observed (e.g., sensors with different sampling periods), because this makes the direct estimation of the parameters describing the relation between these variables impossible (Muteki *et al.*, 2005).

In the case of missing data, it is important to elucidate whether the fact that the variables are missing is related to the unknown values of the variables. The properties of statistical methods for working

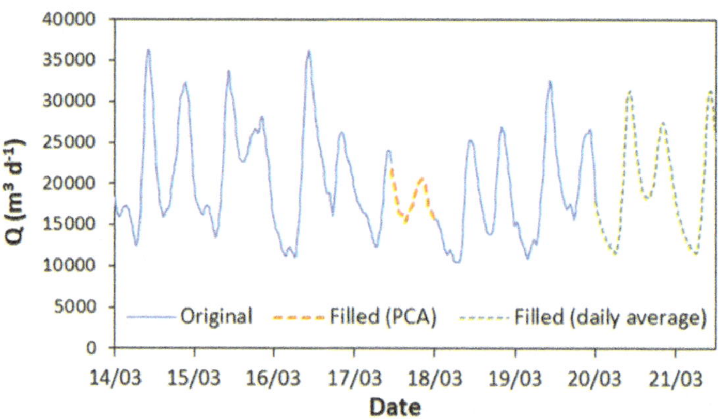

Figure 6.16 Wastewater resource recovery facility influent flow rate along 1 week. Result of filling in the missing data based on a PCA model (day 17 from 11:00, dashed orange line) and based on the average daily pattern (from day 20, dashed green line).

with missing data are conditioned by the nature of the dependencies implied by these mechanisms (Little & Rubin, 2002). The missing data mechanism is important because it affects how much the missing data can bias the imputed data and subsequent data-analytical steps. Three different cases can be distinguished:

(1) Missing completely at random (MCAR) case: when there is no relationship between values of the variables (observed and missing) and the probability that they are missing. The missing positions are just a random selection from the data matrix. This does not mean that the pattern itself is random, but rather that missingness does not depend on the true underlying values. This is a rigid assumption, but reasonable in special cases. For instance, this can be convenient when some variables are very expensive and are measured only for a random subset of the data records, whereas the least expensive variables are measured for all data records (Allison, 2001). Similarly, when the data are absent due to scheduled sensor maintenance protocols not related to the operation of the process (Nelson, 2002), or when there is some random sensor fault or random data overloading problems in the communication with the data storage systems.
(2) Missing at random (MAR) case: when the probability of a data point to go missing depends on other values, which are available, but not on the unknown value of the missing data point (i.e., the propensity for a data point to be missing is not related to the missing data, but it is related to some other data that is actually available). This case is less restrictive than MCAR. The observed data include the necessary information to estimate the missing data. MAR occurs, for example, when some data records are not registered whenever certain other process measurements are outside the safe limits.
(3) Not missing at random (NMAR) or non-ignorable (NI) case: when the probability that a data point is missing depends on the unobserved value of the missing data points. The probability of a data point to go missing is complex to evaluate in this case because it depends on the unknown values of the missing data. This occurs, for example, in censored data: when some measurements are out of some detection limits (either above or below). NMAR is a problematic case, because the distribution of missing data for a set of variables does not follow that of the observed measurements for that same set of variables in other objects. In this case we need quite specialized methods to model the missing data mechanism to obtain good estimates of the parameters of interest.

Ignorability means that we do not need to model the missing data mechanism as part of the estimation process. MCAR is ignorable and MAR can be made ignorable, with the appropriate analysis. In real cases, we do not know the values of the missing data and, thus, we cannot formally prove any of the previous assumptions. Therefore, the mechanism in a particular problem must be selected following intuition and process knowledge. The key question is to distinguish between NMAR (NI mechanism typical of reliability and detection limit analyses) and ignorable mechanisms (MCAR and MAR) usually assumed in most missing data literature and software. The following sections will be focused on MCAR and MAR.

In the next sections, several statistical methods proposed in the literature for handling missing data will be reviewed, ranging from discarding the observations with missing data to relatively complex imputation methods. Some of the methods are quite simple approaches, implemented in most standard statistical packages, that use ad hoc procedures of deletion or imputation to force the incomplete dataset into a complete data matrix (Schafer, 1997). On the contrary, there are other approaches that offer substantial improvements over conventional methods, but at the expense of requiring specialized subject matter knowledge and software. It is important to note that neither method is perfect, and the only good solution to the missing data problem is not having any (Allison, 2001). Thus, there is great value in minimizing the appearance of absent or missing data.

6.6.2.2.1 Complete case analysis

Most statistical packages use, by default, the complete case (CC) method: any observation with absent or missing data is discarded. This is a simple approach, but it has some important drawbacks due to the potential loss of information, that is particularly acute in multivariate settings, where incomplete cases are often a substantial portion of the entire dataset. The loss of information leads to loss of precision (increasing the uncertainty of the estimates) and to suffer bias when the missing data mechanism is not MCAR, but only MAR selection (Little & Rubin, 2002).

6.6.2.2.2 Available case analysis

A natural and less restrictive alternative than the CC approach for univariate analysis is to include all observations where the variables of interest are present. This is the available case (AC) method. With this method the number of data records changes from variable to variable according to the pattern of missing data and this creates practical checking problems if the missing data mechanism is not MCAR.

Kim and Curry (1977) found AC to be superior to CC in simulations based on weakly correlated data, but this is a severe problem with highly correlated data, as noted by Haitovsky (1968). Lopes and Menezes (2005) suggested correcting the correlation matrix as in Higham (2002).

6.6.2.2.3 Imputation-based methods

The inconveniences of the two previous methods (loss of information, bias, and the ill-conditioned estimated covariance matrix) make it advisable to apply methods that in some way complete the data matrix, i.e., filling in the gaps with 'reasonable' values for the absent or missing positions. The objective is, in general terms, to fill in the empty positions of an incomplete data matrix, so that a complete matrix can be obtained. This then allows us to apply data analysis techniques that require that there are no absent or missing data. Ideally, the method that it is used to impute the absent or missing data must respect the correlation structure of the measured variables.

There are very basic and easy-to-apply alternatives, such as interpolating values of the same variable using previous and subsequent observations, averaging values for the variable around the time of measurement (as in the unconditional mean imputation method, which will be discussed later) or repeat values from previous data records.

Other more advanced methods allow filling in data from more or less complex models, exploiting the correlation with other available measured variables.

In the following sections some of these methods are described.

6.6.2.2.3.1 Single-imputation methods A single-imputation method consists of filling in an absent or missing value with a single, plausible value. Imputations are means or draws from a predictive distribution of the missing data and require a method for creating a predictive distribution for the imputation based on the observed data. Several approaches are discussed in this section.

Unconditional mean imputation method: each missing position is filled-in with the mean of the observed values of its column. With this method, the estimates of quantities that are not linear in the data (variances, covariances, percentiles, or measures of shape), are not estimated consistently using standard complete data methods applied to the completed data (Haitovsky, 1968). For example, variances and covariances from the filled-in data are biased towards zero, because the imputed missing values are at the centre of the distribution. Another drawback of this method is that it only uses the information of each variable, that is, it does not consider the multivariate nature of the collected data as the possible cross-correlated variables. The next methods solve this problem in different ways.

Conditional mean imputation (Buck's method): this method (Buck, 1960) starts estimating the mean and covariance matrix from the matrix based on the CCs, and then uses these estimates to impute, in each incomplete data record, the missing values with their least-squares linear predictions on the present variables. Instead of linear regression, a biased regression method, such as ridge regression (RR), principal component regression (PCR), or partial least-squares (PLS) regression, can be used when the correlation structure of the data and the combination of missing variables of a data record imply the inversion of an ill-conditioned matrix.

With this method, reasonable estimates of means are obtained, when the normality assumptions are acceptable. But, as Buck's method does not account for the variance between pairs of variables where both are missing, it underestimates the variances and covariances, although this underestimation is lower than that obtained with the unconditional mean replacement method.

Stochastic regression imputation method: this method is similar to the previous one, but to avoid systematic underestimations of the variability, to the estimated part of each data record is added a random realization of a normal variable with zero mean and covariance matrix that of the estimate of the unknown part of the data record conditioned to the known part of it.

This method draws values from the predictive distribution of the missing values given the observed ones. As in the Buck's method, linear regression can be substituted by a biased regression method to overcome the ill-conditioning problem.

Maximum-likelihood-based imputation method (the expectation maximization (EM) algorithm): the maximum-likelihood problem consists in finding the value for the set of parameters Θ (because we usually assume multivariate normality for the data, the set of parameters Θ is the vector of means and the shape of covariances) that maximizes the probability of obtaining the observed data, that is, to answer the question: for what set of parameters it is most probable to find the observed data?

It can be shown that, under MAR or MCAR, the missing data mechanism can be safely ignored (Schafer, 1997) and all relevant statistical information about the parameters is contained in the observed data likelihood expressed as:

$$L(\Theta \mid \mathbf{X}_{\text{obs}}) \propto P(\mathbf{X}_{\text{obs}} \mid \Theta)$$

The EM algorithm (Little & Rubin, 2002) consists of the following steps:

- Fill in the absent or missing values by initial estimated values.
- Estimate the parameters given the completed dataset.
- (E step) Re-estimate the missing values given the current estimate of the parameters.
- (M step) Re-estimate the parameters given the current, completed data set.
- Repeat E and M steps until convergence.

It should be noted that in step E we can suffer from the problem of ill-conditioning of the covariance matrix of the observed data, and this may require the use of biased regression methods such as RR, PCR, or PLS. See Schafer (1997) for an efficient implementation of this algorithm.

Limitations of the single-imputation methods: After applying any of these imputation methods, the matrix with the imputed missing data appears complete, and this allows analysing the data. This, looking very convenient, can be problematic, as single-imputation methods can distort the multivariate structure of the data. For example, the imputation of means skews the estimated variances and covariances towards zero. In contrast, imputing predicted values from regression models tends to increase the estimated correlations. Ultimately, treating the imputed data as if they were real can be an error because standard errors, *p*-values, and other uncertainty measures, calculated using standard full data methods, can be misleading by not reflecting the uncertainty associated with absent or missing data.

6.6.2.2.3.2 Multiple imputation method Multiple imputation (MI), basically, creates several (say M) values for each absent or missing value that produce a set of data points that reflects the signal data variability (Rubin, 1987). With this method, a number, M, of complete datasets are computed, and each one can be analysed with the standard statistical techniques, estimating the parameters of interest (mean, variance, etc.). As M combinations of the estimated parameters are available, this allows calculating its variances by combining the variability of estimates from within each imputed dataset with the variability of the estimates across M imputed datasets.

MI rests on three main assumptions: a probability model on complete data (observed and missing values), a prior distribution reflecting the uncertainty of the parameters for the imputation model, and the missing data mechanism, which must be ignorable.

The most popular form of the MI method is the data augmentation (DA) algorithm (Tanner & Wong, 1987) that is an iterative process that alternately fills in the data and makes inferences about the unknown parameters, but unlike the EM algorithm, it does so stochastically. DA starts with a random imputation with assumed parameter values (usually from the EM algorithm), and then obtains new parameters from a Bayesian posterior distribution, from the observed and imputed data. This algorithm, in iteration t, alternates between two steps:

I step (imputation) draws $\mathbf{X}_{mis}{}^{t}$ from their conditional distribution given $\mathbf{X}_{obs}$ and $\Theta^{(t-1)}$.
P step (posterior) draws Θ^{t} from their posterior distribution given $\mathbf{X}_{obs}$ and $\mathbf{X}_{mis}{}^{t}$.

This alternating procedure creates a Markov chain that eventually converges in distribution (Schafer, 1997).

In practice, the length of the Markov chain should be long enough to assure stability (convergence) and without depending on initial values. As the algorithm converges to a probability distribution, not a single set of values, this makes it rather difficult to determine whether convergence has, in fact, been achieved (Allison, 2001). Nevertheless, the rate of convergence of the EM algorithm serves as a useful indication of the rate of convergence for DA.

Each run of the DA produces a single, completed dataset. To apply MI, a separate chain of iterations for each of the desired datasets (parallel) or one long chain of DA cycles can be run, taking the imputations separated by a number of iterations to assess the independency in the dataset (because the imputation in each iteration is related to the previous ones and said dependency is reduced with the successive iterations) (sequential).

6.6.2.2.3.3 PCA model building-based methods for high-dimensional datasets PCA models are especially useful in the context of highly correlated datasets with a large number of variables, given its dimensionality and noise reduction power. The PCA model provides us A (latent dimension) latent variables or principal components (PCs) that are independent linear combinations of the K original variables (generally with $A \ll K$) capable of explaining most of the variance of the K original variables. As PCA is an efficient way to encapsulate the correlation structure of a dataset, a method that performs well in building a PCA model for an incomplete dataset should be a good tool to fill in the data. There are two methods that are frequently used to build a PCA model from an incomplete dataset: (a) a generalization of the non-linear iterative partial least-squares algorithm (NIPALS) (Schafer, 1997; Wold *et al.*, 1983) to deal with incomplete observations by performing the iterative regressions using the present data and ignoring the missing data, and (b) the iterative algorithm (IA) (Arteaga & Ferrer, 2005; Walczak & Massart, 2001) that, in essence, consists of filling in the data with the predictions obtained from previous PCA models iterated recursively until convergence.

Folch-Fortuny and co-workers (Arteaga *et al.*, 2020; Folch-Fortuny *et al.*, 2015) proposed a regression-based framework to estimate a PCA model with A PCs from an incomplete dataset and showed that the framework methods overperform the NIPALS adaptation and the IA. The proposed method is an adaptation of the IA. The more general method from the framework is the known data regression (KDR) method, that is equivalent to the EM algorithm, and the other framework methods are the consequence of replacing the linear regression with biased regression methods (KDR with PCR and KDR with PLS).

The IA (Walczak & Massart, 2001) starts by filling in the missing data with initial estimates (zeros or other imputations such as the mean of the known values of each column). This yields a reconstructed dataset from which a PCA model can be fitted. Then the original missing data are replaced by their predictions from this PCA model, yielding a new reconstructed dataset from which a new PCA model can be fitted. This process is iterated until convergence in the predicted values for the missing data.

The regression-based methods framework (Arteaga *et al.*, 2020; Folch-Fortuny *et al.*, 2015) can be seen as an adaptation of the IA where the way to reconstruct the missing positions allows us to differentiate the different particular methods of the framework: KDR method, KDR with PCR method, KDR with PLS method, and trimmed scores regression (TSR) method.

Among the different methods of the framework, the TSR method performed extraordinarily well in most of the data structures and a wide range of missing data percentages (Folch-Fortuny *et al.*, 2015). This data imputation method represents the best compromise solution among prediction quality, robustness against data structure, and computation time. From the other regression-based methods presented, the KDR methods with PCR and PLS offer also good solutions, however, they are more time-consuming, because they imply to fit additional PCR and PLS models.

6.6.3 Filtering and smoothing

6.6.3.1 Purpose

The purpose of smoothing the data in a signal is to highlight the most important trends in the measured variable and to remove the unwanted high-frequency fluctuations (noise).

These techniques are usually used to emphasize and expose those patterns in the data reflecting changes in the measured process variable and reduce or remove fluctuations that are unrelated to the process. A digital filter applies mathematical operations on a sensor signal to reduce or enhance certain components of that signal (e.g., process variations, noise).

6.6.3.2 Exemplary methods

The *moving average* is the simplest and most widely used filter. It is easy to understand, apply, and tune in view of satisfactory results. It is calculated by averaging a given subset of values of a time series to produce their average as output result. Therefore, a new time series is produced with the averages that presents lower variability (and thus lower noise) than the original values. The number of values (N) used in the determination of each moving average (i.e., the window length used by the filter) must be chosen. When the underlying process is relatively constant a large N is preferred (yielding a smoother filtered signal). However, a small window length allows the moving average to follow rapid changes more closely (i.e., it reacts faster to the changes). In turn, this means more noise will remain in the signal.

Figure 6.17 shows an example of moving average filter applied to the dissolved oxygen concentration in a biological reactor, and the effect of using different window lengths of the filter.

The moving average can be calculated using only previous values (e.g., for a window length $n = 3$, three values will be averaged x_{t-2}, x_{t-1}, x_t) allowing a real-time calculation of the moving average. However, for analysing an historical dataset, a centred moving average is usually preferred (for time instant t, the three values averaged would be x_{t-1}, x_t, and x_{t+1}) resulting in a symmetrical averaging when the window length is an odd number. Note that calculating the moving average is similar to applying a rectangular pulse as a filter: for example, for a moving average with $n = 3$, the filter applied would be …0, 0, 1/3, 1/3, 1/3, 0, 0, ….

Because the calculation of the average is sensitive to the presence of outliers, each outlier in the signal to be smoothed will disrupt all the averages produced for the length of the window chosen. For those cases with frequent occurrence of outliers, an effective alternative is the use of the median (instead of the average) as it is a robust estimator of the central position of the window values.

To obtain an even smoother filtered signal, the moving averages or the moving medians can be applied two or more times.

The *local regression smoothing* uses a moving window approach in which each smoothed value is determined by fitting a regression model to the values within the window. For each position of the window, around the window centre a linear or a polynomial function is estimated by minimizing the least-squared function. To avoid distortion in the output smoothed value due to the presence of

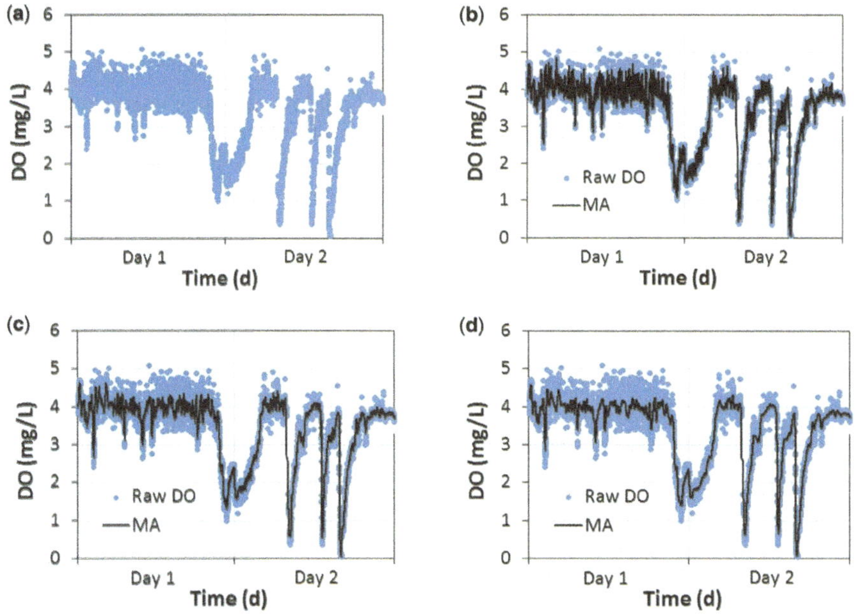

Figure 6.17 (a) DO concentration along 2 consecutive days (one value is recorded for every 20 s) together with a moving average superimposed of window length: (b) $n = 3$ (1 min moving average), (c) $n = 15$ (5 min moving average), and (d) $n = 45$ (15 min moving average).

outliers, robust regression can be used to fit the bulk of the values within the window. When the values of the signal to smooth are clearly non-linear, another option to fit a smooth function and approximate locally the values in the window is to use kernel regression. It is a non-parametric method that uses a kernel (usually the Gaussian kernel) to weight each value of the window and produce the smoothed value for each position of the window. The number of values to consider in the kernel regression is also a trade-off between the smoothing effect and the delay introduced in the produced filtered values.

The *exponential filter* obtains the smoothed output considering all the previous values of the time series and assigning them geometrically decreasing weights over time (Montgomery *et al.*, 2015). The simplest form of this smoother is the first-order exponential filter: $s_t = \lambda x_t + (1 - \lambda)s_{t-1}$, where s_t is the smoothed output at time instant t, x_t the current value of the time series, and λ is the weight of the current value (observation) and it should fall in the range $0 < \lambda < 1$ (although common values that perform well in practice fall between 0.1 and 0.4). Note that the higher the λ the more emphasis is put on the current value and, thus, more closely the smoothed output will follow the behaviour of the time series and reacts faster to its changes. To improve the behaviour of the filter when there is a trend in the values of the time series, the exponential filter can be recursively applied twice, which is known as second-order exponential filter (or double-exponential filter). In a similar way, when there is some cyclic pattern in the values of the time series, the exponential filter can be recursively applied three times, which is known as third-order exponential filter (or triple-exponential filter).

The *Savitzky–Golay filter* is a popular filter that, like the moving average filter, computes the filtered value based on a recent set of data points. It operates by fitting a polynomial (of a given degree) to a recent set of data points by the linear least-squares method. An analytical solution exists if the sequential values of the time series are uniformly spaced in time. This enables fast execution of the filter in real-time applications. The polynomial degree and the window length determine the noise reduction level and the extent of the distortion introduced in the smoothed output. A

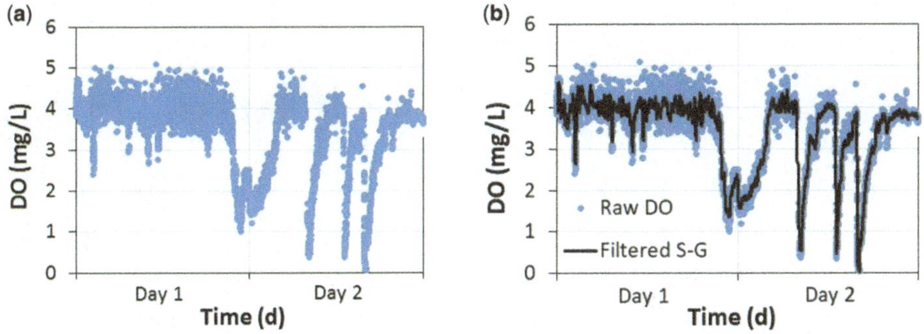

Figure 6.18 (a) DO concentration along 2 consecutive days (one value is recorded for every 20 s). (b) Original signal together with superimposed filtered signal from Savitzky–Golay filter (using a window length of 45 samples and fitting them with a second-order polynomial function).

low-degree polynomial can nicely smooth wide peaks present in the original signal but can hardly represent accurately the shape (height and width) of narrow peaks in the smoothed output (Sousa *et al.*, 2007). Figure 6.18 shows the same example as of Figure 6.17 (DO concentration in a biological reactor) but applying a Savitzky–Golay filter using a window length of 45 samples and fitting them with a second-order polynomial function.

Wavelets (small oscillating waves) can be used for de-noising, signal filtering, compression, and feature extraction. A wavelet is used to decompose the original signal into different scales of decreasing levels of detail (or resolution) using a mother wavelet which is successively scaled and translated in time. Different families (types or shapes) of wavelets are available and many of them have proven to be useful (e.g., Haar, Daubechies, Symlets, Meyer, etc.). Usually, the wavelet choice is empirically selected to have a shape characteristic of the behaviour that the original signal exhibits. In this way, less of the expected behaviour in the original signal will be smoothed out.

In practice, discrete wavelet transform (DWT) is implemented as a sequence of complementary high-pass and low-pass filters, decomposing the signal into details and approximation parts. Each decomposition level includes filtering the input signal (or its approximation from the previous level) through a pair of complementary filters followed by a down-sampling operation in which one in every two values of the filtered outcome is discarded. This procedure (see Figure 6.19) is repeated up to the

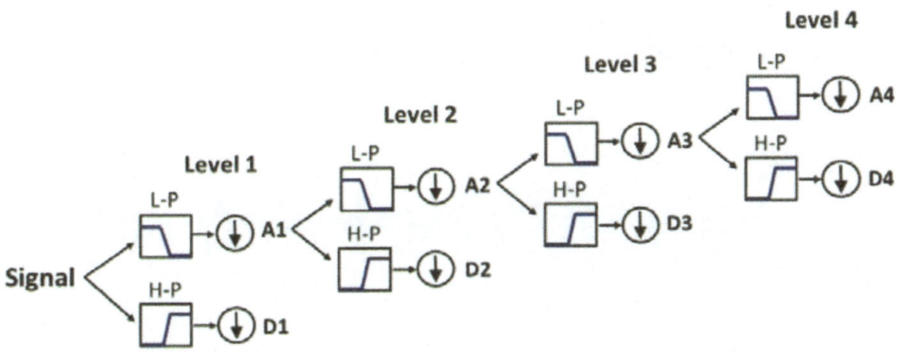

Figure 6.19 Scheme of signal decomposition up to level 4. A: approximation, D: details, H-P: high-pass filter, and L-P: low-pass filter.

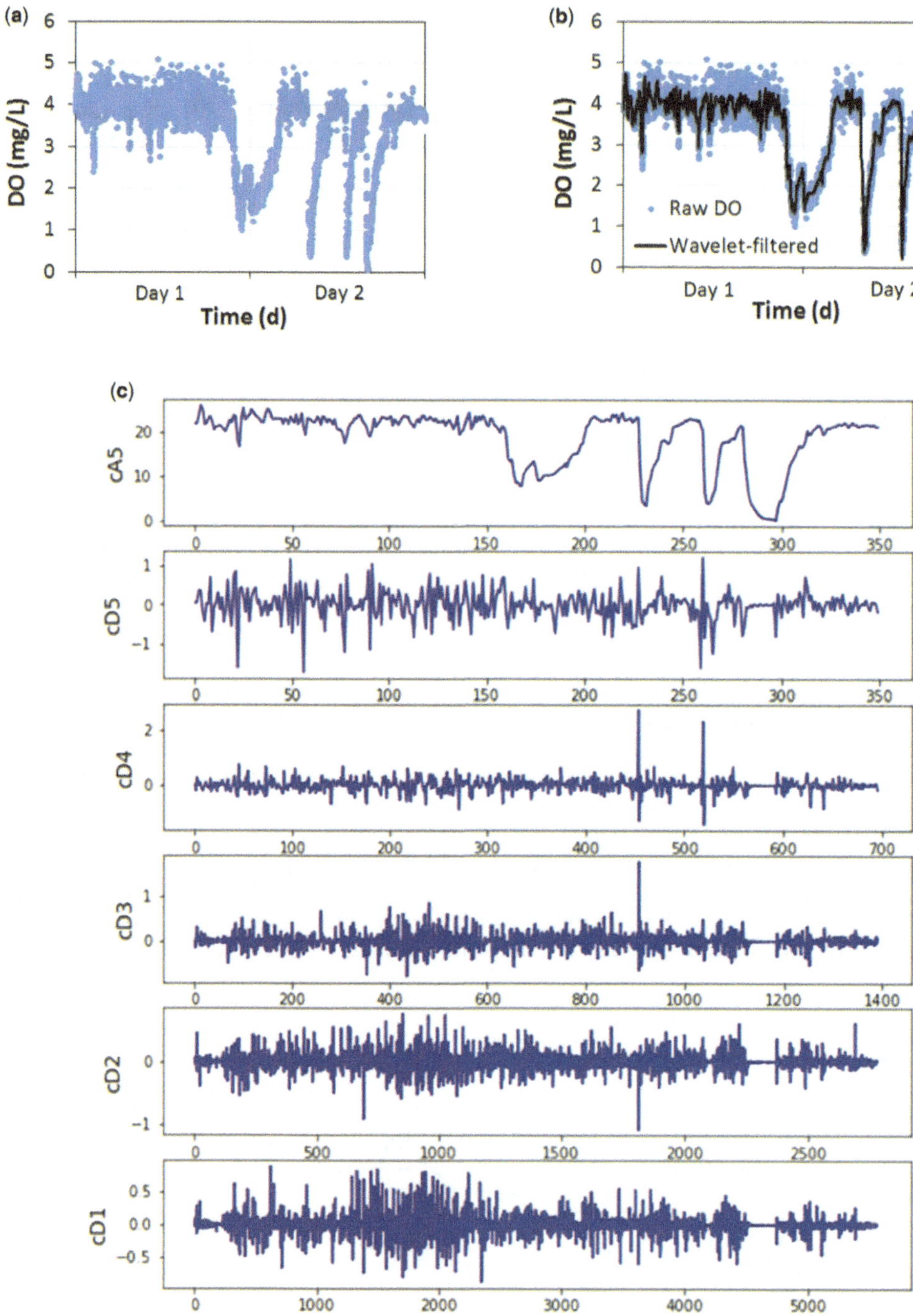

Figure 6.20 (a) DO concentration along 2 consecutive days (one value is recorded for every 20 s). (b) Original and wavelet filtered signal (after five levels of decomposition with wavelet Daubechies 2 – db2), (c) approximation (cA) and detail coefficients (cD) of each level of decomposition.

desired level of precision. The decomposition level can be empirically chosen based on the nature of the processed signal or via automatic criteria such as entropy (Mallat, 1989).

DWT is an orthogonal base decomposition; thus, it is an invertible operation (i.e., the original signal can be reconstructed reversing the steps). The details represent the high-frequency part of the input signal, whereas the approximation (the output of the low-pass filter) is a smoothed version of the input signal at the previous resolution. At a given level of decomposition, the details are mainly fluctuations due to unwanted noise, which can be removed (zeroed via thresholding) without distorting the important patterns in the original signal. Although the DWT is based on elaborated mathematics, the strategy to remove high-frequency noise is simple: (1) use DWT to decompose the original signal; (2) set detail coefficients higher than the given threshold to zero, and (3) use the inverse DWT to compute the smoothed signal.

Figure 6.20 shows the same example as the previous ones, the DO signal from a biological reactor, but applying the wavelet filtered signal after five levels of decomposition with wavelet Daubechies 2.

6.6.4 Feature engineering

6.6.4.1 Purpose

The purpose of feature engineering is to transform the primary data into relevant features (new variables) to improve subsequent data analytical steps, primarily by suppressing irrelevant information and computing variables that are considered high relevant to subsequent tasks.

Using relevant (i.e., informative, meaningful, discriminating) features is essential to succeed in data-driven modelling (pattern recognition, classification, or regression). In feature engineering, technical knowledge (of the process or of the data) is used to extract or create features (i.e., variables or attributes or descriptors) from the data that are relevant for the particular use case, that is, new promising features that highlight information that is relevant to the subsequent task. Process knowledge is used to generate or emphasize or extract aspects of the primary data that matters for the data-driven modelling purpose at hand. The extracted features from the original recorded variables can be significantly more relevant than the recorded variables themselves as Aguado *et al.* (2022) showed when developing a predictive data-driven model as soft sensor. This fact highlights the importance of the feature engineering process in the development of useful artificial intelligence solutions.

6.6.4.2 Exemplary methods

Combining several existing features into a new feature is more informative for the empirical modelling purpose at hand. For example, the DO concentration in the aerobic reactor of a WRRF is important, but its ratio with respect to the DO set point could be more informative when analysing the aeration system performance. The daily total energy consumption of a WRRF (kWh) is important, but for comparison purposes its ratio with respect to the daily wastewater flow rate (kWh/m^3), or the population equivalent (kWh/PE) can be more informative.

Compute the difference between two sensor signals. Clearly, the computed difference should be a meaningful operation from the technical knowledge of the process. For example, Ruano *et al.* (2012) demonstrated that the pH difference over the aerobic reaction zone of a continuous WRRF conveyed information on the nitrification process in the reactor and even this pH difference was used for control purposes. Thürlimann *et al.* (2018) used the pH difference between the first and last aerated tanks of a biological treatment line to detect ammonia peak load events and these authors proposed an aeration rule-based controller in combination with qualitative trend analysis (used to detect the relative changes in the calculated pH difference) that was successfully tested on several full-scale municipal facilities.

Add lagged features to incorporate process dynamics into the empirical modelling. This pre-processing technique creates new features from the existing ones, shifting the feature values along time. This technique is widely used in data-driven modelling for time-series analysis and forecasting. When a given feature x_t is lagged, new features are created simply by shifting the original feature for the desired number of time units (n): x_{t-1}, x_{t-2}, x_{t-3}, x_{t-n}, such as a moving window of size n. In this way, an empirical model can capture how the current situation/behaviour (at time instant t) is influenced by the previous values of the features. This allows different modelling possibilities to represent the dynamics of the process depending on which features are lagged and the number of lags considered. Initially, the number of lags can be selected based on process knowledge or performing a sensitivity analysis trying different options and evaluating them.

Encode qualitative variables to convert them into a numerical form. Values contained in a data point can be qualitative (e.g., the on/off state of a pump or a blower), and some empirical learning algorithms can only work with numerical variables. To incorporate this qualitative information into the modelling, one-hot encoding can be applied when there is no natural ordering between categories of the variable to be encoded. With this encoding technique a new binary variable (known as dummy variable in statistics) is created for each possible category of the variable to be encoded. For example, if there are two categories (on/off) for the state of a pump, two binary variables would be created and coded as [1, 0] and [0, 1]. For each timestamp only one of the binary variables will be 1 and the other 0. When there is a natural ordering between categories of the qualitative variable to be coded, an integer value can be assigned to each category. This coding technique is referred to as ordinal encoding.

Discretising continuous features can sometimes be used to help the empirical model detect patterns in the data that could not be so evident when using the continuous values of the feature. The discretization converts a continuous variable into a series of discretized or nominal intervals. For example, this technique can be used to discretize the dissolved oxygen concentration in an aerobic reactor tank into three categories (low, medium, high), which could be later encoded to be able to incorporate them into the empirical model using one-hot encoding as [1, 0, 0], [0, 1, 0], and [0, 0, 1], respectively. Note that discretization usually induces a substantial loss of information.

Extract information from the timestamp that can be useful for the modelling purpose at hand, for example, hour of the day, weekday or weekend, public holiday or not, season of the year, etc. These date–time based features are interesting for predicting variables that exhibit a typical pattern (e.g., temperature, influent flow rate, etc.).

Compute the derivatives of the primary data, which can provide useful information in different situations. The first derivative of a sensor signal is the rate of change of the signal with time, and it is usually interpreted as the slope (or the tangent) to the signal at each time instant. The second derivative is the derivative of the first derivative, and it is a measure of the curvature of the signal (i.e., the rate of change of the slope of the signal). Higher-derivative orders can be computed similarly (e.g., jerk: 3rd derivative; snap: 4th derivative). The derivatives can be used to detect peaks in primary data. Derivatives have been used in analytical spectroscopy for several purposes (Mark & Workman, 2007): (1) for spectral discrimination (i.e., to accentuate small structural differences between nearly identical spectra); (2) for the correction for irrelevant background absorption; and (3) for spectral resolution enhancement (i.e., to increase the apparent resolution of overlapping spectral bands to facilitate the determination of the number of bands and their wavelengths). Care should be taken with numerical differentiation because it usually degrades the signal-to-noise ratio. Because differentiation of a signal increases the relative amount of noise (signal-to-noise ratio of a derivative is almost always poorer than that of the original signal), use of differentiation in combination with an adequate smoothing is essential for successfully applying this technique, to optimize the signal-to-noise ratio. Derivatives of the primary data from sequencing batch reactors (SBRs) can be calculated to detect important breakpoints (ammonia valley, nitrate knee, nitrate apex, DO break point, etc.) (Dutta & Sarkar, 2015) along the time-evolution profiles of low-cost sensors (oxidation reduction potential (ORP), pH, DO, etc.) because they are closely related to the biological reactions that are taking place (nitrification, denitrification, organic matter removal, etc.). This enables a real-time control of the length of the SBR phases providing better flexibility to process deviations and variations in the influent composition (Marsili-Libelli, 2006) than when classical SBR control is used, which is based on a fixed time schedule that do not allow for the adaptation of neither the length of the stages nor the cycles. In continuous

WWTPs, sensor signal derivatives have also demonstrated to be useful, especially where sensors are prone to drift and signal contains information in its derivatives (Ruano *et al.*, 2012; Thürlimann *et al.*, 2018; Wang *et al.*, 2021).

6.6.5 Transformation

6.6.5.1 Purpose

Transformations are non-linear conversions of the signal data such as the log-transformation and the squared root transformation, that allow correcting departures from normality in the distribution of the raw data. By applying a mathematical function to the values of a variable it is possible to reshape its distribution. The main motivation for applying a transformation to the data is to convert a skewed distribution into a normal one (or at least into a more symmetric distribution) (Eriksson *et al.*, 2013).

Normality of the data is a very common assumption in many statistical analyses, especially in parametric statistics. This assumption allows the development of confidence intervals and to perform hypothesis tests. Deviations from the normal distribution will reduce the reliability of these tests and also will reduce the efficiency of the data analysis. However, these deviations of normality do not completely invalidate the results nor the conclusions of the statistical tests unless it is a severe violation of normality. Anyway, applying a transformation to a given variable, it is possible to make its distribution more symmetric.

Sometimes the presence of extreme values in the dataset can unduly influence data-driven model development. These extreme values can skew the distribution of the signal data, making it deviate strongly from normality. Measurement errors, data corruption, and rare events can be some of the reasons for the presence of these extreme values in the collected dataset. Note that transformations (such as logarithmic transformation or squared root transformation) may mask outliers taking high extreme values. A WRRF domain expert could identify which extreme values can be considered outliers and only then they could be removed from the dataset. The extreme values that are not outliers and make the distribution of the variable non-normal, may be pre-treated so the distribution is reshaped into a more symmetric one.

6.6.5.2 Exemplary methods

The *logarithmic transformation* is very often used because variables with a log-normal distribution are relatively frequent in nature (e.g., concentrations, flow rates), exhibiting right-skewed distributions (tails on the right - positive skew). This type of transformation is interpretable as a change in a logarithmic scale imply proportional relative changes on the original scale. If the original value of the variable is x_i, the transformed value (x_i*) would be $x_i^* = \log(x_i)$. Note that because log(0) is undefined – as is the log of any negative number –, when using a log-transformation, a constant should be added to all values to make them all positive before transformation, that is, $x_i^* = \log(\text{constant} + x_i)$.

Figure 6.21 shows the histogram for 2-year laboratory measurements of chemical oxygen demand (COD) concentration, together with its logarithmic transformation. As can be seen in this figure, the transformation has converted the positively skewed original distribution of the measured COD values into a more symmetrical one.

The *square root* and the cube root are also used to transform variables that exhibit right-skewed distributions. These transformations are also known as *power transformations* as they can be expressed as x_i^p, where p is the root power (1/2 for square root, 1/3 for cube root). With their application, high values of the variable get compressed and low values of the variable become more spread out, but the compression and spreading are smoother than when the logarithmic transformation is applied.

The *Box-Cox* is a useful family of transformations that includes all the aforementioned transformations. This family has a parameter (λ) that determines the type of transformation that will be applied to the data. When the value of λ is 0 ($\lambda = 0$), the applied transformation is the natural

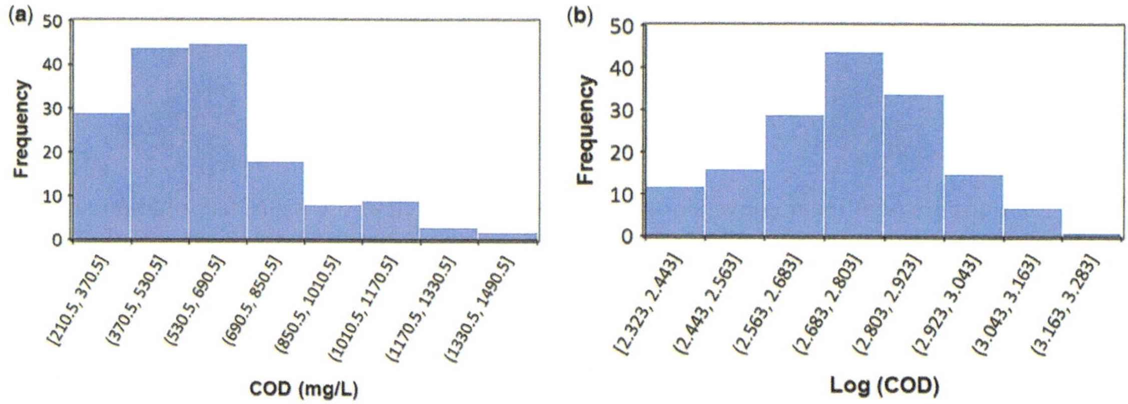

Figure 6.21 (a) Histogram of the COD concentration at the influent of urban wastewater resource recovery facility (158 values, corresponding to 2 years of lab-measurements). (b) Log-transformed COD values to make its distribution more symmetric. Note that plotting the histogram of a variable is a visual and easy way of assessing its normality.

logarithm, but if $\lambda \neq 0$, a power transformation is used. In the special case of $\lambda = 1$ there is no change in the shape of the distribution and the transformed data are just shifted downwards:

$$x_i^* = \log(x_i) \text{ when } \lambda = 0 \text{ and } x_i^* = (x_i^{\lambda} - 1)/\lambda \text{ when } \lambda \neq 0$$

It should be mentioned that to allow for negative values of the variable (x_i) a slight modification of the Box-Cox transformation was proposed by Bickel and Doksum (1981):

$$x_i^* = \log(x_i) \text{ when } \lambda = 0 \text{ and } x_i^* = (\text{sign}(x_i)/x_i/^{\lambda} - 1)/\lambda \text{ when } \lambda \neq 0$$

Finally, for the less-frequent case in which the collected variables exhibit a left-skewed distribution (i.e., the tail extends clearly to the left – negative skew), **squared root**, **cube root**, and **exponential** transformation could be applied to convert them into more symmetric distributions.

6.6.6 Scaling

6.6.6.1 Purpose

The purpose of scaling is to change the range of a given variable, that is, modify its numerical values but retaining their relative positions/proportions. Thus, the shape of the distribution is not affected.

Usually, empirical models are used to uncover the underlying patterns in the data to support decision making and increase process knowledge with monitor, control, and predictive applications. Thus, frequently many variables from the process are simultaneously analysed to be able to take into account their correlations and interactions. These variables may have really different numerical ranges (i.e., pH, DO, and flow rate). In addition, the numerical range of a signal is affected by the chosen unit of measurement. Scaling can be used to avoid that some variables dominate over others due to their higher numerical range or to ensure that analytical results are insensitive to the original unit of measurement. Many empirical algorithms are sensitive to the scale and do not perform well when the input variables have very different scales. The scaling of the variables allows correcting this situation.

Note that the scaling can regulate the impact that a given variable has on an empirical model. A variable that it is known to be relevant for a particular case can be assigned a higher weight in order to increase its contribution to the further empirical modelling. Thus, prior knowledge of the relevance of a variable can be incorporated via the scaling. Importantly, many analytical tools are sensitive to the ranges of the data inputs, meaning that any change in scale will result in a change of the estimated values or parameters.

6.6.6.2 Exemplary methods

Scaling to unit variance, it is the usual and recommended scaling procedure when no prior knowledge on the data is considered and the variables do not share the same measurement unit. Each value of the variable is divided by the overall standard deviation, so that each variable has unit variance. In this way, all the variables have the same chance to contribute to the further data analysis. Note that the resulting values are not bounded to a specific range.

Together with the unit variance scaling, another pre-processing technique named mean-centring is often applied to raise the interpretability of the later empirical model. Mean-centring converts all values in fluctuations around zero instead of around the mean, thus focusing on the fluctuating part of the data and contributing to remove the offset (if there is). When this combination is used (named auto-scaling or standardization), for each value of the variable, the mean value is subtracted and then divided by the standard deviation. This approach results in scaled data with zero-mean and unit-variance (see for instance Figure 6.22).

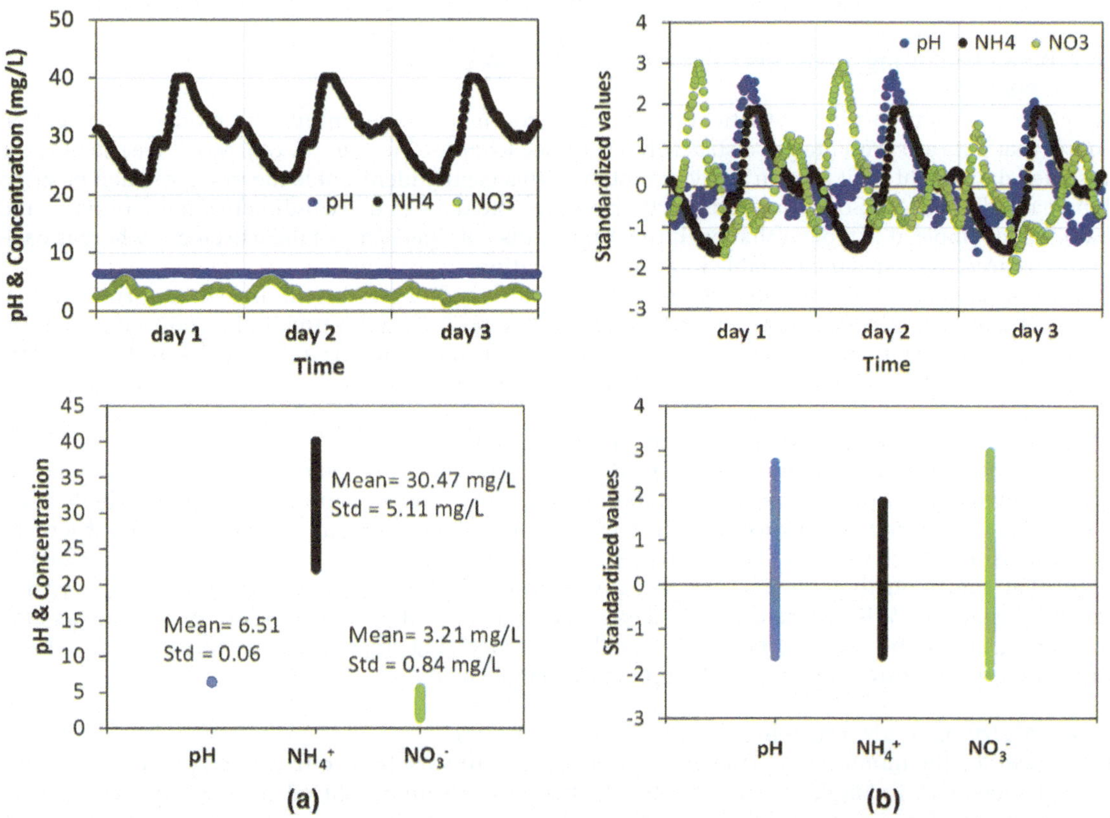

Figure 6.22 pH, ammonia, and nitrate concentration in the anoxic reactor of a wastewater resource recovery facility during 3 consecutive days: (a) recorded values (i.e., the mean value of every 10 min) and (b) after mean centring and scaling to unit variance.

Min–max scaling (also known as normalization) is another common technique to achieve the same scale of the variables. In this case, for each value of the variables, the minimum value is subtracted and then divided by the range (i.e., the maximum minus the minimum). Therefore, the scaled values lie in range from 0 to 1. This scaling approach is frequently used with variables that will be used in neural network models (because they often expect input values in the range of 0–1). Note that this scaling procedure is more sensitive to outliers.

Block scaling is employed to avoid that a group of naturally related variables dominate over other variables in the later empirical modelling just due to the high number of variables in the group (Eriksson *et al.*, 2013). For example, if for an empirical model we are considering the influent flow rate, the DO in the aerobic reactor and the spectra produced by a spectrophotometric sensor (e.g., 221 wavelengths), we could use block scaling for the spectra to allow the other two variables (flow rate and DO, that could be scaled to unit variance) have possibilities of contributing to the developed empirical model. In the block-scaling approach all the variables of each block are scaled so that the sum of their variances is 1. To achieve this, each value is multiplied by the inverse of the square root of the number of variables of the block.

There are also other scaling approaches less frequently employed in the WRRF context. For example, the *Pareto scaling*, which uses the inverse of the square root of the standard deviation as a scaling factor and some approaches that use a size measure (the mean or the square root of the mean) instead of a dispersion measure as a scaling factor, such as the *level scaling* or the *Poisson scaling*.

6.6.7 Data structuring: methods to handle unstructured data

6.6.7.1 Purpose

Many tools for data mining, statistical process control, machine learning, and model identification require that data are presented in a structured way (see Chapter 2 for definition). More specifically, the presented data ought to fit into an array or matrix with a clearly identifiable meaning for each position along each dimension. Some methods further require the data to be two-dimensional (columns and rows). For example, it is typical that each row represents a unique time of measurement whereas each column represents a specific variable in a specific location.

Unfortunately, data are not always presented in this kind of structured array-like format. When so, the data are classified as unstructured data (see Chapter 2 for definition and examples). A natural approach to analyse unstructured data is to transform it into structured data, prior to further data analysis. In what follows, we discuss such data pre-processing methods by means of a few examples.

6.6.7.2 A note on non-reversibility of data structuring

In general, data structuring cannot be reversed, meaning that the original unstructured data cannot be reconstructed exactly from the structured data alone. Thus, data structuring almost invariably leads to some loss of information. Consequently, care must be taken to (a) store the original unstructured data if considered relevant for future purposes and (b) to fine-tune the data structuring process in a way that trades off the benefits (e.g., effective and efficient use and storage of data) against the loss of information. Where feasible and relevant, irreversible data pre-processing steps, such as data structuring, should be executed before reversible pre-processing steps (such as centring and scaling). Data structuring is often one step in data pipeline management.

6.6.7.3 Methods to structure particle measurement data

It is typical for the number of particles in water samples to vary from one water sample to the next. As explained with Example 2.4 in Chapter 2, this leads to unstructured data when properties of individual particles are measured. To enable analysis, these unstructured data can be transformed through (a) binning, (b) density function estimation, and (c) distance computation. This is illustrated with data from Example 2.4 from Chapter 2. As explained in Chapter 2, these two data records contain measurements of four variables, four light intensities measured in two water samples (named forward scatter count – FSC, side scatter count – SSC, fluorescein isothiocyanate, and peridinin chlorophyll protein) for 56,358 and 164,810 particles. The top panels in Figure 6.23 shows the measurements for two variables (FSC and SSC). Note how these measurements are presented in log-scale, thus implying a transformation step (see Section 6.6.5) prior to structuring.

Binning consists of splitting the measurement range into separate bins. The location of these bins is the same for every data record. Binning is equivalent to the generation of a conventional histogram

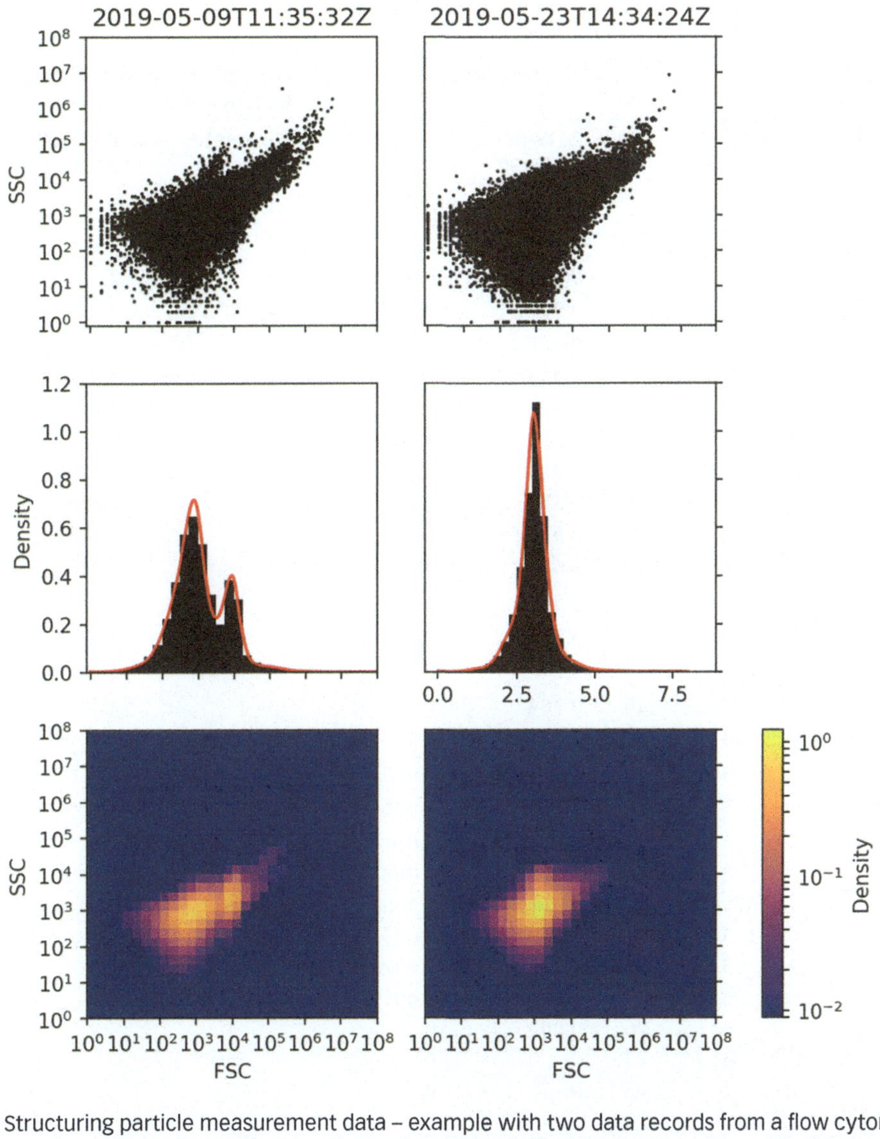

Figure 6.23 Structuring particle measurement data – example with two data records from a flow cytometry signal. Left: first data record; right: second data record. The timestamp is displayed on top. Top: original measurements for two light emission intensities (FSC and SSC). Middle: normalized histogram (black bars) and log-spline density (red line) for the FSC variable. Bottom: two-dimensional normalized histogram for the FSC and SSC variables, displayed here as a heatmap.

when the bins are equal in size, as is typical. The histogram can be normalized by dividing each count by the total number of particles in the data record and by the bin width. Examples for the FSC variable are shown in the middle panels of Figure 6.23 and in Table 6.1. These histograms are computed with 32 equal-sized bins. When executed for all 137 data records, this leads to a 137 × 32-dimensional array. These structured data can then be analysed easily by conventional data analytical methods.

Binning can easily be generalized to a multivariate case. For example, the measurement range of both the FSC and SSC variables can be split into 32 equal-length intervals. This leads to the generation

Table 6.1 Structuring particle measurement data.

Histogram			Normalized Histogram			Log-Spline Density Function Coefficients		
Bin	2019-05-09 11:35:32	2019-05-23 14:34:24	Bin	2019-05-09 11:35:32	2019-05-23 14:34:24	Parameter	2019-05-09 11:35:32	2019-05-23 14:34:24
1	17	45	1	0.001	0.001	1	−3.14	4.174
2	71	105	2	0.006	0.003	2	−2.196	−2.64
3	52	102	3	0.005	0.003	3	−2.095	−2.705
4	106	216	4	0.009	0.006	4	−0.595	−0.177
5	245	381	5	0.021	0.01	5	0.362	−0.044
6	392	762	6	0.034	0.02	6	1.84	2.169
7	766	1,284	7	0.067	0.034	7	2.747	2.54
8	1,307	2,456	8	0.114	0.066	8	3.68	4.781
9	2,572	4,762	9	0.224	0.127	9	1.279	3.416
10	4,317	8,922	10	0.376	0.239	10	3.621	1.394
11	6,579	16,302	11	0.573	0.436	11	−0.201	0.781
12	7,418	27,761	12	0.646	0.742	12	0.129	−0.599
13	6,088	41,849	13	0.53	1.119	13	−1.035	−1.639
14	3,711	24,245	14	0.323	0.648	14	−2.937	−1.167
15	2,302	9,248	15	0.2	0.247	15	−0.322	−7.481
16	4,404	5,183	16	0.384	0.139	16	−28.572	−0.946
17	3,479	2,653	17	0.303	0.071	17	−19.163	−16.196
18	812	1,325	18	0.071	0.035	18	8.666	−16.099
19	424	795	19	0.037	0.021	19	−48.311	−2.078
20	318	505	20	0.028	0.014			
21	209	298	21	0.018	0.008			
22	156	175	22	0.014	0.005			
23	94	123	23	0.008	0.003			
24	38	85	24	0.003	0.002			
25	35	36	25	0.003	0.001			
26	19	4	26	0.002	0			
27	2	3	27	0	0			
28	1	1	28	0	0			
29	0	0	29	0	0			
30	0	0	30	0	0			
31	0	0	31	0	0			
32	0	0	32	0	0			

Example with two data records from a flow cytometry signal. The table shows the structured data, i.e., vectors of uniform length when computing (a) a histogram, (b) a normalized histogram, or (c) the coefficients of a log-spline density function.

of $32 \times 32 = 1024$ separate bins. The normalized two-dimensional histograms are shown in the bottom panels of Figure 6.23. Through this binning, each of the 137 data records is converted into a sequence of numbers of uniform size. In the example, it leads to 1024 bin counts for every water sample, thus leading to a 137×1024-dimensional array. The histograms discussed above can be interpreted as a piece-wise constant density function. A more refined function can be obtained by using a piece-wise linear binning approach. For examples, we refer to Walther *et al.* (2009) (Equation (6.1)) and Russo *et al.* (2022) (flow cytometry case).

As an alternative to binning, one can also consider *estimating a (multivariate) density function* to the data for each water sample separately. Importantly, the density function must be parameterized, meaning that the same number of density model parameters is used to describe the data for every water sample. To this end, one can use semi-parametric models, such as log-spline density models with a fixed spline basis (see, e.g., Ramsay & Silverman, 2005). Note that non-parametric density models, such as kernel density estimation, cannot be used for this purpose as the number of parameters changes as a function of the number of particles. The middle panels of Figure 6.23 show two log-spline density models using the same cubic spline basis. The fitted cubic spline function describes the log-density by means of 16 polynomial segments of degree 3. In turn, these polynomials are described by 19 function parameters, which are known as spline coefficients. These spline coefficients are listed in Table 6.1. In this case, fitting this kind of density function to each data record leads to structured dataset containing 137 rows containing 19 coefficients each. Density function estimation can be used in multivariate cases as well (not shown). Practical identifiability issues may arise however in large-dimensional cases (e.g., six variables and higher).

A final and third method to structure particle measurements consists of computing a distance between each pair of data records. In the presented case, this will lead to a symmetric 137×137-dimensional distance matrix. Defining and computing a distance metric for pairs of data records with varying dimensions is not trivial. The Wasserstein distance is a popular approach to do so however (Bigot, 2020; Frogner *et al.*, 2015). The Wasserstein distance is also known as the earth mover's distance (EMD) because it can be interpreted as the minimum effort needed to move one pile of dirt to match the shape of another pile (Orlova *et al.*, 2016; Rubner *et al.*, 2000). In general, computing the EMD requires solving a large-dimensional linear programme for every pair of data records. This can be computationally prohibitive (Bigot, 2020; Frogner *et al.*, 2015). However, a simple solution is available in the univariate case, leading to efficient computation of the EMD. The left panel of Figure 6.24

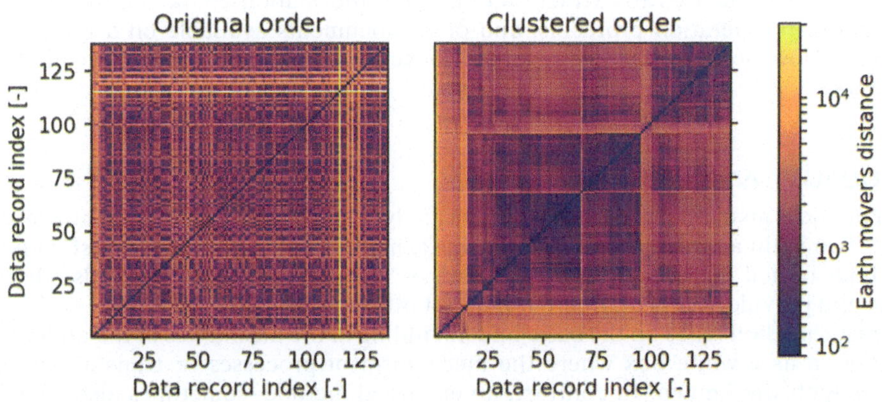

Figure 6.24 Structuring particle measurement data with the EMD – example with 137 data records from a flow cytometry signal. Left: pair-wise EMD for every pair of data records, in the original order (note the 114th data record appears to be very dissimilar to every other data record (in yellow)). Right: pair-wise EMD after ordering the data records according to clustering distance.

shows the pair-wise EMD based on the FSC variable for the 137 data records in the exemplary dataset. The 114th data record (timestamp: 2019-06-11 03:44:12) appears to be very dissimilar to every other data record. This turns out to be a data record with an extremely low number of particles (i.e., 70, see Figure 2.6 in Chapter 2). This means that both the number and the distribution of the FSC measurements are substantially different. The right panel shows the same pair-wise distances after the data records are ordered according to clustering distance, following hierarchical clustering (using Ward linkage). The block-diagonal structure of the resulting image suggests that the data records can be grouped into meaningful groups. Importantly, this kind of clustering requires that the data are structured through distance computation, as illustrated here, or through the binning or density estimation method described above.

6.6.7.4 Methods to structure text data

Several methods exist to transform unstructured text into a structured data format. One simple approach consists of the bag-of-words approach (Goldberg & Hirst, 2017). In this case, one counts the number of times a word appears in a document and reports the frequency of each word. The reported counts results in a histogram for every document being analysed, with word frequencies reported for every word in any of the analysed documents. This leads to a structured data format given that the number of considered words is constant. This bag-of-words representation can then be used as an input for conventional tools for data mining. A more refined approach is known as term frequency × inverse document frequency (Rajaraman & Ullman, 2011). This method aims to account for the fact that certain terms, especially articles and pronouns, are common across all documents and are therefore less informative about the contents of a particular document.

In recent years, alternative machine learning models have been developed that also account for the order in which words appear in a text. Some of these models are based on embeddings. Roughly speaking, such embeddings are numeric representations of words or phrases in the text that are internal to the model. As these embeddings are numeric, this enables quantitative analysis and identification of highly performant machine learning models for natural language processing (NLP) tasks, including translation. Practical examples include embeddings from language models (Peters *et al.*, 2018) and bidirectional encoder representations from transformer models (Devlin *et al.*, 2019).

The above techniques belong to the highly specialized field of NLP. This is currently a very active field of research with applications in document translation and sentiment analysis. In principle, this type of techniques can be used to extract and navigate information contained in manuals, reports, procedures, standard operation protocols, and other documents available on a typical wastewater treatment plant. However, to our knowledge, the above-mentioned techniques have not been used in the water sector.

6.7 DATA QUALITY ANALYSIS

This section provides an overview of the main data analytical techniques with potential use for WRRF applications. Two main approaches are presented: (1) methods rooted in mechanistic understanding and (2) methods rooted in statistics and data analysis (data-driven). Both methods rely on different types of data and provide access to different types of information.

A mechanistic method relies on the basic understanding of the mechanics of the underlying system. They can be seen as a white box where the mechanism of processes or transformations of input variables to outputs are known. In contrast, an empirical method, sometimes also called statistical method, relies on observations (e.g., sensor signals) rather than theory. They can be seen as a black box where correlations are obtained from the analysis of the data. In this case, the fundamentals and physical mechanisms underlying the systems are not explicitly considered.

6.7.1 Methods rooted in mechanistic understanding (mechanistic methods)

6.7.1.1 Purpose and application for data quality assessment

Mechanistic models seek to describe causation and to establish a mechanistic relationship between inputs and outputs. In this regard, they can be used to detect gross errors and improve data quality by exploiting the redundancy in the form of conservation laws for mass, electrons, and energy. Data reconciliation by mass balancing aims to enhance the quality of raw measured data by ensuring consistency of the modified data with known mass balances.

6.7.1.2 Methods

Sanity checks and mass balances are typical examples of mechanistic relations. This section briefly presents the main sanity checks that should be performed and discusses the application of data reconciliation based on linear mass balances to improve data quality, dealing with absent or missing data, and detecting gross errors. The methodology is illustrated for a wastewater treatment process.

6.7.1.2.1 Univariate approach: sanity checks

Sanity checks are quick and basic tests that allow assessing whether an analytical determination or the result of a calculation is reasonable, that is, if the value could be possible (real) or not. For example, when working with concentrations and flow rates, the values cannot be negative. Thus, the basic check consists in comparing the recorded value with the numerical criteria. If the recorded value meets the criteria a quality label would be flagged as *valid* otherwise the quality label for the data would be *invalid*.

It is advisable to perform this type of checks before starting the main data analysis. Considering the definitions and analytical determinations several plausibility checks can be performed:

- Concentration of any compound ≥ 0
- Flow rates ≥ 0
- Total suspended solids (TSS) $\geq$ volatile suspended solids (VSS)
- TSS in return AS $\geq$ TSS in AS tank
- Total COD $\geq$ biochemical oxygen demand ($BOD_{ultimate}$) $\geq BOD_5$
- Total COD $\geq$ soluble COD
- Total nitrogen (TN) $\geq$ soluble nitrogen
- Total phosphorus (TP) $\geq$ soluble phosphorus
- Total kjeldahl nitrogen (TKN) $\geq NH_4$-N
- Total nitrogen = TKN + NO_3-N + NH_2-N

Whenever a check fails, the reason for the error or inconsistency should be investigated and the corresponding value be corrected (e.g., using the true value if it was found, or if the analytical determination repeated) or deleted (in case no explanation or correction is possible).

Additionally, to further assess the collected data, typical ratios between variables can be calculated and checked whether they fall within the typical range of a WRRF of the same type or not (e.g., municipal WWTPs). Thus, advanced sanity checks are performed to assess whether the results fall within a plausible range. These ratios can be calculated and compared at different points of the water treatment line: raw wastewater, primary effluent, AS tank, etc.: TN/total COD; TP/total COD; total BOD_5/total COD; TSS/total COD; VSS/TSS. The values of these typical ratios can be found in IWA (2012).

6.7.1.2.2 Multivariate approach: mass balances

Mass balances, especially COD and nitrogen mass balances, are essential not only in the WRRF design phase, but also for process modelling, monitoring plant performance, and follow-up of plant products such as biogas production, P-recovery, nutrient recovery, and side-stream treatment (Rieger *et al.*, 2010). For these tasks, it is crucial to close all mass balances (flow and substances) within defined accuracy (e.g., 5–10%) and to explain all the gaps/losses in the mass balances, in order to detect and remedy inconsistencies in the dataset and to point out potential measurements/sensors containing errors. Within this approach, if the key process variables of primary data fit all the constraints of the mass balances set, they are therefore considered reliable and valid, otherwise new estimates must be carried out using the available set of mass balances to obtain proper secondary data.

Data reconciliation is a proven technique to evaluate the consistency of collected data (Crowe, 1996; Ozyurt & Pike, 2004). It involves a procedure of optimally adjusting estimates for variables such that these estimates satisfy the conservation laws and other constraints (Crowe, 1996) and are therefore more accurate than the original values. In this section, linear mass balances serve as the constraints based on which data reconciliation is performed. More specifically, mass balance-based data reconciliation re-estimates the value of measured variables (with true values x) taking into account their measured values y and the constraints $h(x)$ involving those variables. The newly estimated variables denoted $\hat{x}$ are such that the weighted difference between measured variables and estimated values is minimized. The weights are grouped in the matrix $\boldsymbol{W}$, which contains the measurement variance, possibly complemented with additional expert knowledge regarding the uncertainty of the measurement. As such, the mass balance-based data reconciliation problem is translated into an optimization problem, consisting of a set of equations which must be solved analytically or numerically (Equation (6.1)):

$$\left.\begin{array}{r} \min SS(\hat{x}) \\ \text{s.t. } h(\hat{x}) = 0 \end{array}\right\} \tag{6.1}$$

with $SS(\hat{x}) = (y - \hat{x})^T W (y - \hat{x})$.

Data reconciliation is applied to denoise measurements and to detect gross errors. Key variables may be measured or not; their identification means that improved estimates of their values ($\hat{x}$) are obtained through the calculation from other, measured variables, given the prevailing constraints $h(x)$. The modified measurements meet all the constraints (i.e., fit all mass balances) and are therefore considered more reliable and accurate compared to the original values. When key variables are not measured, it may be possible to compute optimal estimates through the same approach. This occurs when the unmeasured variables are part of a set of independent mass balance equations whose number is equal to or larger than the number of unmeasured variables.

Typical examples of key process variables in a WRRF that need to be known with high accuracy are influent and effluent mass flow rates of the AS process (biological reactor) as well as the waste AS mass flow rate. The oxygen requirements for COD and nitrogen removal are also important process variables and therefore typically need to be estimated.

The most commonly used constraints for wastewater treatment process are mass balances of conservative quantities such as total mass flows (or volumetric flows, in case density can be assumed constant) or mass flows of total COD ($mCOD$), total nitrogen (mTN), total phosphorus (mTP), and dry solids. Mass balances can be set-up over individual and/or combined unit processes. This requires that the resulting mass balances contain all key variables.

Important prerequisites for successful data reconciliation are to satisfy the redundancy and steady-state conditions. Redundancy of variables means that their measured values can also be calculated from other (measured) variables. However, for many WRRFs, sufficient measured data are often not initially available and additional measurements typically need to be carried out to ensure the required degree of redundancy for data reconciliation. Le *et al.* (2018) proposed an experimental design procedure for WWTPs to determine sets of additional measurements, which guarantee that key variables can be identified through subsequent data reconciliation, by satisfying the required degree of redundancy.

The data reconciliation procedure further assumes that all measurements are normally distributed around a nominal value and have a known or estimated measurement uncertainty. Because the constraints are usually steady-state constraints, an important prerequisite to apply data reconciliation is that the plant should be in steady-state operation during the period for which data are used. For instance, there should not be any heavy rain or a change in the waste sludge flow. At least 2 weeks of data or a period covering one time the SRT is recommended.

Before drawing conclusions from the reconciled data, data validation is required. To this end, data reconciliation is often accompanied by statistical tests for gross error detection. The most commonly used ones are the global test, the measurement test, and the nodal test. For a more detailed explanation of the data reconciliation procedure and the tests, the reader is referred to Le *et al.* (2022).

6.7.1.2.2.1 Illustrative example The wastewater treatment plant under study (Le *et al.*, 2022) is described by a set of 16 mass balances that were set-up around individual and combined unit processes (Table 6.2). The flow balances were set-up around main unit processes; the boundaries for COD, TP, TN, and TKN balances were set-up over combinations of unit processes as indicated by the dashed line in Figure 6.25. The measurements are depicted in Table 6.3. Data reconciliation could balance 22 out of 32 measured variables, meaning that their reconciled values fulfilled the constraints (mass balances), and their precision was improved. Moreover, data reconciliation allowed to calculate nine unmeasured variables, four of which are not even unmeasurable as such. The latter are related to COD and nitrogen removal around the activated units: they are denitrified nitrogen (46-DENI), nitrified nitrogen (47-NITR), required oxygen for COD removal (48-OCcod), and the total required oxygen

Table 6.2 Set of mass balances describing the wastewater resource recovery facility under study.

No.	Unit	Bilinear Mass Balance	Unit
1	Primary settler (PS)	$Q_4 + Q_5 - Q_7 - Q_{28}$	m^3/d
2	Activated sludge units (AS)	$Q_7 - Q_{17} - Q_{26}$	
3	Secondary clarifier (CL)	$Q_{26} - Q_{27} - Q_{37}$	
4	Waste sludge thickener (ST)	$Q_{28} - Q_{31} - Q_{39}$	
5	Primary sludge thickener (PT)	$Q_{27} + Q_{31} - Q_{34}$	
6	Digester (DIG)	$Q_{34} - Q_{36} - Q_{38}$	
7	Dewatering centrifuge (DEW)	$Q_{37} + Q_{38} + Q_{39} + Q_{40} - Q_5$	
8	Primary settler (PS)	$mTP_4 + mTP_5 - mTP_7 - mTP_{28}$	kg/d
9	Activated sludge units (AS)	$mTP_7 - mTP_{17} - mTP_{26}$	
10	Secondary clarifier (CL)	$mTP_{26} + mTP_{28} - mTP_5 - mTP_{36}$	
11	Primary settler (PS)	$mCOD_4 + mCOD_5 - mCOD_7 - mCOD_{28}$	
12	Activated sludge units (AS)	$mCOD_7 + mCOD_{45} - mCOD_{17} - mCOD_{26} - OC_{cod} - 2.86 \times DENI$	
13	Secondary clarifier (CL)	$mCOD_{26} + mCOD_{28} - mCOD_5 - mCOD_{36} - mCOD_{43}$	
14	Activated sludge units (AS)	$mTN_7 - mTN_{17} - mTN_{26} - DENI$	
15	Activated sludge units (AS)	$mTKN_7 - mTKN_{17} - mTKN_{26} - NITR$	
16	Wastewater resource recovery facility	$OC_{net} - OC_{cod} - 4.57 \times NITR$	

Q = flow; mTP = mass flow of total phosphorus; mCOD = mass flow of total COD; mTN = mass flow of total nitrogen; mTKN = mass flow of total kjeldahl nitrogen. The number in the first column indicates the stream number; DENI = denitrified nitrogen; NITR = nitrified nitrogen; OCcod = required oxygen for COD removal; OCnet = total required oxygen. The mass flow term is the product of measured flow and corresponding concentration.

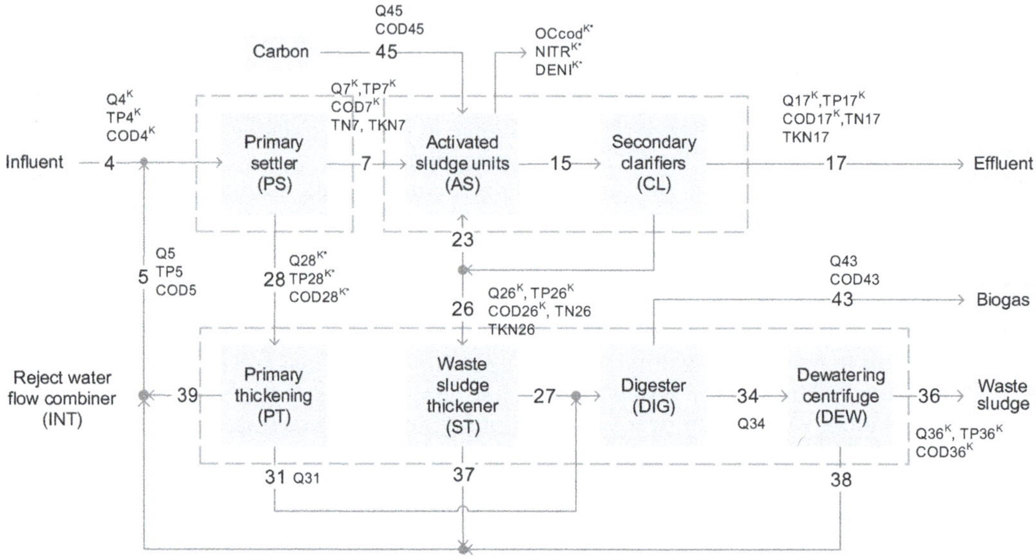

Figure 6.25 Simplified process diagram of wastewater resource recovery facility under study (from Le, 2019). The variables indicated in the figure are measured in terms of flow and concentrations. The dashed lines represent the boundaries for COD, total phosphorus (TP), and nitrogen mass balances (TN and TKN).

(49-Ocnet). Note that, although these variables may also have been calculated based on the mere mass balances, the application of data reconciliation resulted in a higher precision.

The ***W*** matrix can be directly calculated from the collected data as standard deviation of the mean of the measurement. In some cases, expert knowledge of the measurement uncertainty could be used as an input to form the ***W*** matrix. An example of how to form a ***W*** matrix from data of a monitoring campaign can be found in Forouzanmehr *et al.* (2021)

All three gross error tests could pick up potential errors in the measured data (Table 6.4). The global test, which evaluates the term $SS(\hat{x})$ (from Equation (6.1)), provided an overall signal that there may be a gross error in the dataset. The measurement test pointed out two specific variables of TP_{26} and TP_7 as suspicious ones. The nodal test indicated that the mass balance around the AS unit $mTP_7 - mTP_{17} - mTP_{26} = 0$ potentially had a problem. The reconciled values of mTP_7 and mTP_{26} indeed significantly deviated from the reported measured values (Table 6.3). Gross error elimination strategies to pinpoint the actual erroneous measured variables were applied (Narasimhan & Jordache, 2000). In this case, this led to the detection of an error in the measurement of total phosphorus in waste AS stream (mTP_{26}), in its turn causing a suspicious value of mTP_7 (due to the smearing effect).

After eliminating this error, the measured and the reconciled values were consistent. This indicated two important outcomes: (1) the flow sheet of the plant was correctly set-up and understood and (2) the measurements were reliable and could be further used for other purposes such as modelling and plant optimization.

6.7.1.3 Potential and perspectives for application of data reconciliation for wastewater resource recovery facility data

Although data reconciliation has been widely studied in (bio)chemical engineering for decades (e.g., Madron *et al.*, 1977; van der Heijden *et al.*, 1994), this concept so far has received relatively little attention in wastewater treatment process engineering. Still, it is clear that WWTPs also have a lot to gain from acquiring more reliable data for further application in process evaluation, optimization,

Table 6.3 Measured data and data reconciliation results.

Number in Process Diagram and Stream Name		Q m-³/day	mTP kg/day	mCOD kg/day	mTN kg/day	mTKN kg/day
4	Wastewater resource recovery facility influent	57,008 ± 3,421 56,943 ± 1,819	446 ± 41 441 ± 33	30,836 ± 3,084 30,796 ± 2,493		
5	Reject water	6,288 ± 441 6,371 ± 285	332 ± 34 332 ± 34	9,076 ± 908 9,076 ± 908		
7	Settled influent	58,101 ± 3,487 58,177 ± 1,819	556 ± 56 434 ± 40	21,461 ± 1,932 21,477 ± 1,796	3,182 ± 287	3,178 ± 286
15	Inflow clarifiers CL	119,575 ± 8,371 119,635 ± 4,089				
17	Wastewater resource recovery facility effluent	56,906 ± 2,846 56,895 ± 1,819	112 ± 12 117 ± 11	1,882 ± 189	665 ± 67	134 ± 13
23	Return activated sludge	61,475 ± 4,304 61,459 ± 3,846				
26	Waste activated sludge (WAS)	1,285 ± 78 1,283 ± 77	193 ± 56 317 ± 40	11,120 ± 2,224 11,099 ± 2,014	971 ± 195	959 ± 144
27	Thickened WAS	133 ± 8 133 ± 8				
28	Primary sludge	5,194 ± 364 5,137 ± 283	338 ± 61	18,450 ± 3,750		
31	Thickened primary sludge	182 ± 10 182 ± 9				
34	Digested sludge	315 ± 23 315 ± 11				
36	Wastewater resource recovery facility waste sludge	48 ± 9 49 ± 9	312 ± 57 324 ± 34	8,910 ± 3,030 8,949 ± 2,471		
37	Centrate WAS thickening	1,150 ± 77				
38	Centrate dewatering	267 ± 14				
39	Overflow primary thickener	4,956 ± 283				
37	Centrate WAS thickening	1,150 ± 77				
43	Biogas	6,372 ± 199		11,470 ± 230 11,470 ± 230		
45	External carbon source			1,780 ± 89		
46	Denitrified nitrogen (DENI)				1,547 ± 353	
47	Nitrified nitrogen (NITR)				2,086 ± 321	
48	Oxygen for COD removal (OCcod)			5,817 ± 3,121		
49	Total oxygen required (OCnet)			15,346 ± 3,447		

Numbers in black = measured values (y), which are calculated as mean ± standard error of the mean of the measured data for the period of measurement campaign, green = reconciled values ($\hat{x}$); blue = calculated values of unmeasured variables; red = measurements potentially containing an error. The mass flow term is the product of measured flow and corresponding concentration.

Table 6.4 Result of gross error detection.

Test	Residue From	Degree of Freedom	Critical p-Value	Indication
Global test	$SS(\hat{x})$ (from Equation (6.1))	11	0.05	Gross error detected with p-value = 0.0027
Measurement test	$y - \hat{x}$ (for each individual measured variable)	18	0.0014	mTP7 and mTP26 with p-value = 0.00045
Nodal test	$h(y)$ (on each constraint containing measured variables)	21	0.0012	mTP7 − mTP17 − mTP26 = 0 with p-value = 0.00044

modelling, and benchmarking. The need and possible added value of data reconciliation is increased by the current digitalization trend, creating a massive flow of data. Moreover, the paradigm shift from WWTPs focusing merely on sanitation and effluent quality, into true resource recovery facilities further increases the potential benefits of automated data reconciliation, because reliable information is needed regarding multiple process objectives.

Several important steps have been taken towards an automated data reconciliation tool for wastewater treatment processes. Structured approaches for experimental design and for data reconciliation have been set-up by Le *et al.* (2018). Their application to several full-scale wastewater treatment plants has been proven successful (e.g., Forouzanmehr *et al.*, 2021). The corresponding software packages have been written in MATLAB and were found flexible enough to handle most of the problems related to data reconciliation, albeit for offline use, that is, not directly connected to the SCADA system of the plant.

To spread the application of experimental design and data reconciliation and open up their benefits for all wastewater treatment plants, further developments are needed towards a standalone software package for online application, which can be coupled to or integrated in the existing SCADA/ information system of WWTPs. The focus should be on the data pre-processing steps, for which a module should be tailored to collect, pre-process, automatically organize input information, and make them ready for the reconciliation procedure. Because the flow measurements of a well-instrumented plant are usually recorded with high frequency, online data reconciliation for the flow measurements seems most easily achievable. The reconciled values of the flow rate are paramount in balancing mass flows and concentrations of components because flow rates appear in all the related balances.

6.7.2 Methods rooted in data analytics and statistics (empirical methods)

6.7.2.1 Purpose and application for data quality assessment

These methods (also called empirical methods) seek to establish statistical relationships and correlations between inputs and outputs. In contrast to mechanistic models their structure is driven by the data, and they can be applied to process data from WWTPs in order to assess the quality of the data, forecast relevant parameters of interest, and provide valuable additional information, that can be used for different purposes such as modelling or within control strategies at the unit process level.

These methods rely on a dataset, that can include sensor data, laboratory data, and metadata encoding plant information. As a coarse simplification two main tasks can be considered:

- *Prediction*: the dataset consists of input variables and output targets, and the goal is to assign an output to an unseen input. Several algorithms for prediction have been developed (neural networks, kernel methods, decision trees, among others).
- *Exploration*: the dataset consists only of input variables, and the goal is to discover some interesting properties of the data records, such as taxonomies (clusters), relevant subspaces (projections), and distributions. Again, many algorithms are available (linear/non-linear dimensionality reduction methods, mixtures of Gaussians, self-organizing maps, among others).

Independent of the method and the goal, a training dataset (inputs and outputs for prediction, and only inputs for exploration) is used to build and train the model. Based on the type of learning used within the training phase, most of the methods are categorized as *supervised* or *unsupervised* as illustrated in Figure 6.26. The supervised approach uses a labelled dataset with the real anomalies and clean data for training the model, whereas the unsupervised set-up does not require any labels. Supervised methods make use of labelled inputs and outputs to train a model to predict a class (classification) or the value of a continuous variable (regression). Unsupervised models work with unlabelled datasets to discover hidden patterns in the data. Table 6.5 summarizes the main differences between both the supervised and unsupervised approaches.

As soon as the model is built, the model is validated and eventually tested using an independent validation dataset and a testing dataset. The validation dataset must be labelled in all cases. The quality of the learning process and the performance of the method is then evaluated using some performance metrics (further described in Section 6.7.4).

This section provides an overview of different empirical methods used for both exploration and prediction within the wastewater treatment context.

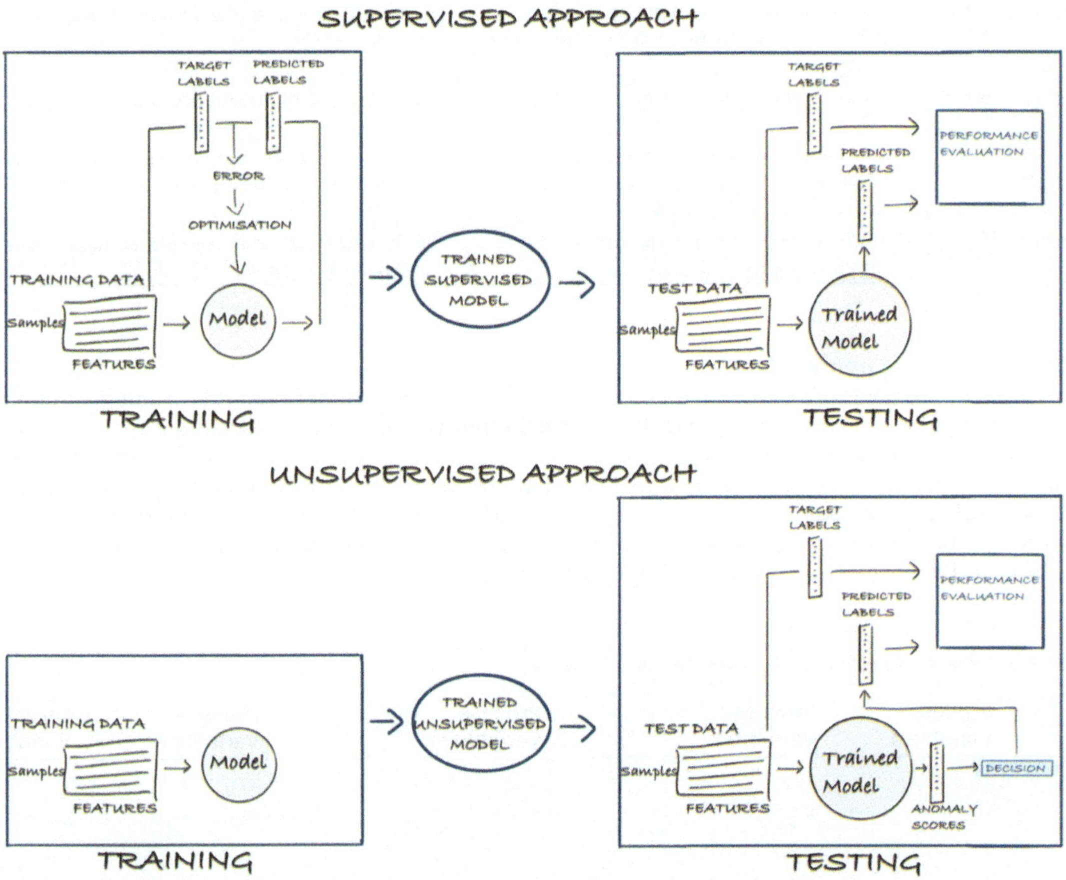

Figure 6.26 Supervised and unsupervised approaches for data quality according to the availability of labels in the dataset. In these figures, *samples* correspond to data records and *features* to variables.

Table 6.5 Key differences between supervised and unsupervised approaches for automated data quality assessment.

	Supervised	Unsupervised
Labelled data	Make use of labelled inputs and outputs to train models	Work with unlabelled datasets
Problem to be solved	Classifying data from learned patterns or predicting outputs	Discover hidden patters and relationships in the data
Human intervention	Often requires human intervention to label the data appropriately	Work independently. Human intervention is required for validating output variables
Accuracy	Generally higher due to information captured in labels	Generally lower due to absence of labels
Complexity	Simpler but resource-intensive due to the need for labelled data. Can be time-consuming to train and may require frequent update	Computationally complex, large training set to produce the intended outcomes
Data required	Inputs and outputs variables	Only input variables
Learning	Learning based on expert-provided labels occurs offline or with some delay	Learning can take place in real time
Algorithms	Classification, regression	Clustering, dimensionality reduction, control charts
Anomaly detection	User must provide examples on the training data of every single type of anomaly to be detected	Able to detect any type of anomaly, including ones which have never been seen before
Main challenge	Build a complete and meaningful labelled dataset for training	Determining what is considered acceptable or typical (ground truth)

6.7.2.2 Supervised approaches

Supervised approaches are based on a collection $D = \{(\boldsymbol{x}_n, y_n)\}_{n=1}^{N}$ of data records (Table 6.6) indexed by n. Each pair $(\boldsymbol{x}_n, y_n)$ consists of an M-dimensional input vector $\boldsymbol{x}_n$ and associated target y_n. Usually, the components of the input vector $\boldsymbol{x}_n = (x_{n,1}, x_{n,2}, \ldots, x_{n,m}, \ldots, x_{n,M})$ are the values available for easy-to-measure process variables, whereas the target y_n are the values available for a hard-to-measure variable, either provided by a domain expert or obtained from expensive and time-consuming laboratory analysis. Two common settings of supervised learning, namely classification and regression will be described below and illustrated in Figure 6.27.

Table 6.6 Collection D of N input–target data points $(\boldsymbol{x}_n, y_n)$.

	Process Variable 1	Process Variable 2	...	Process Variable *m*	...	Process Variable *M*	Target Variable
$\boldsymbol{x}_1$	$x_{1,1}$	$x_{1,2}$		$x_{1,M}$		$x_{1,M}$	y_1
$\boldsymbol{x}_2$	$x_{2,1}$	$x_{2,2}$		$x_{2,M}$		$x_{2,M}$	y_2
	...	...		...		...	...
$\boldsymbol{x}_n$	$x_{n,1}$	$x_{n,2}$		$x_{n,M}$		$x_{n,M}$	y_n
	...	...		...		...	...
$\boldsymbol{x}_N$	$x_{N,1}$	$x_{N,1}$		$x_{N,M}$		$x_{N,M}$	y_N

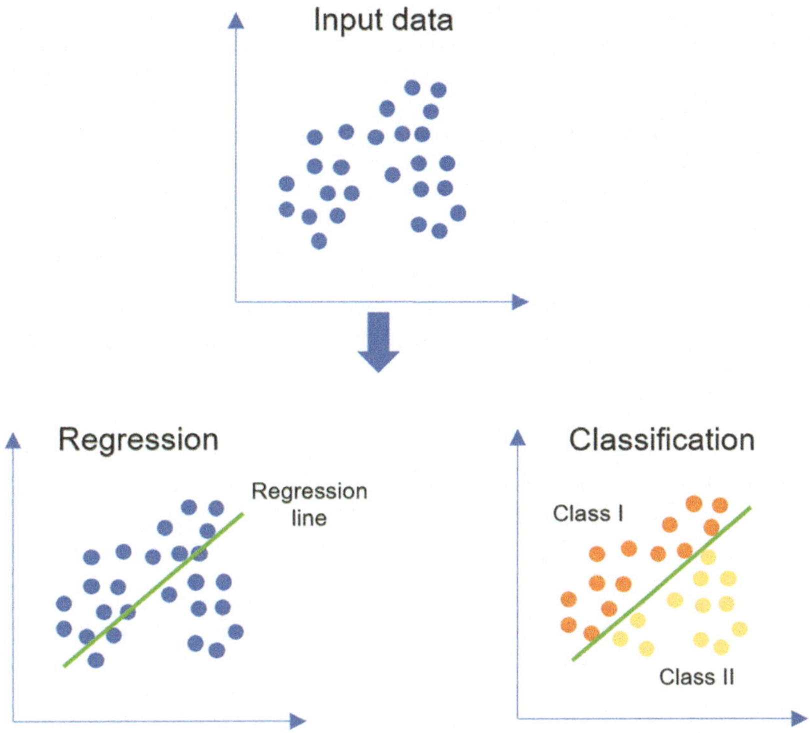

Figure 6.27 Examples of supervised methods.

6.7.2.2.1 Classification

In classification, output targets y_n can only take on discrete values. Distinct target values are used to label different classes: the notion of class is used to denote data records, pairs of inputs and output, with common characteristics. The labels can be used to represent different operating regimes of a plant or a unit (e.g., high and low loads), as well as other characteristics such as their status (e.g., normal and abnormal conditions). Classification is then about modelling the relation f between the inputs (measurements) and the target (e.g., operating regime) and using it to predict the target based on the inputs only. In this sense, quality data assessment of new data boils down to determining whether a new data point belongs to a specific class or not.

Many classification approaches are available (Friedman *et al.*, 2001). From a practical point of view, the choice between these models should be based on the application and on the properties and availability of the data. More than often, however, the choice is conditioned by the familiarity of the user with a particular method and by the availability of software implementations. Some of the most popular ones have been implemented in WWTPs. A few examples that are worth mentioning are naive Bayes (NB; Carvajal *et al.*, 2015), logistic regression (LR; Salman & Salem, 2012), k-nearest neighbours (kNN; Kim *et al.*, 2016), random forest (RF; Vitorino *et al.*, 2014), support vector machines (Fan *et al.*, 2004; Tian *et al.*, 2012), and ANNs (Boger, 1992; Hanbay *et al.*, 2008).

NB is a generative model aiming at estimating the probability that a certain input–output pair belong to a certain class, given that the inputs are observed. LR, kNN, RF, and ANN are discriminative models designed to learn the, typically non-linear, function that assigns a data record to a specific class. According to Ng and Jordan (2002), discriminative models are expected to perform generally better than generative ones (a) when the assumptions in the generative models are not valid and

(b) when large and representative datasets are available. ANN models are well-known for their flexibility and have proven their ability to outperform other models, especially in the big-data regime. Importantly, however, flexibility may lead to models that are over-specialized to a specific set of data, a phenomenon often denoted overfitting.

When deployed for use in real-time applications for safe and long-term operation, classification models must be robust to measurement and labelling errors. They may also require frequent updates to represent newly labelled data. That is because new data reflect changes in the environmental conditions (e.g., changes in the wastewater composition or seasonal patterns) and changes in controlling the units, and thus changes in the input–target relation. An important factor to consider when selecting a classification model is its computational complexity. Simple models such as LR and NB are usually easy to learn and update. Conversely, learning and maintaining more complex models such as ANNs is often more demanding, because their learning procedure associates with solving a non-convex optimization.

Together with the classical methods mentioned above, more recent methods based on deep neural networks, deep learning (Goodfellow *et al.*, 2016) methods, are becoming popular. These methods come with the promise of being capable of learning non-linear dynamics and complex interactions between the variables in wastewater systems (Harrou *et al.*, 2018; Mamandipoor *et al.*, 2020; Zhang *et al.*, 2018). It is worth mentioning that these methods have not yet found their way into practice.

6.7.2.2.2 Regression

In regression, output targets are allowed to take on continuous values. In contrast to classification, a regression problem consists of determining the input–target function under the more explicit assumption that target values are affected by errors. Similar to discriminative methods for classification, the function is parameterized, and learning consists of determining the values of the parameters that best describe the data. Fitting a line that best describe a set of at least three data points is the simplest regression problem. The function to be determined is, in this case, the equation of a line parameterized by its slope and intercept. Their values are usually obtained via least-squares minimization (Kay, 1998). More sophisticated regression approaches generalize this idea to more complex, non-linear, functions, with multiple inputs. Again, once learned, the input–output function can be used to predict the target values directly from input variables, instead of measuring or determining them directly. Predicted values can then be compared with the current measurements to detect abnormal variations and anomalies within the target variable. Figure 6.28 shows an illustrative example of actual values together with the expected values according to the regression model fitted with the original values.

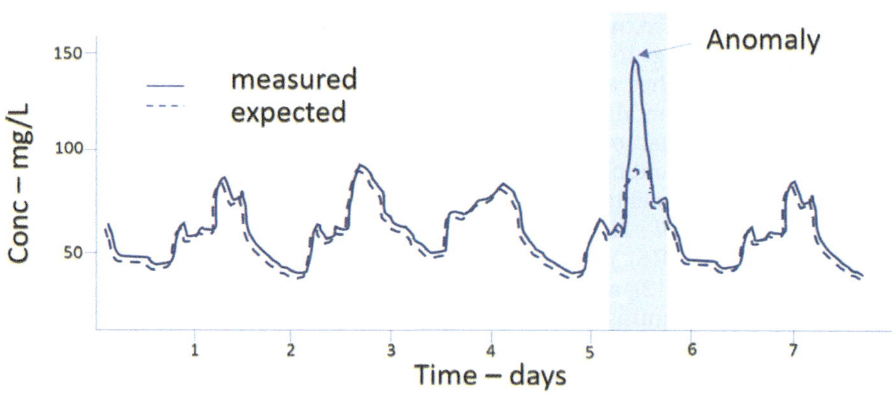

Figure 6.28 Example of regression methods to detect anomalies in a key process variable.

Differences between actual and expected values out of the thresholds previously defined allow inconsistencies or anomalies to be detected.

Some of the most popular regression techniques include linear and non-linear regression methods, regression trees (RTs, a flowchart-like structure that encodes the function as a tree to be navigated from observations, the branches, to conclusions, the leaves), random forests (a combination of decision trees, where each tree is used for regression before their solutions are combined to produce the final prediction), artificial neural networks, kernel regression (a linear regression problem formulated in a high-dimensional space), and Gaussian process regression (GPR, a kernel method that estimates the most probable random function that represents the data, given the measured input–output pairs) and PLS (a multivariate projection method on latent structures; Wold *et al.*, 2001). We refer the reader to Friedman *et al.* (2001) for a more exhaustive characterization of these techniques.

Some results of these approaches in environmental processes can be found in the research literature. For example, Granata *et al.* (2017) estimated the main wastewater quality indicators using RTs and support vector regression. Zhou *et al.* (2019) performed a daily wastewater inflow prediction using RF, considering weather features and domestic water usage patterns.

Unlike supervised approaches that provide optimal values of the function parameters, GPR is a data modelling technique that aims at estimating the input–output function as a collection of random input variables capable of incorporating the uncertainty existing in the output measurements (Rasmussen & Williams, 2005). GPR has been applied in environmental processes, including monitoring and fault detection in WWTPs (Samuelsson *et al.*, 2017; Zambrano *et al.*, 2019).

Similar to the majority of regression techniques, the data $D = \{(x_n, y_n)\}$ $(n = 1)^N$ modelled with GPR are assumed to be such that the inputs are quantities assumed to be known without errors, whereas the output measurements are modelled as $y_n = f(x_n) + \text{noise}_n$. When the noise processes have a Gaussian distribution and are independent for each data point, the output measurements jointly have a Gaussian distribution which is completely specified by a mean function $\mu(x)$ and a covariance function $k(x, x)$ selected by the user in such a way that its elements $k(x_n, x_{n'})$ express the similarity between pairs of input data. The covariance function determines the structure of a GPR model: a prior knowledge of the process can be valuable when selecting an appropriate covariance function. Whenever specific data properties need to be captured, a combination of different covariance functions should be considered (see Lloyd *et al.*, 2014). A selection of commonly used covariance functions can be found in Samuelsson *et al.* (2017), in which GPR is extensively used for process monitoring and fault detection using measurements from flow rate and ammonium sensors in WRRFs.

The covariance function is represented by a matrix $\boldsymbol{K}(X)$ of size $N \times N$ whose elements $k(x_n, x_{n'})$ are by a vector of unknown coefficients or hyperparameters p that characterize function k. A typical choice for the elements of K is the squared-exponential kernel function in which $k(x_n, x_{n'}) = p_1 \exp[-p_2(||x_n - x_{n'}||)]$ and $p = (p_1, p_2)$. A simple approach to select the hyperparameters from data is to choose the vector p whose likelihood is maximized, given the output measurements and corresponding inputs in the training dataset. In regression we are interested in estimating the output y for a previously unseen input x: this produces a distribution which, in GPR, is a Gaussian distribution whose mean and variance are easily determined using matrix operations.

6.7.2.3 Unsupervised approaches

Unsupervised techniques use a dataset $D = \{\boldsymbol{x}_n\}_{n=1}^N$ of N data records, each consisting only of M-dimensional input variables $\boldsymbol{x}_n \in X^M$, without corresponding target values or labels. These approaches can be used to explore and discover certain properties of the data, primarily the existence of groups of similar records or of low-dimensional and yet meaningful projections, without the help provided by class or other output annotations (Bishop, 2006). First, an overview of the simplest unsupervised approaches and their use for univariate monitoring is given, and second, some more advanced feature extraction approaches based on artificial intelligence techniques are described such as the well-known PCA (Jackson, 1980), and its use in multivariate monitoring.

6.7.2.3.1 Univariate monitoring methods

Typical univariate ($M = 1$) data monitoring methods apply algorithms and data models to data records of a single-process variable at a time. The main goal of these techniques is to determine whether a change has occurred in that variable. To detect changes in the data, these methods have been applied successfully over the course of almost 100 years in many industries as part of product quality control, fault detection in parts manufacturing, for improvement of business and banking process, and for general anomaly detection in the continuous (chemical) process industries (Baklouti *et al.*, 2018; Corominas *et al.*, 2011, 2018; Newhart *et al.*, 2019; Sánchez-Fernández *et al.*, 2018; Von Sperling *et al.*, 2020). The details of how these methods are applied in practice are often taught as part of introductory statistics courses in many undergraduate programmes in engineering.

The main premise of applying these methods is to first establish what good operation may be and then to detect a change away from good operations and label it as *anomalous* or *faulty*. The detection ability is obtained by applying simple models to signal data. The models for detecting anomalous events are often built on the assumption that (1) the signal data mean is constant with time, and (2) that its spread, or variability, also do not change over time (Montgomery, 2019). A change is then detected when (1) either the signal data mean shifts, drifts, or otherwise moves away from its assumed fixed point, (2) the variance of the signal data changes over time, or (3) both the mean and the variance change. These basic models are the focus of the following discussion. More advanced models that are not constructed on this assumption (e.g., models that allow for time-varying means) are also readily available, though often overlooked in basic applications of these technologies (Box *et al.*, 2015).

The changes detected in a signal data are found using algorithms that compare the values to prescribed limits that are set-up for the purpose of anomalous event detection. Data are often plotted, or charted, to aid the use of such algorithms by process operators and engineers: such charts are called control charts, because the user of the chart is assuming a stable process signal data, with a fixed mean and unchanging measurement variability. When changes occur, the user of the control chart can infer that a fault has occurred so that an investigation can be undertaken to first substantiate if a change is indeed a fault and then to identify the likely cause of the presumably degraded operation that fostered the change in the signal data. For example, a sensor in a plant may drift over time: the drift can be detected on a control chart that compares the signal data to laboratory data and the user of the chart can take action to recalibrate the sensor (Thomann *et al.*, 2002). Whenever visualization and charting are not needed, nor they are a priority, these methods can be fully automated using software.

The general use of a control chart is shown in Figure 6.29. For this illustration, the control chart has one upper control limit. When a fault or event occurs, time passes before the event is detected on the

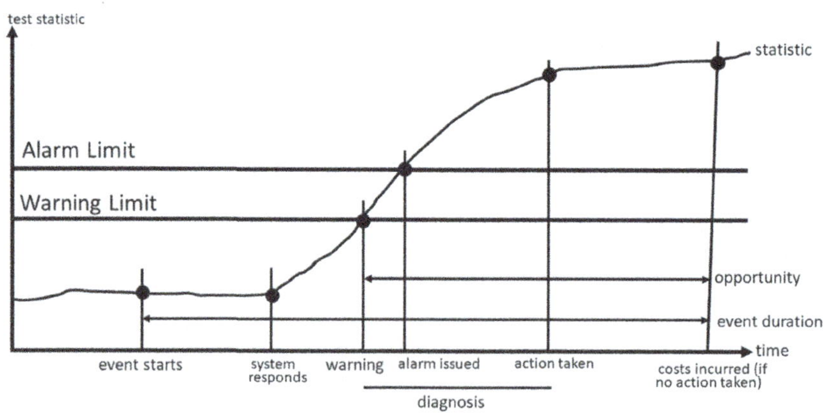

Figure 6.29 Basic control chart use (*source*: Miletic *et al.*, 2008).

chart as violation of the control and warning limits. Once the detection of abnormal data is achieved, an action can be promptly taken, hopefully before costly economic or environmental damage occurs.

This use of the chart can be applied to any control charting methodology, regardless of the algorithm or the model assumptions that are used to construct the event detection system and thresholds. For example, a control chart may have two limits (often, an upper and a lower threshold), two warning limits, and the chart may be applied in practice with an algorithm (e.g., the run-rules) that counts the successive observations so that alarms can be issued accordingly. Many variants and embellishments of the basic run-rules algorithm are available: notably, the very popular Western Electric rules and related methods have been developed largely by industrial practitioners and not academic researchers (Mitra, 2008). As such, these approaches to managing the responsiveness to faults are not focused on theoretical constructs but on the practicalities of implementation and use by non-experts.

Univariate monitoring methods should be understood as computational systems that employ fault detection models to identify anomalies and to foster follow-up actions to analyse the causes initiating the violations. The ability to analyse faults in a signal data (e.g., to stipulate whether a fault stems from a sensor malfunctioning, an upset of the biological process in a WRRF, or a combination of these) will depend on the process model that forms the basis for the control chart (e.g., a fixed mean control chart). Such models can be developed using expert knowledge in a supervised way, in such a way the label of good operation is either set by a process expert or by an algorithm via an (almost entirely) unsupervised bootstrapping manner. When automatic, labelling process data can be thought as a learning mechanism in which the algorithm iteratively discovers about the anomaly patterns from historical data, it is important to note that iterative approaches require the successive retraining of the control charts. This activity is coupled with the examination of alarm rates and process examination by operators until the model (the control chart) reflects a desirable target operation (Miletic *et al.*, 2008).

To operate under a reliable data acquisition and logging system, the practical use of such systems will inevitably require periodic model maintenance (i.e., control chart re-training) and various upgrades of the data communication system. These activities should be continuously checked and validated in relation to the engineering objectives of applying a fault detection system based on control charts. User training and interface design can also be of concern in many practical applications.

6.7.2.3.1.1 Shewhart control chart The most commonly used control chart is the one developed by Walter Shewhart of Bell Labs in 1931 (Newhart *et al.*, 2019). This method is considered to be the oldest statistical process monitoring technique. In Shewhart's charts, data records x_n or their running averages $\overline{x_n}$ over a short time-window are plotted together with a centre line (CL) corresponding to the grand mean of the data (i.e., the mean, μ_0, of the averages $\overline{x_n}$). The middle line is assumed to correspond to a fixed and normal operating point, and the interest remains in understanding how data records deviate from it: large deviations are then assumed to be related to anomalous conditions. This is achieved by adding the upper control limit (UCL) and lower control limit (LCL) to the plot, two quantities that are calculated by adding and subtracting a certain number k (commonly 2 or 3) of the variable's standard deviations (σ) to the mean: that is, $\mu_0 \pm 2\sigma$ (Montgomery, 2019). Whether a data record x_n (or the running average $\overline{x_n}$) is found above the UCL or below the LCL, a statistically significant change has most likely occurred. Thus, a decision of data quality assessment is made comparing actual data with control limits, exceeding control limits will indicate anomalous data. For instance, Figure 6.30 shows an example of the application of Shewhart control chart to DO signal data in a WRRF.

The empirical modelling approach underlying Shewhart charts depends strongly on the control limits used. They are assumed to stem from a variable with a normal distribution and whose data records are not autocorrelated in time and stationary. Although restrictive, this assumption can be easily relaxed to accommodate processes that do not produce normally distributed data. Such methods often apply non-parametric statistical techniques that do not refer to any specific data distribution and that produce control limits that are entirely empirical (Box *et al.*, 2015). Such approaches are especially useful for applications in WRRFs where the normality and time-correlation assumptions are often violated.

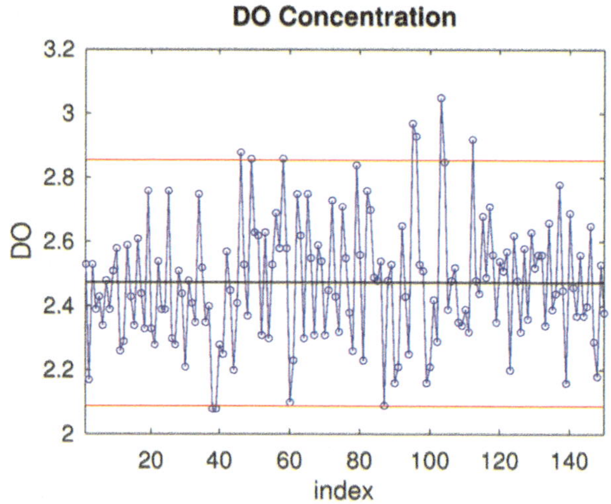

Figure 6.30 Example of a Shewhart control chart for DO concentration (in blue) measured in a large European wastewater resource recovery facility. The UCL and LCL are depicted in red, and the CL is in black. The chart uses Student's *t*-distribution as a reference with 0.025 as each tail area (95% confidence interval).

6.7.2.3.1.2 Cumulative sum control chart A cumulative sum (CUSUM) control chart is a modification of the Shewhart chart. This technique first appeared in the 1950s and it was used to detect small persistent process shifts. This and other modifications of the Shewhart chart are mainly of academic interest and they are rarely found in practical implementation of fault detection systems. Modifications to Shewhart's basic chart that are based on run rules (e.g., the one by Western Electric) and data-filtering techniques are applied more often. As a typical textbook example, an overview of the CUSUM chart is provided here.

This CUSUM chart plots the cumulative sum of previous and present deviations of the data records from the grand mean (μ_0): that is, $C_t = \sum_{n=1}^{t} (x_n - \mu_0)$. This chart can be represented in two ways but the most common is the tabular CUSUM form. Notice that the tabular form discussed in textbooks is primarily a teaching device and not a practical means of application of this technology. The tabular CUSUM is based on two statistics: C_t^+ and C_t^- are the sum of deviations above and below the grand mean μ_0, respectively. They are referred to as the one-sided upper and lower CUSUMs:

$$C_t^+ = \max\left\{0, x_t - (\mu_0 + K) + C_{t-1}^+\right\} \tag{6.2}$$

$$C_t^- = \max\left\{0, (\mu_0 - K) - x_t + C_{t-1}^-\right\} \tag{6.3}$$

The initial values are $C_t^- = C_t^+ = 0$. The constant K is referred to as the reference value and it is often chosen to be $K = x_1 - \mu_0/2$, where x_1 represents the out-of-control value which is aimed to be detected quickly (Ferrer-Riquelme, 2009). The CUSUM control chart is powerful at detecting persistent changes in data records, particularly shifts from in-control values μ_0 to another out-of-control value, x_1.

Both statistics C_t^+ and C_t^- are plotted in the control chart for each data record, whenever either of the statistics lie beyond a decided threshold (control limit), the process is considered to be out-of-control. A common value of this threshold is chosen to be five times the standard deviation. Both parameters, the threshold and K values, are crucial for a proper CUSUM control chart performance.

For instance, Figure 6.31 shows an example of the application of CUSUM control chart to DO signal data in a WRRF.

6.7.2.3.1.3 Exponentially weighted moving average control chart The EWMA control chart is another modification of the Shewhart chart, which was established in the 1950s (Roberts, 1959). By updating the UCL and LCL to account for some non-stationary characteristics, the EWMA method also detects small persistent process shifts. This chart plots the most recent and previous data records by multiplying the most recent one by a forgetting factor λ (commonly $0.05 \leq \lambda \leq 0.25$) and the EWMA by $1 - \lambda$: that is, $\text{EWMA}_i = \lambda \overline{x}_i + (1-\lambda)\text{EWMA}_{i-1}$. By weighing data records in exponentially decreasing order, the most recent values contribute more to the computation of the moving average than the oldest ones. If the parameter λ is small, then the EWMA control chart performs as a CUSUM control chart; if close to zero, then EWMA performs more like a Shewhart control chart.

The CL of the EWMA method corresponds to the grand mean (μ_0) and the control limits; UCL and LCL are calculated by the following expressions discussed by Montgomery (2019):

$$\text{UCL} = \mu_0 + L\sigma\sqrt{\frac{\lambda}{2-\lambda}\left[1-(1-\lambda)^{2i}\right]} \tag{6.4}$$

$$\text{LCL} = \mu_0 - L\sigma\sqrt{\frac{\lambda}{2-\lambda}\left[1-(1-\lambda)^{2i}\right]} \tag{6.5}$$

Here, the factor L is the width of the control limits, whose recommended values are around 3.

Figure 6.32 shows an example of the application of EWMA control chart to DO signal data in a WRRF.

This type of chart is found often in fault detection systems because its underlying models can also be used as low-pass filters for data and to make one-step ahead predictions of signal data that have non-stationary means. Practical applications of these methods include the training of the chart model

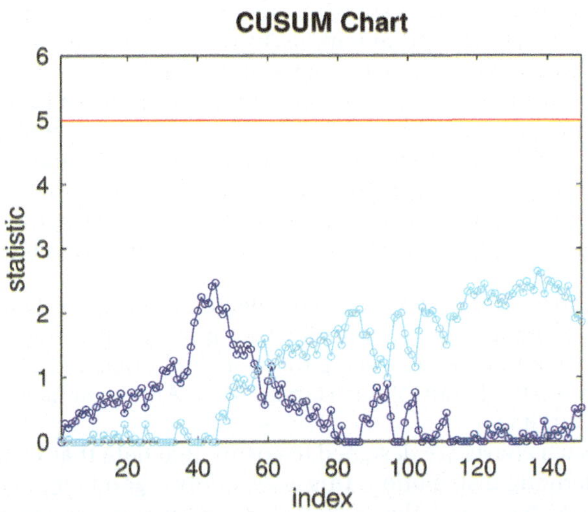

Figure 6.31 Example of a CUSUM control chart for DO concentration (large European wastewater resource recovery facility) showing C_t^+ in light blue and C_t^- in dark blue. Chart parameters are: $K = 0.01$ and $h = 5.0$. The selection of the parameters was made to display the values of the statistics well. In practice, these values are tuned for fault detection capability and not aesthetics.

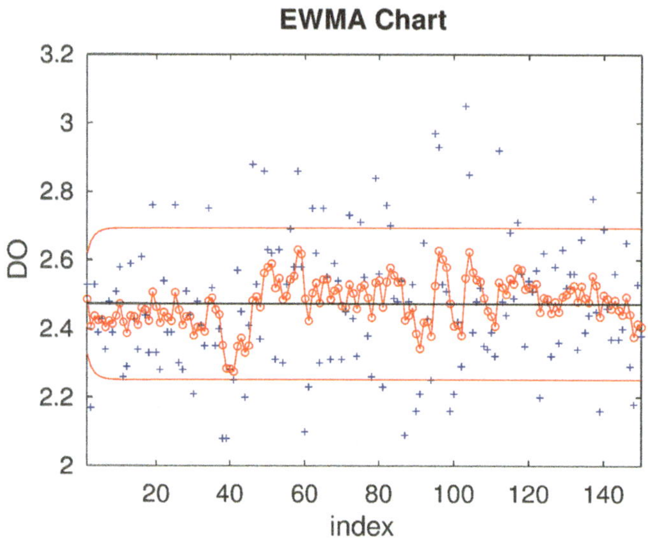

Figure 6.32 Example EWMA chart for DO concentration (large European wastewater resource recovery facility) showing UCL and LCL in red and CL in black. Data are in blue. The EWMA model values are in red with circles. Parameters: $L = 3$ and $\lambda = 0.25$ on a subgroup size of one (i.e., a chart for individual data points).

to determine the λ parameter from a training dataset, or to set/configure it via user selection. The selection of the λ parameter in effect determines if the chart will behave as a low-pass filter or as a fault detector of some capability to identify mean changes of various sizes.

6.7.2.3.2 Multivariate monitoring methods

Multivariate monitoring methods are a direct extension of the univariate methods (Section 6.7.2.3.1). In contrast to the univariate techniques that handle the data records of only one variable ($M = 1$) at a time, the multivariate methods deal with data comprising multiple variables ($M > 1$) simultaneously. The main practical aspect in the use of multivariate methods is the ability to take advantage of the correlations that arise among measured variables to solve engineering and production problems via exploration of data structures. Multivariate methods are applied, for example for online monitoring of industrial processes (Kourti & MacGregor, 1996; Kourti *et al.*, 1996; Miletic *et al.*, 2008; Qin, 2012).

Advanced feature extraction approaches include the transformation of the original multivariate space into a feature space (such as PCA) to reduce dimensionality and clustering techniques (Figure 6.33).

6.7.2.3.2.1 Multivariate control charts One example demonstrating the benefits of the multivariate approach is in control charting. In typical control chart applications, single variables are plotted on a time-series plots with UCL and LCL (e.g., using three-sigma limits) and are used to trigger an alarm when a process is out of statistical control (Oakland, 2003). An alarm flag is raised whenever a data record violates its control limits.

Because the limits for such charts are designed to ensure that data that violate the control limits are in the far tails of the underlying distribution, this is certainly a good engineering practice. However, typical control charts ignore, by design, the existence of interactions between process and production variables. Such interactions can be measured and quantified by computing correlations between variables. When these correlations are ignored, a rich source of information about the process being monitored is lost, limiting the capability of the monitoring system to detect real faults or events occurring in the process.

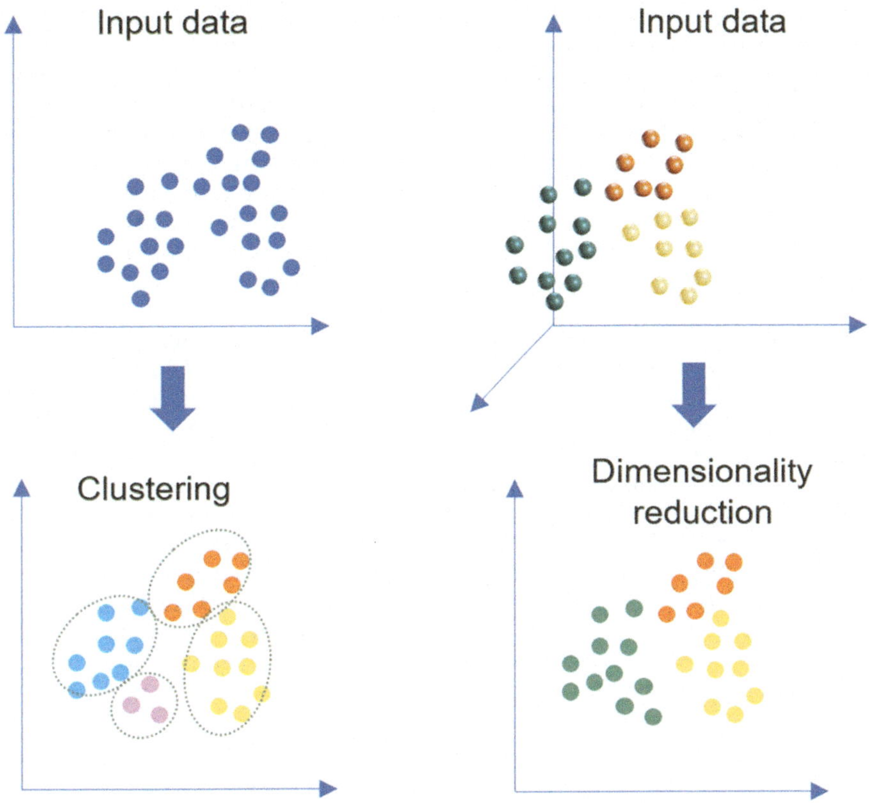

Figure 6.33 Example of unsupervised multivariate methods.

In all practical situations, where real wastewater treatment processes are monitored, it could be expected that signal data of a single-process unit are related to each other. These correlations stem from the connection each variable has to other variables, via the underlying process they are all measured on. The extent of these interactions will dictate the ability of the univariate charts to detect alarms. The application of multivariate methods to fault detection via control charting provides a clear advantage in most practical situations because correlations are modelled directly.

The multivariate charts can be constructed using Hotelling's T^2 statistic, the multivariate analogue to the familiar Student's *t*-statistic. The T^2 statistic (Equation (6.6)) is used to summarize the correlations among the process variables via the $M \times M$ data covariance matrix $\mathbf{S}_x$ (see Mitra, 1998):

$$T^2(\mathbf{x}_n) = (\mathbf{x}_n - \bar{\mathbf{x}})^T \mathbf{S}_{\mathbf{x}}^{-1} (\mathbf{x}_n - \bar{\mathbf{x}}) \tag{6.6}$$

For each data record $\mathbf{x}_n$, Equation (6.6) results in the scalar value $T^2(\mathbf{x}_n)$, which measures the squared distance between $\mathbf{x}_n$ and the expected value $\bar{\mathbf{x}}$. This distance is weighted by inverse covariances, and it quantifies how similar $\mathbf{x}_n$ is to the normal conditions $\bar{\mathbf{x}}$. Because this statistic can be computed for each observation of the vector $\mathbf{x}_n$, T^2 values are easily charted as time series. It is useful to plot T^2 over time together with statistical control limits that can be looked up in statistical *F*-tables once the degrees of freedom are defined or calculated as empirical percentiles. As such, the T^2 chart provides a clear indication that is related to the breach of the typical in-control pattern of a process.

In wastewater treatment, using the T^2 statistic has been proposed, for instance, for monitoring sequencing batch process to support optimization of the nutrient removal performance (Villez *et al.*, 2010). In addition, T^2 statistic is often employed when using unsupervised techniques such as the PCA for monitoring the treatment processes (see, for instance, Rosén, 2001).

The use of multivariate control chart is illustrated in Figure 6.34. Two variables (x_1 and x_2) taken from a process are charted using univariate Shewhart charts with three sigma control limits. Neither of the individual charts on the variables x_1 and x_2 show alarms of any kind along the entire time series of the data. Note however, that these variables are highly correlated. Conversely, multivariate control charts, using the T^2 statistic, that are designed to account for the correlation between the monitored variables provide a clear indication that the process is out of statistical control.

6.7.2.3.2.2 Principal component analysis PCA is a multivariate method that has been widely studied and applied in the wastewater sector and there are many case studies reviewed, for instance in Aguado *et al.* (2006); Haimi *et al.* (2013); and Newhart *et al.* (2019). It is considered a mature technique for practical applications (Corominas *et al.*, 2018). PCA has been used in the wastewater sector, for instance, for monitoring treatment processes, instrumentation fault detection and isolation, visualizing variations in collected datasets, analysing composition of wastewaters, and as a data pre-treatment technique for various modelling tasks. This approach aims at reducing the dimensionality of a multivariate dataset when there are many correlated variables. PCA thus projects the variables from the original dataset into the much smaller number of new informative variables (latent components).

PCA models (Jackson, 1980) are built from $\mathbf{X} = (\mathbf{x}_1, \ldots, \mathbf{x}_n)^T$, the $N \times M$ data matrix, where each of the N rows $\mathbf{x}_n = (x_{n,1}, x_{n,2}, \ldots, x_{n,m}, \ldots, x_{n,M})$ collect the values of M process variables for n data records. Again, the variables refer to either signal data (an automated instrument) or laboratory data.

The PCA factorizes the data matrix $\mathbf{X}$ into two separate matrices $\mathbf{T}$ and $\mathbf{P}$, plus a residual matrix $\mathbf{E}$:

$$\mathbf{X} = \mathbf{T}\mathbf{P}^T + \mathbf{E}, \tag{6.7}$$

where $\mathbf{T}$ is a $N \times A$ data summary, or *score*, matrix and $\mathbf{P}$ is a $M \times A$ *loading*, or projection, matrix. The residual, or error, matrix $\mathbf{E}$ has size $N \times M$. The quantity $A \ll M$ denotes the number of retained PCs. Several methods are used to select the number of PCs: cross-validation and heuristic techniques are commonplace (see Camacho & Ferrer, 2014; Jackson, 1993; Wold, 1978).

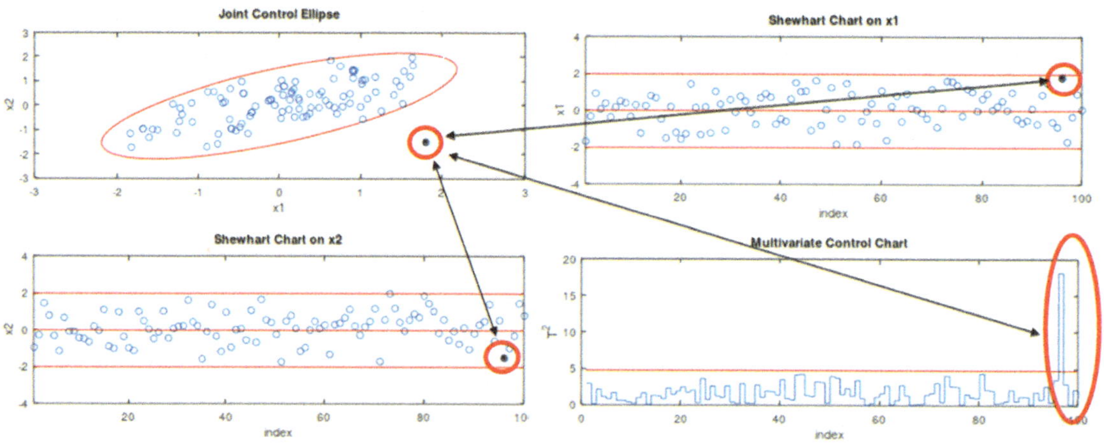

Figure 6.34 Univariate and multivariate control chart examples (*source*: Miletic *et al.*, 2019*b*).

The columns of the **T**-matrix represent independent explanatory variables that are often related to the chemical, physical, and biological driving forces of the plant or process unit. Mathematically, these variables are linear combinations of the original process variables (usually correlated) that are collected in the plant. These linear combinations (i.e., weighted sums) result in a single value called the ***t***-score whose value is stored in the rows of the **T**-matrix with each column of the **T**-matrix corresponding to one linear combination of variables. The weights for these combinations that relate the values in the **X**-matrix to the **T**-matrix are stored in the **P**-matrix. This matrix projects the variables from the original *X*-space into the much smaller and possibly informative *P*-space. PCA thus constructs a mapping of the original process data onto the underlying (chemical, physical, and biological) driving forces that are of interest for operating the process.

In applications of PCA, two summary statistics, the already presented Hotelling's T^2 and the squared prediction error (SPE), can be used to detect process faults, and to gauge model adequacy and data quality. The SPE statistic is a scalar measure of the quality of the PCA model: it is computed using matrix methods to provide a single value for model diagnostics via control charts on this metric. These statistics can themselves be decomposed to reveal, for example, which process variables are related to a fault, or an outlier in the dataset via a visualization method called *contribution plots* (Westerhuis *et al.*, 2000). These powerful fault diagnostic techniques are useful also for automated fault diagnostics, in applications where process data are analysed repeatedly over time (Miletic *et al.*, 2004, 2019a).

Use of SPE and T^2 statistics for outlier detection and contribution plots for diagnosis is illustrated in Figure 6.35. In this real-world example data from a large municipal WRRF in the USA are analysed: COD, TSS, and TSS/COD ratio were identified as the most likely outlier sources (inCTRL Solutions, 2018). Thus, a decision of data quality assessment is made plotting SPE and T^2 statistics of actual data in a control chart with their corresponding control limits. Exceeding these control limits will indicate anomalous data.

It is important to notice that wastewater treatment systems are dynamic and non-linear processes by nature. For that reason, adaptive and non-linear extensions of the standard PCA method have also been proposed for WRRF applications (see e.g., Haimi *et al.*, 2016; Rosén, 2001). Typically, the standard PCA approach is used first and if its performance is not satisfactory, the model complexity is increased with adaptive or non-linear extensions that are described in Section 6.7.2.3.3.3.

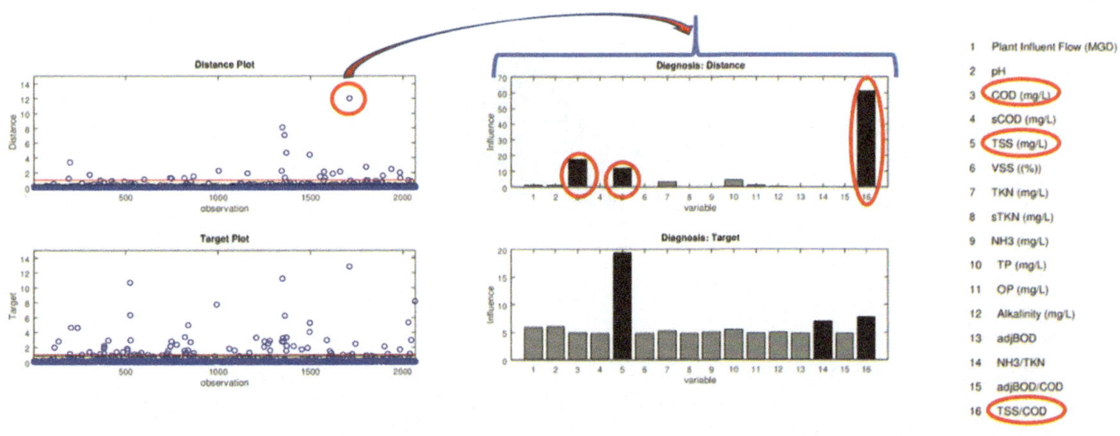

Figure 6.35 Example of outlier detection and diagnosis in a wastewater resource recovery facility using contribution plots (*source*: inCTRL Solutions, 2018). Note that for discussions with the client *Distance Plot* refers to SPE and *Target Plot* refers to T^2.

Furthermore, note that PCA can be used in the analysis of data from batch processes. This is possible if the input data (the **X**-matrix) are properly pre-processed prior to being used for modelling. These techniques are useful in many applications, including SBRs in wastewater treatment (Aguado *et al.*, 2007). The approaches, often named *multiway PCA*, include, in addition to batch number, an extra dimension of, for example, time or extent of reaction in the data matrix. Multiway PCA methods require a reshaping of the data for the **P**-matrix in Equation (6.7): a computation often named *unfolding* (Nomikos & MacGregor, 1995) which is readily executed using most computer programming languages.

6.7.2.3.2.3 Clustering Clustering techniques aim at grouping process data records into meaningful subsets, or clusters. Each cluster consists of data records similar to each other but different from data records in other clusters. Unlike classification techniques (Section 6.7.2.2.1) that group data in classes based on pre-labelled data, clustering techniques are unsupervised methods for investigating patterns in a dataset with no pre-existing labels. Thus, a decision of data quality assessment is made by defining each cluster, and analysing whether actual data are out or in the confidence region of each cluster. There are several different clustering techniques, which are described in detail in well-known statistical textbooks (e.g., Hastie *et al.*, 2009).

Clustering is primarily used in WRRFs for monitoring, visualizing, and diagnosing the process operating conditions. The most popular clustering techniques proposed in wastewater treatment applications are based on the *K*-means algorithm as well as on fuzzy clustering algorithms (Corominas *et al.*, 2018; Haimi *et al.*, 2013). Clustering has often been combined with dimensionality reduction obtained by means of PCA (see e.g., Aguado & Rosén, 2008; Maere *et al.*, 2012). Particularly, clustering the PCA scores provides a tool for automated diagnosis of process faults and deviations. An example of clustering in a case study of a European municipal WRRF is presented in Figure 6.36

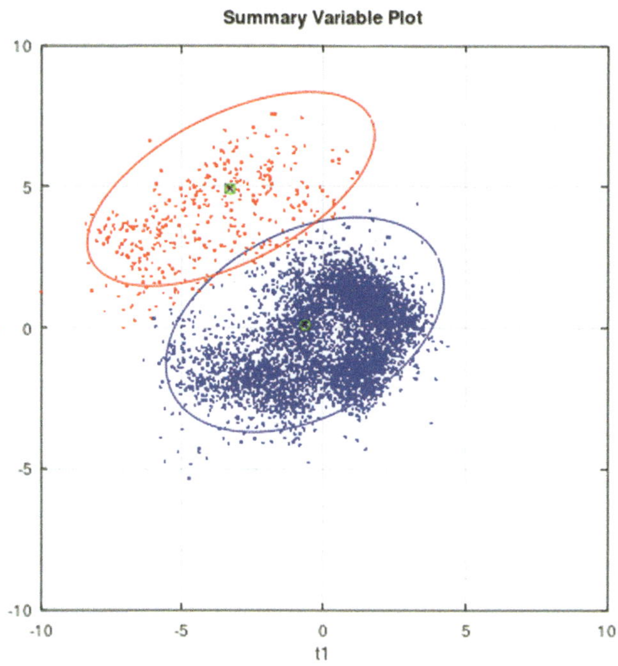

Figure 6.36 Example of using clustering for grouping wastewater resource recovery facility process data pre-treated with PCA (*source*: inCTRL Solutions, 2020).

(inCTRL Solutions, 2020). Data in the $t_1 - t_2$ PCA score space are grouped in two clusters, each associated with a different operational situation: ellipses show non-parametric confidence regions around the clusters.

In addition to PCA, clustering methods have been combined with unsupervised ANN methods called self-organizing maps (SOMs) (Kohonen, 2001). The potential of using SOMs and clustering has proven itself to be useful for wastewater process monitoring applications (see e.g., García & González, 2004).

6.7.2.3.3 Common challenges for application of the methods

Modern automation systems are composed of efficient in-line devices which are capable of measuring, performing computation, and storing large amounts of process data, at very fast sampling times. As a result of our improved ability to control the processes, available data do not only reflect intrinsic mutual dependences among highly instrumented facilities, but also their time-dependence or, roughly speaking, autocorrelation. Time-dependence, due to the causal nature of the underlying physical system, is emphasized by the increased sampling frequencies. Additionally, as the processes are often operated under diverse regimes and subjected to modifications and ageing, the correlation and autocorrelation structure displayed by the data may also vary with time, a property referred to as non-stationarity.

A number of procedures that can be used to adapt standard techniques to autocorrelated and non-stationary processes will be shown. A number of statistical principles is also presented that can render these techniques robust to abnormal data, or outliers, and a principle for generalization to non-linear data.

It is considered again a dataset consisting of $N \times M$ data matrix, where each of the N rows $\boldsymbol{x}_n = (x_{n,1}, x_{n,2}, \ldots, x_{n,m}, \ldots, x_{n,M})$ collects the values of M process variables for n data records. Unsupervised techniques for dimensionality reduction such as PCA aim at explaining the covariance structure of the data with a small number of component directions in the data space. The PCs, defined to be linear combinations of the original variables and constrained to be pair-wise orthogonal, correspond to directions along which the projected data records show the largest variance. The underlying assumptions of PCA are often violated because of autocorrelation and non-stationarity, whereas outliers compromise the estimation of the model parameters (i.e., the centre and covariance matrix of the assumed Gaussian distribution). An overview on this classical multivariate technique will be described, with the understanding that principles and procedures can be extended to structurally similar techniques and to univariate methods, which may be highly affected to the presence of outliers.

6.7.2.3.3.1 Sensitivity to outliers Real-world datasets usually contain data records that behave differently from the bulk of the data. These data records are referred to as *outliers* and their presence may be due to measurement or recording errors, or they may result from a data acquisition campaign performed under exceptional conditions. Whatever the origin, outliers may compromise statistical analyses by *masking* out data records that anomalously deviate from the bulk of the data, potentially leading to a distortion of the main sources of variability present in the data. These effects are observed in classical measures of location and scatter, such as the mean and the covariance, and they may hinder the reliability of the wide range of statistical analyses that build up on those estimators. Commonly used unsupervised techniques such as the already mentioned PCA and k-means models, as well as supervised techniques such as linear regression models and quadratic discriminant classifiers are thus fragile to outliers.

Robust statistics aims at computing fits to these models, as if the outliers were absent. Some standard robust procedures for accommodating outliers are surveyed. The classical estimators of location and scatter are the mean vector $\bar{\mathbf{x}} = N^{-1}\mathbf{1}^{\mathrm{T}}\mathbf{X}$ and covariance matrix $\mathbf{S}_X = (N-1)^{-1}(\mathbf{X}-\bar{\mathbf{x}})^T(\mathbf{X}-\bar{\mathbf{x}})$. Both estimators have a *breakdown value* of 0%, indicating that even a small number of outliers is sufficient to compromise their inference. In the univariate case, commonly used robust estimators of location and scatter are the *median* and the *median of all absolute deviations from the median*, respectively. In the multivariate case, commonly used robust estimates are based on the *minimum covariance determinant* (MCD) criterion. In MCD, the subset of data records whose covariance matrix has the smallest determinant, thus minimum spread, is identified and used to calculate the empirical mean vector and covariance matrix. The number of elements in the subset is user-defined. Computing the MCD estimators is non-trivial because of the combinatorial nature of subset searches. Efficient alternatives and other robust estimators exist (Maronna *et al.*, 2019).

In PCA, the centre of the data, the covariance matrix that is decomposed, and the variance that is maximized are inferred using classical estimators, thus rendering this technique vulnerable to outliers. A number of variants of robust PCA are available; the simplest ones replace fragile estimators with robust counterparts, whereas more sophisticated alternatives develop on projection pursuit (Hubert *et al.*, 2005) and low-rank decompositions and sparse matrices (Candès *et al.*, 2011). Outliers have similar effects also on unsupervised techniques for clustering such as k-means. This is because k-means aims at grouping the data by minimizing the sum of squared distances between data records and the centres of a user-defined number k of clusters. Being the centres estimated by data means, this technique is also vulnerable to outliers. Trimmed k-means methods aim at robustifying this approach to clustering (Gallegos & Ritter, 2005). The fragility of k-means is also observed when softening with a mixture of Gaussian distributions; here clusters' location and scatter are again estimated from data. Corona *et al.* (2013) use robust PCA when developing soft-sensors in a municipal wastewater treatment plant, but robust approaches to clustering used in the water and wastewater sectors are scarce.

6.7.2.3.3.2 Sensitivity to autocorrelation and non-stationarity *Dynamic PCA* (Ku *et al.*, 1995) extends PCA to autocorrelated data based on the intuition that adding L-lagged data records are equivalent to modelling an L-order autoregressive process. Dynamic PCA practically reduces to applying PCA to the augmented data matrix $\widetilde{\mathbf{X}}_L = [\mathbf{X}_n \dots \mathbf{X}_{n-l} \dots \mathbf{X}_{n-L}]$ with $\mathbf{X}_{n-l}$ the l-time back-shifted matrix $\mathbf{X}$. For example, by letting $L = 1$ we would assume the existence of a vector of latent variables $\mathbf{z}_n$ of dimension $R < D$ and dynamics $\mathbf{z}_n = \mathbf{B}\mathbf{z}_{n-1} + \mathbf{w}_n$, for some matrix $\mathbf{B}$ and independent Gaussian variables $\mathbf{w}_n$. L is a user-defined hyper-parameter. The use of dynamic PCA in fault analysis and statistical process control is widespread and demonstrated on real-world and simulated studies (see Yoo *et al.*, 2004 and Villez *et al.*, 2005). Current developments of interest aim at extending quality metrics such as T^2 and Q statistics to models based on dynamic PCA (Ferrer, 2014).

Because the number L of time-lags is selected only once and then kept fixed, dynamic PCA is non-adaptive and thus not adequate for data from non-stationary processes. In situations where stationarity does not hold, it is necessary to equip these techniques with mechanisms to automatically adjust its parameters. Techniques such as recursive PCA (Li *et al.*, 2000) and moving window PCA (MWPCA, by Wang *et al.*, 2005) are based on the intuition that, as time passes, new data records should be included into the model, whereas old data records should be given increasingly less relevance. In MWPCA, this intuition is taken to the extreme. At each time point, a fixed number of old data record is dropped out the model while an equal amount is added, the model's parameters are then re-computed anew on a sliding-window basis. Window size and number of dropped out data records can be fixed or adjusted adaptively. Again, applications in fault analysis are commonplace and often based on signal data of diverse nature: Haimi *et al.* (2016) used MWPCA with an adaptive window length for fault analysis in a municipal wastewater treatment plant, whereas Awhangbo *et al.* (2020) used fixed-window MWPCA for monitoring an anaerobic digester with data from near-infrared spectra. The fault detection performances of recursive and classical PCA are compared on simulated and full-scale applications by Gernaey *et al.* (2004), whereas Rosen and Lennox (2001) studied similar problems using recursive and multiscale PCA.

6.7.2.3.3.3 Non-linear process data Techniques such as *kernel PCA* (KPCA; Schölkopf *et al.*, 1998) mitigate the fact that PCA cannot handle non-linear data. KPCA diagonalizes a positive semidefinite kernel restricted to the data records in $\mathbf{X}$. Under the assumption that there exists a function φ that maps the data records into some arbitrarily high-dimensional *variable space* (with new features), KPCA performs PCA in the extended variable space. Explicit knowledge of φ is however not necessary, as all computations can be performed directly from the positive semidefinite kernel $k\left(\mathbf{x}_n, \mathbf{x}_{n'j}\right)$ when treating it as the Gram matrix $\varphi(\mathbf{x}_n), \varphi\left(\mathbf{x}_{n'}\right)$. A popular choice for the kernel is the Gaussian bump $k\left(\mathbf{x}_n, \mathbf{x}_{n'}\right) = -\exp\left(-\| \mathbf{x}_n - \mathbf{x}_{n'} \|^2 / \sigma^2\right)$, classical PCA is recovered when the linear kernel $k\left(\mathbf{x}_n, \mathbf{x}_{n'}\right) = \mathbf{x}_n^T \mathbf{x}_{n'}$ is used. The kernel extension of robust PCA (*kernel ROBPCA*, Debruyne & Verdonck, 2010) is used when relations among variables are assumed to be non-linear. Kazor *et al.* (2016) compares static, dynamic, adaptive, and adaptive-dynamic versions of PCA, KPCA, and other manifold learning techniques; Klanderman *et al.* (2020) extend this work for a fault analysis platform.

6.7.3 Guidelines for method selection

With basis on the methods presented in Sections 6.7.1 and 6.7.2, this section aims at providing general considerations to support the selection and application of methods for anomaly detection.

6.7.3.1 Framework for application of data quality assessment

Data quality assessment in practice requires us to consider different steps in order to provide an efficient, reliable, and meaningful evaluation. As a good practice, domain experts and data end-users should always be involved in setting-up the data validation framework and also in assessing the performance of the validation (as specified later in Section 6.7.4). Method selection and application should involve the following steps as illustrated in Figure 6.37 where links with different chapters in the STR (C: chapter, S: section) are also included:

(1) Identify the needs of the end-users: in the short (e.g., real-time visualization and flagging) and long terms (e.g., estimation of concentrations and loads. This can help to determine what type of faults or anomalies would be the most important to detect, and therefore the most suitable methods. See Section 3.3 in Chapter 3 for causes of degraded data quality or other anomalies that are of interest.

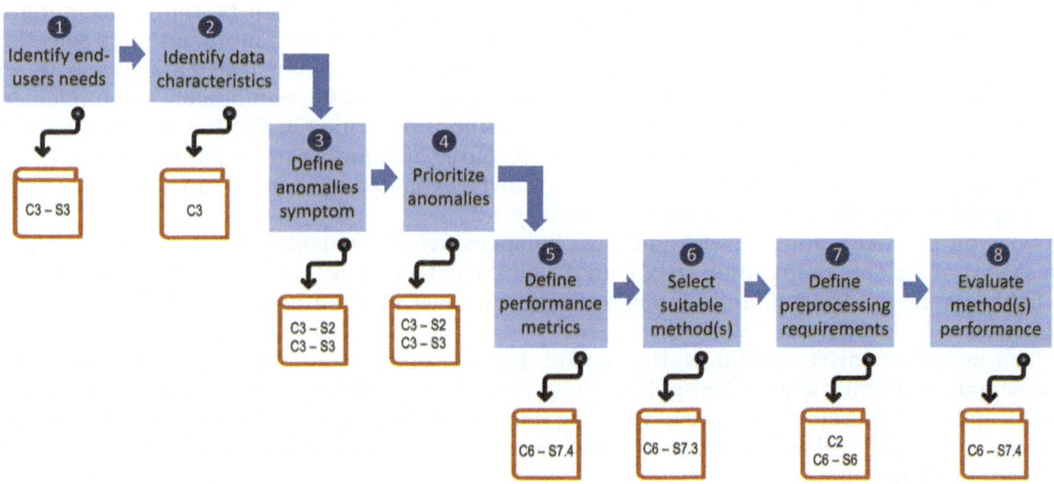

Figure 6.37 Practical framework for selection of data quality assessment methods.

(2) Identify data characteristics: temporal characteristics such as seasonal changes, effect of weather events (e.g., rainfall), non-stationary phenomena (e.g., drift), and potential correlation between the measured variables. This encompasses also whether historical data are available and at what frequency, and whether metadata are already available, such as annotation indicating perceived issues with data quality (see Chapter 3).
(3) Define the symptoms of the anomalies listed in step 1: see Sections 3.1 and 3.3 in Chapter 3 to link the anomalies of interests to symptoms that may be detected in the data.
(4) Prioritize the type of anomalies: according to the importance of the detection and consequences of anomalies being undetected. This information will also influence the evaluation of the method performance.
(5) Define one or more performance measures by which algorithmic methods will be judged (Section 6.7.4). Clearly define how these measures are evaluated. Where relevant, include a description of how data are split between calibration, validation, and testing to avoid biased evaluations of performance.
(6) Select the most suitable method: both characteristics of the data and the user needs are key factors to determine the most suitable method to be applied for data quality assessment. In practice, different algorithms will detect certain types of anomalies better than others. This may lead to using an ensemble of methods as the best approach.
(7) Define the requirements for data pre-processing based on the structure of the available data (see Chapter 2) and the chosen tools. This includes data selection, centring, scaling, and more (see Section 6.6). Apply pre-processing steps consistently when comparing different methods and ensure that the split between calibration, validation, and testing is maintained.
(8) Calibrate, validate, and test the methods of choice on a representation segment of your data.

Metadata as described in Chapter 3 play an important role in the data quality assessment process. Any transformation applied to the raw data (as described in Section 3.2.4, Chapter 3) must be carefully reported. Metadata about the quality of the data (as described in Section 3.3, Chapter 3), as a result from the data quality assessment process must also be described.

6.7.3.2 Selection of the method(s)

The current practice for selection and implementation of the anomaly detection method relies on an ad hoc trial-and-error approach, which apart from being subjective it can be time-consuming and costly. Ideally, a more systematic decision-making process could provide a more optimum selection.

A first decision point is made according to use of prior information as illustrated in Figure 6.38, what implies whether previous mechanistic knowledge of the process (known redundancy and fundamental relationships) is known and therefore can be used. Methods described in Section 6.7.1 belong to this category. If this information is unknown (or time/resource consuming to obtain through, e.g., extra experiments), the information from historical data can be exploited by implementing methods that are driven by the data as the ones described in Section 6.7.2.

As a good practice, sanity checks and mass balances (when available) should always be applied, even as part of the data preparation step, with the aim of detecting gross errors within the data before any further analysis. Sensors and variables of interest must then be involved in the mass balance which need to be known exactly.

When using data-driven methods, the second decision point should be made about the specific implementation. From the practical point of view, the choice will be mostly conditioned to (1) the available data and therefore the type of learning that is possible (supervised – from labelled datasets, unsupervised – from unlabelled datasets), and also to (2) the desired behaviour/output of the data validation implementation.

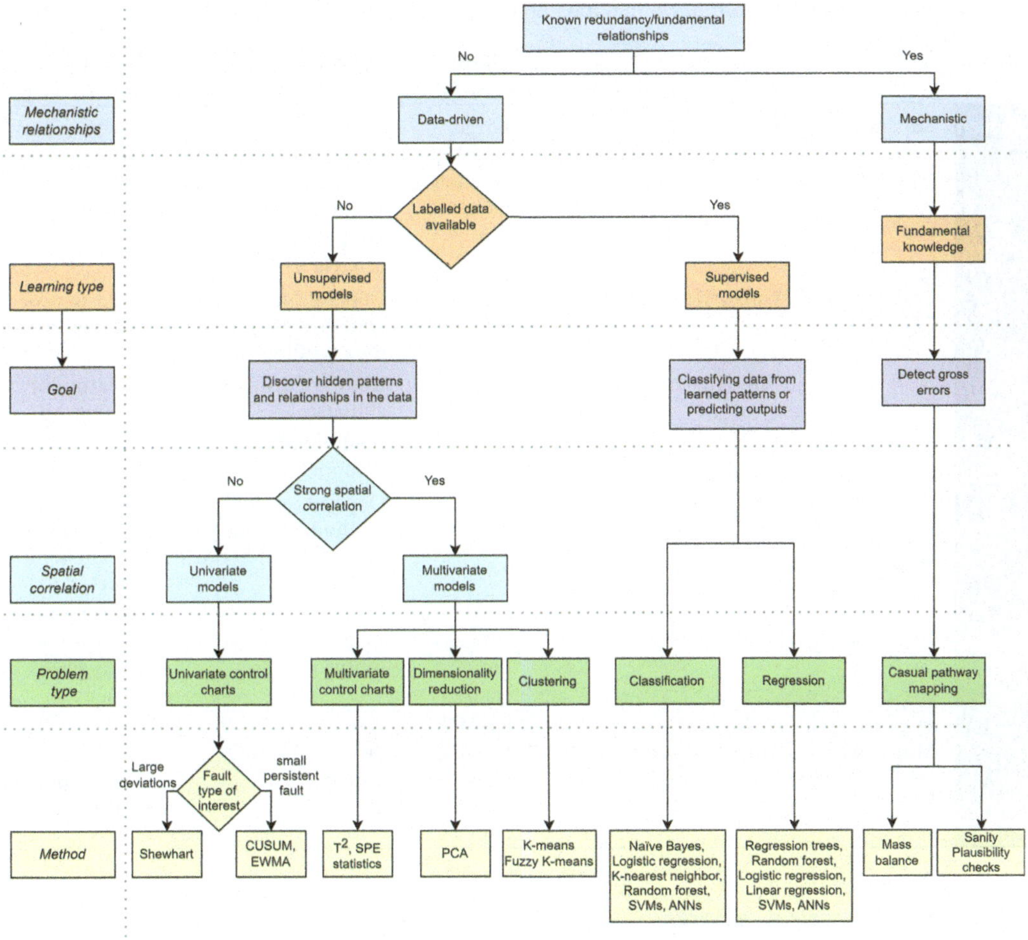

Figure 6.38 Guide to data quality analysis methods as described in Sections 6.7.1 and 6.7.2.

Table 6.7 provides a simple overview of the data requirements (inputs) and possible outputs depending on the method implementation. General data-driven methods will require large datasets with a large variability for training and evaluation purposes. Independently of the anomaly detection approach, the output of the methods can be represented as (a) a label indicating whether an instance is an anomaly or not, and (b) a score or confidence value indicating the degree of abnormality (Markus & Uchida, 2016). Scores can be converted into a label using a suitable threshold (Alferes *et al.*, 2013).

In practice, the final choice of the data quality assessment method will involve the points discussed previously and also the available resources for implementation. As an example, within the data-driven approach, unsupervised methods have been preferred and more extended because having a fully labelled training and test dataset is complex (e.g., anomalies may be not known in advance, or may occur spontaneously as novelties during the test phase), demanding a lot of effort and technical expertise.

Table 6.7 Summary of inputs and outputs within the implementation of the data quality assessment methods.

Method				Inputs	Relevant Parameters	Outputs
Data-driven	Supervised	Classification		Large training labelled datasets	Depend on the method, e.g., number of neighbours (kNN)	Whether an instance is an anomaly or not, e.g., labels
		Regression			Depend on the method, e.g., splitting criteria (reg tree)	
	Unsupervised	Univariate	Control charts	Large unlabelled datasets	Control limits, mean, standard deviation, data window size	Violations of control limits
		Multivariate	Control charts		Limit for control statistic	Violation of multivariate control statistic, contribution plots
			PCA		Number of components	
			Clustering		Depend on the method, e.g., number of clusters (*k*-means)	Data that does not belong to any cluster are detected as anomalies
Mechanistic	Univariate	Sanity checks		Process data, smaller dataset	Plausible ratios, limits, ranges	Whether the sanity check has passed
	Multivariate	Mass balances		Process data, smaller dataset	Stoichiometric parameters	Check data consistency and identify gross errors

6.7.3.3 Additional practical remarks

Independently of the chosen approach, it is important to take into account that both mechanistic and data-driven methods have their own strengths and limitations as summarized in Table 6.8 and they should not be considered as exclusive methods. Current efforts are actually directed to compensate the limitations of each other to enable a synergy between them (Bikmukhametov & Jaschke, 2020, among others). Hybrid methods include the use of mechanistic models to train and test data methods, reduce data dimensionality, reduce data-model complexity, reduce computation time, augment data, and reduce empirical uncertainty, among others.

In practice, the most extended methods for anomaly detection in the WRRF context are the well-known univariate control charts introduced in Section 6.7.2.3.1 due to their easier implementation and use. However, not all types of anomalies will be detected, especially for complex time series. Table 6.9 highlights the main strengths and limitations of each chart. From the practical point of view, it is recommended to combine EWMA or CUSUM charts with Shewhart charts to gain the benefits of both types of charts, so that both small and large shifts in the mean or the process standard deviation (σ) can be detected fast enough.

As introduced in Section 6.7.2, independently of the selected method, a proper data split is required to provide the right information for a two-stage process: first, a subset of data for construction and calibration of the method, and second, a subset of data for validation and refining of the method. Also, although mechanistic methods require deeper knowledge of the system compared with empirical

Table 6.8 Strengths and limitations of data-driven and mechanistical approaches.

	Strengths	Limitations
Data-driven	• Easier construction and set-up. • Parametrization effort is relatively low. • Specially used when it is difficult, costly or impossible to know the mechanistic information of the system. • An empirical method can provide reliable results when it is based on a substantial amount of test data with enough variability.	• Output may be not enough transparent. • Large amount historical data is needed. • In most cases, they are applicable only to particular operating conditions (similar to the training data). • In general, lower accuracy, less preferred for precise analysis. However, accuracy increases with the amount of data provided, the selection of an adequate algorithm and the identification of inputs and outputs of interest.
Mechanistical	• Higher accuracy. • Lower input data points (observations) are required. • Extrapolation to different operating conditions is possible within the calibration domain. • Output of the method is transparent.	• Often mechanistic information is not available or requires time consuming and complex calculations. • Oversimplified assumptions can lead to wrong conclusions. • Requires experimental data that may be difficult to obtain. • Significant investment in method development, testing, and application.

Table 6.9 Practical considerations while using control charts.

	Strengths	Limitations
Shewhart	• Simple and intuitive. • Effective for detecting large changes in process parameters. • Widely used due to their relative ease of implementation and interpretation.	• May take a long time to detect a small persistent shift in the process parameter.
CUSUM	• More efficient in detecting small shifts in the mean of a process (2σ or less). • It is easy to identify visually the shifts in process mean. • Provide the fastest out-of-control detection of sustained shifts for almost all shift sizes and estimation errors.	• Establishing and maintaining of CUSUM charts are more difficult. • CUSUM charts are slower in detecting large process mean shift. • Harder to detect and interpret patterns.
EWMA	• Thanks to the weighted factor, the EWMA chart is much more sensitive to smaller shifts in the process mean (0.5 to 2σ). • Also used to smooth the effect of known, uncontrollable noise in the data. • Can identify trends, which allows us to identify whether a process changed positively or negatively. • Because it is based on a digital filter, it can be used for forecasting and feedback/feed-forward control.	• EWMA charts are also slower at detecting larger shifts. • Inertia effect. If an upward shift happens when the EWMA statistic is in the lower part of its range (LCL), it first must come back up to the centre line before it can continue up to the UCL. This delays the detection of the shift in the mean. • No explicit procedure to find the most suitable value for the forgetting factor λ.

methods, process knowledge will be required in any implementation for a proper method set-up and validation of the method's output to avoid misleading results (see Section 6.7.4).

A sequential or parallel combination of methods could provide a more accurate and reliable result. Such implementation could encompass, for example, the application of simpler methods (sanity checks, control charts) for a basic data quality validation, followed by a more complex assessment. Univariate and multivariate methods can also be combined to detect faults affecting individual sensors and out of statistical control situations affecting the whole process.

6.7.4 Performance of the methods for anomaly detection

This section provides guidance for assessing the performance of anomaly detection methods. First, the step-wise process (workflow) for performance assessment is outlined and guidance is provided for each of the steps (Section 6.7.4.1). Next, a working example is described to illustrate the usefulness of the process. Therein, four anomaly detection methods are compared on real WRRF data as an example of how the different performance metrics can be used to select anomaly detection method (Section 6.7.4.2). Finally, merits and pitfalls of performance assessment are discussed (Section 6.7.4.3).

6.7.4.1 Workflow

The goal with assessing the performance of anomaly detection methods is two-fold. First, the performance assessment should answer whether the performance meets the practical requirements and end-use needs. Second, based on such assessment, the best method which distinguishes abnormal (non-typical) from normal data is identified. The five basic steps for a detailed performance assessment include:

(1) Quantitatively define the goal and targeted performance of the anomaly detection problem upfront.
(2) Select metric/s that reflects the goals in step 1.
(3) Identify or produce datasets that will be used for evaluation.
(4) Compute and assess metric/s and benchmark the anomaly detection methods.
(5) Interpret and place the results into context.

6.7.4.1.1 Step 1: define the needed performance

The first step is to define the lowest acceptable detection performance, which still fulfils the end-user needs. Detecting subtle anomalies such as a minor sensor drift, with high demands on no false alarms will be more difficult, and likely more costly to develop, compared to methods that provide rough indications and support, for example, for prioritizing sensor maintenance. An early definition of the anomaly detection scope is recommended as it may reveal impossible tasks early, instead of after completion of an extensive assessment study.

6.7.4.1.2 Step 2: select metrics

Performance assessment metrics can be divided into two categories: dynamic and static (Kurtoglu *et al.*, 2008). Static metrics quantify the anomaly detection method performance with respect to a collection of data instances independent of their ordering in time, whereas dynamic (and hybrid) metrics do consider such ordering. Table 6.10 classifies the main metrics (static and dynamic/hybrid) applicable to the water resource recovery field for anomaly detection.

6.7.4.1.2.1 Static metrics As Table 6.10 illustrates, a wide collection of static performance metrics for anomaly detection can be found in the literature. The most popular metrics are based on the confusion matrix. The confusion matrix contains four numbers:

- True positives (tp) or correctly detected anomalies
- False positives (fp) or incorrectly detected anomalies

Table 6.10 Main metrics applied to the water resource recovery field for anomaly detection.

Classification	Metric Type	Metrics	References
Static	Confusion matrix derived	True positive rate/hit rate/ sensitivity/recall False positive rate/false alarm rate False negative rate/miss rate True negative rate/specificity Negative-predictive value Precision/positive-predictive value Accuracy F-score/F-measure	Fawcett (2006); Frank *et al.* (2019); Schneider *et al.* (2019)
	Explicitly probabilistic*	ROC graphs AUC/area under the ROC curve Partial area under curve Kappa coefficient/Cohen's kappa/ kappa statistics Matthews correlation coefficient (MCC)/phi coefficient	Fawcett (2006); Frank *et al.* (2019); Samuelsson *et al.* (2019)
	Regression	RMSE R^2 score (coefficient of determination) ME and MAE MSE Mean square derivative error (MSDE) PBIAS Mean absolute error relative (MEAR) Sum of residuals	Hauduc *et al.* (2015)
Dynamic/ hybrid	Dynamic	Mean time to detection Detection of stability factor	Kurtoglu *et al.* (2008); Fellini *et al.* (2019)
		Detection speed (ARL1)	Baklouti *et al.* (2021)
	Hybrid	Weighting: anomaly detection evaluation indexes Anomaly detection index (I_{FD})	Corominas *et al.* (2011); Alferes *et al.* (2017); Carlsson and Zambrano (2016)

Different names for the same metrics are separated with a '/'.
*Probabilistic metrics but also based on the estimates from the confusion matrix.

- False negatives (fn) or missed anomalies
- True negatives (tn) or correctly detected normal data

From this basic set, the metrics accuracy, precision, recall, and $F1$-score can be obtained. Accuracy is the ratio between correctly identified data records and total number of data records:

$$\text{Accuracy} = (\text{tp} + \text{tn})/(\text{tp} + \text{tn} + \text{fp} + \text{fn})$$

Precision is the proportion of actual anomalies correctly detected by the method:

$$\text{Precision} = \text{tp}/(\text{tp} + \text{fp})$$

Recall expresses the ability of the method to find what proportion of actual anomalies is correctly detected:

$$\text{Recall} = \text{tp}/\text{tp} + \text{fn}$$

Generally, there is an inverse relationship between these last two measures, which is called the precision-recall trade-off. For this reason, a popular score used to measure method performance for unbalanced classification problems (as in the case of anomaly detection, where anomalies belong to the minority class) is the $F1$-score, computed as:

$$F1 = 2 \times \text{precision} \times \text{recall}/(\text{precision} + \text{recall})$$

An example of the application of confusion matrix-derived performance metrics is presented in Section 6.7.4.2.

Explicit probabilistic metrics are also based on the estimates from the confusion matrix and allow for systematic performance quantification of anomaly detection methods usually from a specific threshold value. For instance, the area under the curve (AUC) indicates the performance of a binary classifier and summarizes the method performance into a single measure which ranges from 0 to 1. Excellent methods, with a good measure of separability, have an AUC ~1; a random guess has an AUC of 0.5. Frank *et al.* (2019) provide more details. The AUC can be visualized in the form of the receiver operating characteristic (ROC) curve (Bradley, 1997) by plotting the total positive rate as a function of the false positive rate. The above-mentioned AUC metric represents the area calculated from the integration of the ROC curve (Russo *et al.*, 2021).

Different regression metrics can also be applied to characterize the performance of methods for anomaly detection, including specific ones and others borrowed from more general classification fields. Characteristics of these metrics vary in the way they are computed (Hauduc *et al.*, 2015), such as: (i) metrics that are averaged over the number of data records on which they were computed, allowing to compare metric performances obtained on datasets of different sizes (e.g., root mean square error, or mean and mean absolute errors; or (ii) relative metrics divided by data records allowing for comparison across different variables (e.g., per cent bias).

6.7.4.1.2.2 Dynamic and hybrid metrics Dynamic performance metrics requires the analysis of the dynamic target and predicted labels (in the case of supervised methods). This might imply the analysis of two binary signals (e.g., 0 means an anomaly is absent, and 1 an anomaly is present) or the multivariate time-series analysis of two signals.

For instance, Kurtoglu *et al.* (2008) defined different dynamic metrics for anomaly detection, such as time to detect and detection stability factor. These indicators are commonly applied to systems where a rapid response is needed to ensure continued safe operation, for example, time from anomaly incipience to first detection is commonly used in aerospace applications (see e.g., Vachtsevanos *et al.*, 2006). Hence, such dynamic metrics are also of interest in WRRF applications for which rapid response is a key requirement. Indeed, temporal metrics should be regarded as an interesting approach for all those processes where anomaly detection outputs may be reviewed continuously because time is critical.

For example, rapid detection of an anomaly in a DO sensor might save energy consumption and ensure the system operation can switch to a fall-back strategy which ensures safe WRRF effluent quality and read more on fault tolerant control (Abbaspour *et al.*, 2020). An example of a more complex dynamic metric is proposed in Corominas *et al.* (2011), which presents a hybrid time-based anomaly detection evaluation index of anomaly detection methods. This index takes into account several characteristics, such as false alarms, false acceptance, and undesirable switching from correct detection to missed detections during an anomaly event (code for using the performance index is available in the supporting information of the same paper). Specifically, the index proposed by Corominas *et al.* (2011) penalizes fps at a flat rate but penalizes fns using an exponential decay function covering a maximum penalty value over time. The resulting metrics from Corominas *et al.* (2011) are the following: G (penalization points) obtained by the evaluated method, $G_{\max}$ (maximum penalization points) when completely wrong detection is assumed, J (best is 1) overall reliability, J_{FAC} (best is 1) reliability of the method against false acceptance, and J_{FAL} (best is 1) reliability of the method against false alarms.

6.7.4.1.3 Step 3: identify datasets

Selecting a good dataset for performance assessment (here denoted evaluation dataset) has similar challenges as selecting a good dataset for training (training dataset). One difference is that the evaluation dataset needs to contain labelled data from both classes, that is, typical and anomalous data. Unsupervised methods, on the contrary, can be trained with simply unlabelled data.

A ground truth as label is ideal, although it may be practically infeasible to obtain. Instead, subjective expert annotations can be used to manually distinguish typical from anomalous periods in a dataset. It should be noted that anomalies are rare by nature, which reflects their presence in data. This concerns real data as well as the evaluation dataset. Although it may be desired in theory to have a balanced proportions of anomalous and typical data in both training and evaluation data, this is not achieved in practice for several reasons; for example, let us assume an anomaly detection imbalanced scenario with 1% anomalies: for us to balance the evaluation data, 98% on the majority class should be thrown away, risking to remove important data points necessary to evaluate the model. This technique, called undersampling, is not recommended for the evaluation dataset. Explaining this further: it is commonly assumed that the majority class is sampled from the same population, but this might not be the case: there might be some instances in the majority class for which the model does not perform well. Removing such data during evaluation will prevent us from understanding how the model is performing in a general scenario and if more actions are needed such as retraining of the model with different hyperparameters. Further, deciding which instances of the majority class should be removed creates an additional hyperparameter that would need testing. Instead, it is important to select appropriate performance measures that consider the imbalance in either class, such as $F1$-score, precision recall curves, etc. In this case, no undersampling is needed, rather a careful consideration of the meaning of such measures.

The need for labelled data also depends on the complexity in selected performance measures. For example, dynamic metrics additionally requires knowledge of when the anomaly was initiated, that is, the transition from typical to an anomaly in time. The static metrics, by contrast, can be evaluated independently of time ordering. From the viewpoint of annotations, it is a much simpler task to annotate the time for when a fault is resolved, rather than when the fault begun. Therefore, it is easier to obtain evaluation data for static measures where a fault followed by a mitigating action is sufficient, by contrast to the dynamic measures where the opposite ordering is required. Dedicated experiments with purposely and precisely introduced anomalies may be needed to evaluate dynamic metrics unless the anomaly is a frequent and clear problem (and if so, the term anomaly should be revised).

Regardless of whether static or dynamic measures are considered, the dataset should be representative of the identified application and expected conditions. Representative here refers to process and influent variations, noise levels and process operations that are in line with the conditions where the anomaly detection application is intended. Representativity also refers to how well the anomaly is expressed akin to a signal-to-noise ratio. For example, evaluating the performance of a bias detection method on data with a 300% bias is limited as best if the goal is to detect a 10% bias.

Finally, a general rule (and probably the most important one) is to *always* use separate datasets for method training as for evaluation to avoid biased performance results.

6.7.4.1.4 Step 4: evaluate the performance metrics

We describe a classic method performance evaluation workflow in Figure 6.39. On the left side, we refer to method training as the process by which a mathematical method is identified by means of a data selected for this task. During method training, the method's parameters are identified based on the training dataset. If ground truth labels are available in the dataset, the method is trained in a supervised fashion and subsequently optimized on the training labels. If labels are not available, unsupervised, or one-class approaches are used for method training (for more information, we refer the reader to Section 6.7.2). Note that labelled data are needed for performance assessment regardless of whether an unsupervised or supervised method is used.

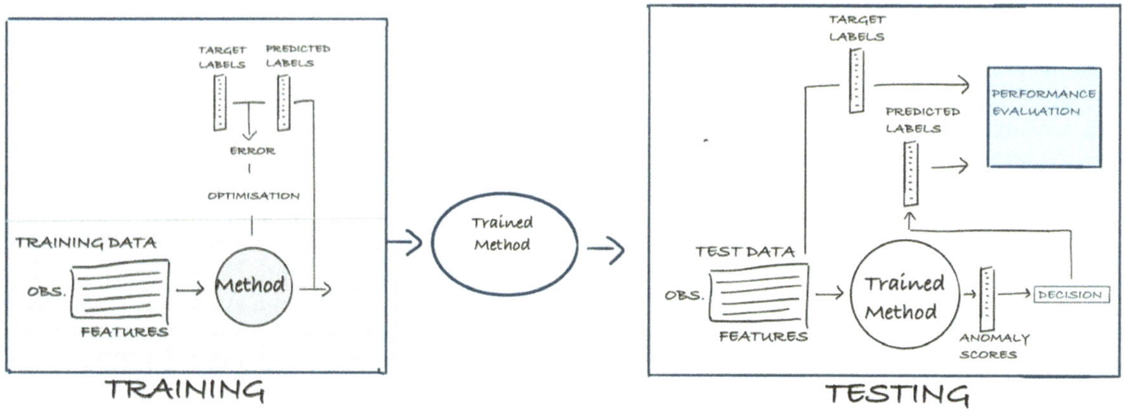

Figure 6.39 Method training and evaluation workflow. Left: the supervised method uses a labelled dataset for training. If labels are partially or not available, unsupervised or one-class approaches are used for training. This is represented in the grey area on the training box. Right: once the method is trained, it is used on the test data to compute the performance metrics. Note that the test data have to be labelled in all cases.

On the right side of the workflow, once the method is trained, the test data are fed to the method. Here, the method will return an anomaly score. Afterwards, a decision function is implemented on the anomaly score to predict a binary label for each unseen test data point. The predicted label is compared with the target label given for the test data and the results are used to compute performance metrics. Note that the target labels are provided by a human expert (subjective) or ground truth (with reference measurements) and are needed to perform method evaluation.

6.7.4.1.5 Step 5: interpret and place the results into context

The final step is to interpret the metrics by placing the results into context. This includes revisiting the goal (step 1), considering deviations between reality and the used datasets (step 3) as well as comparing the results for the studied methods with benchmark methods. We recommend two benchmarks.

First, a simple method that can be developed and implemented without special data science skills and serve as a baseline measure of the added value for a more advanced method (probably the studied one). A simple method has several benefits such as ease of education, implementation, and usage. More elaborate methods should show better performance as they likely have higher costs.

Second, a random detection method should serve as the lower performance bound. Any method should perform better than random guess. By using such benchmarks, it becomes easier to interpret the performance of the studied method, in the context and comparison of the benchmarks.

6.7.4.2 Working example

Here, we illustrate how the performance metrics come in play by replicating part of the results shown in Samuelsson *et al.* (2018). First, the experimental data are outlined, next the basic confusion matrix with related measures is given. Then, hybrid measures combining static and dynamic metrics are illustrated. Two methods A and B are compared with a simple approach (pragmatic engineering approach) and a coin toss detector (random selection when to raise an alarm or not).

The example is based on data from fouled and clean DO sensors in a 350,000 PE WRRF (Bromma, Sweden). There, a DO sensor's bias magnitude was recorded during successive fouling for a period of 7 months.

Step 1: Define detection goal

The goal is to detect poor DO readings due to fouling. A bias of 0.5 mg/L was defined as the targeted detection limit to be useful for sensor maintenance actions. It is desired to reach the detection goal with a maximum of 10% false alarm rate. Achieving a low false alarm rate is a priority compared to a high detection rate.

Step 2: Define performance metrics

The key metric is to compute the confusion matrix, which is easy to interpret and is further used to obtain ROC curve, accuracy, precision, recall, and *F*1-score (Figure 6.40). The hybrid metrics in Corominas *et al.* (2011) as well as mean-time-to-detection were included to cover the time (dynamic) aspects, as a complement to the static metrics.

Step 3: Identify datasets

The data in Samuelsson *et al.* (2018, 2019) were produced with the purpose of evaluating fault detection methods. Hence, the dataset contains clear annotations for typical and anomalous observations, as well as five periods for training data. Details about the data are given in the original publications.

Step 4: Compute and evaluate metrics

Static metrics: In this case, for method A, the matrix shows that out of 1262 typical and 1126 anomalous data records, the method correctly classifies most of them (1195 and 983 respectively).

Method B, on the other hand, correctly classifies about half of typical instances (632) and 802 anomaly instances. These results can be further extrapolated in the form of accuracy, precision, recall, and *F*1-score (Figure 6.40).

The above metrics are reported in Figure 6.40 for both methods. It can be noted that precision, recall, and *F*1-score present similar values for typical and anomaly classes. In fact, these metrics are usually used in the case of imbalanced datasets only (in this case instead, typical and anomaly classes are balanced). Still, it is always important to differentiate the metrics *per class*: in fact, although method B presents a total accuracy of 0.60 which might seem acceptable, its recall performance on the typical class is not better than a random guess of 0.50.

The statistics from the confusion matrix are also commonly visualized by graphs. The receiver operating characteristics (ROC) is obtained by plotting the detection rate (also: true positive rate) as a

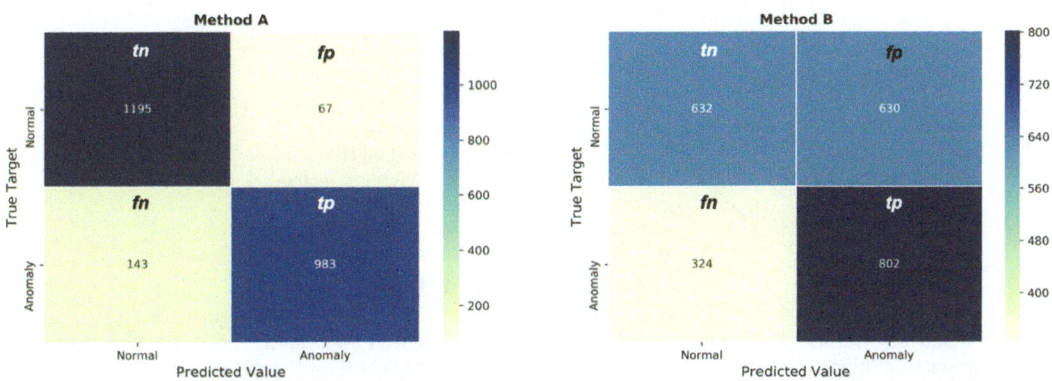

Figure 6.40 Confusion matrices for methods A (left) and B (right). Note that high values in the diagonal entries are representative of a model with high precision and recall.

function of the false alarm rate (also: false positive rate) (Kay, 1998) and is illustrated in Figure 6.41 for this specific example. Many machine learning studies instead include the equivalent precision–recall curve, which is preferred over ROC when there is data imbalance and lacking anomalous data records as indicated by Ozenne *et al.* (2015).

Hybrid and dynamic metrics:

In this section, the dynamic performance metric proposed in Corominas *et al.* (2011) is also used. Several parameters need to be tuned for the method. PFAC,0, PFAC,sat, and tFAC values for false acceptance were 20, 50, and 3 data records (45 min), respectively, and kswitch was equal to 1. PFAL,0 for false alarms was set to 20. The evaluation system is set-up that a perfect detection case receives 0 penalization points (G) and the completely wrong detection case receives the maximum penalty G_{max} (8.0×10^4 for this example). Accordingly, the perfect detection case would provide 100% reliability (J) and completely wrong detection provides 0% reliability (J). If the decision of the anomaly detection system results in false alarms, delayed detections, delayed with inertia, or intermittent detections, there is an increase in the penalization points compared to the perfect detection case. In general terms, method A is more reliable (88%) than method B (64%). Yet, method A is not much better than the pragmatic method (92%). The random detection results in the worst performance by far (35%).

When only evaluating the reliability of the system against false acceptance (J_{FAC}), method B is less reliable (71% reliable) as compared to method A (88% reliable) because it is less capable of identifying the anomaly events. Yet, the pragmatic approach is more reliable as compared to method A. When focusing on false alarms, method B gets more penalization points as compared to method A (G_{FAL} of X and Y points, respectively), obtaining a 50 and 95% reliability against false alarms (J_{FAL}). Method A is much more reliable against false alarms. The random method is 90% reliable, and the pragmatic method 86%.

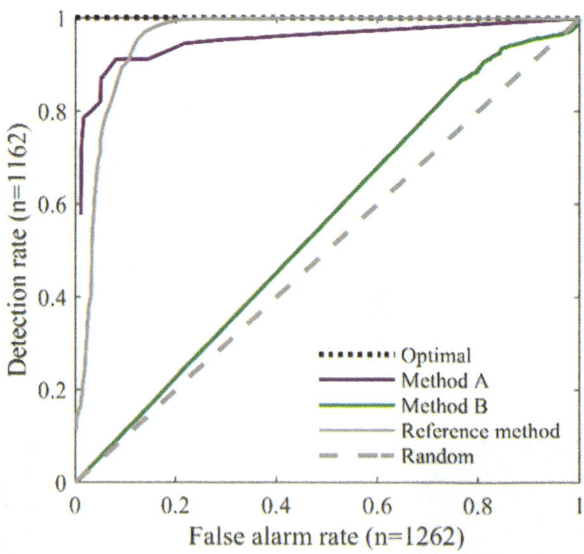

Figure 6.41 ROCs for method A (purple solid line) and method B (green solid line). The optimal performance is shown for comparison (black dotted line) with a constant 100% detection rate and zero-false alarm rate. The simple reference method (grey solid line) contrasts the poor performance of method B. The lowest acceptable performance is indicated with diagonal grey dashed line, which provides the result for a coin flipping detector (equal probability of raising an alarm as not raising one).

Overall, method A performs better than method B, but does not outperform the pragmatic approach. Even if method B is less performant, it is still better than the random approach.

Step 5: Interpret and place the results into context

From Figure 6.40, both methods A and B perform better than the random detector which is the first evidence that they are actually useful. However, method A, which performs better than B, is not superior to the reference method. This indicates that the more advanced method only provides minor added value and if implementation is difficult, the reference method would be a better choice. Another way of interpreting the result is that the data are easy to analyse, without a need for advanced anomaly detection methods.

The performance metrics in Table 6.11 supports the results in Figure 6.41. In addition, the needed detection goal of 10% false alarms is achieved by method A, which is therefore suited for implementations and fine-tuning. The performance of the methods obtained by the hybrid metric is similar to the one obtained by the static metrics. The hybrid metric includes weighting parameters which can be adjusted to give more importance to the dynamic or the static component of the hybrid metric.

6.7.4.3 Merits and pitfalls of performance assessment

The implementation of the step-wise procedure described above and illustrated with an example might not be that easy when facing an evaluation of an anomaly detection algorithm. In this section, we provide merits and pitfalls which might help end-users bring the proposed approach into practice.

Vague names: Many static performance measures have vague names such as sensitivity and accuracy, which seemingly can indicate a good performance. It is essential to fully understand how the metric at hand was obtained to interpret the results.

Imbalanced datasets: Balanced datasets, that is, equal amounts of anomalous and typical data, are desired for most metrics. A dataset with mainly typical data records can easily show close to 100% accuracy for a method constantly avoiding alarms.

Table 6.11 PFAC,0, PFAC,sat, and tFAC values for false acceptance were 20, 50, and 3 samples (45 min), respectively, and kswitch was equal to 1.

	Method A	Method B	Pragmatic	Random
False alarm	**0.053**	0.50	0.13	0.10
Precision	**0.94**	0.56	0.87	0.46
Recall	0.87	0.71	**0.99**	0.094
*F*1-score	0.90	0.63	**0.91**	0.16
Accuracy	**0.91**	0.60	**0.91**	0.52
Mean time to detection (days)	1.60	0.94	0.42	18.71
G (penalization points) obtained by the method	9.1×10^3	2.9×10^4	6.1×10^3	5.2×10^4
G_{max} (maximum penalization points) when completely wrong detection is assumed	8.0×10^4	8.0×10^4	8.0×10^4	8.0×10^4
J (best is 1) overall reliability	**0.88**	0.64	0.92	0.35
J_{FAC} (best is 1) reliability of the method against false acceptance	**0.88**	0.71	0.97	0.09
J_{FAL} (best is 1) reliability of the method against false alarms	**0.95**	0.50	0.86	0.90

PFAL,0 for false alarms was set to 20. Bold text indicates best performing method.

Representative data: It should always be remembered that the method performance in a real application will only be similar to the performance assessment results when reality is represented in the performance assessment data. Using a dataset with obvious anomalies can produce great results as measured by a certain metric but may fail in practice if the anomaly is less pronounced.

Using only sophisticated and flexible methods: Flexible methods such as deep neural networks with many parameter methods are generally likely to overfit the training data and are much more complex to maintain when deployed. Always compare complex methods to simple benchmark methods to judge the needed sophistication.

Good, and good-enough performance: Commonly, the expectations for confidence in method predictions are higher than what can be obtained from the available accuracy in data. It is therefore important to decide early what is good enough for the desired application to avoid wasting resources on applications where data availability of quality is lacking.

As a summary, the general take home message is to:

(1) Clearly define the needed detection performance at an early stage of the assessment.
(2) Select metrics that are applicable given the data at hand for the specific application, and representative for the real problem. Otherwise, data and metadata collection needs to be revised to, for example, include annotations about anomalies and known errors.
(3) Put the performance assessment into context. Assure that data used for evaluation reflects reality.

6.8 TAKE-HOME MESSAGES

Data quality assessment techniques are increasingly demanded to be implemented in WWTPs/WRRFs to really improve the decision-making process for smart operation. To this aim, data processing pipeline has been defined including the main steps of initial review, data pre-processing, and data quality assessment techniques. A significant number of techniques has been briefly introduced from basic to advanced ones, including both mechanistic and empirical methods. Useful criteria have been defined for proper technique selection:

- Data processing steps included in a data pipeline are specific to the case and purpose.
- Data pre-processing technique selection depends on the raw data characteristics and the final data quality technique to be applied.
- If redundancy/fundamental relationships are known, mechanistic techniques should be applied.
- If redundancy/fundamental relationships are mostly unknown, empirical methods should be applied.
    - When available, sanity checks and mass balances should always be a part of the data processing pipeline.
- The selection of a data-driven technique is conditioned by the available data, labelled/unlabelled, and the aim of the data validation implementation.
- At the time of writing, hybrid methods (mechanistic and data-driven) are being studied more intensely to compensate the limitation of both approaches.
- Assessing the performance of anomaly detection methods strongly depends on the selected metrics (dynamic and static), which must be carefully selected given the data at hand for the specific application.

REFERENCES

Abbaspour A., Mokhtari S., Sargolzaei A. and Yen K. K. (2020). A survey on active fault-tolerant control systems. *Electronics*, **9**(9), 1513, https://doi.org/10.3390/electronics9091513

Aguado D. and Rosén C. (2008). Multivariate statistical monitoring of continuous wastewater treatment plants. *Engineering Applications of Artificial Intelligence*, **21**(7), 1080–1091, https://doi.org/10.1016/j.engappai.2007.08.004

Aguado D., Ferrer A., Ferrer J. and Seco A. (2006). Comparison of different predictive models for nutrient estimation in a sequencing batch reactor for wastewater treatment. *Chemometrics and Intelligent Laboratory Systems*, **84**, 75–81, https://doi.org/10.1016/j.chemolab.2006.03.009

Aguado D., Ferrer A., Ferrer J. and Seco A. (2007). Multivariate SPC of a sequencing batch reactor for wastewater treatment. *Chemometrics and Intelligent Laboratory Systems*, **85**(1), 82–93, https://doi.org/10.1016/j.chemolab.2006.05.003

Aguado D., Noriega-Hevia G., Ferrer J., Seco A. and Serralta J. (2022). PLS-based soft-sensor to predict ammonium concentration evolution in hollow fibre membrane contactors for nitrogen recovery. *Journal of Water Process Engineering*, **47**, 102735, https://doi.org/10.1016/j.jwpe.2022.102735

Alferes J., Tik S., Copp J. and Vanrolleghem Peter A. (2013). Advanced monitoring of water systems using in situ measurement stations: data validation and fault detection. *Water Science & Technology*, **68**(5), 1022–1030, https://doi.org/10.2166/wst.2013.302

Alferes J. and Vanrolleghem P. A. (2016). Efficient automated quality assessment: dealing with faulty on-line water quality sensors. *AI Communications*, **29**(6), 701–709, https://doi.org/10.3233/AIC-160713

Alferes J., Copp J., Weijers S. R., Cussonneau G., Fay G., Dembele A. and Vanrolleghem P. A. (2017). Validating data quality for water quality monitoring: Objective comparison of different data quality assessment approaches. ICA 2017.

Allison P. D. (2001). Missing Data 2001. Sage: Thousand Oaks.

Arteaga F. and Ferrer A. (2002). Dealing with missing data in MSPC: several methods, different interpretations, some examples. *Journal of Chemometrics*, **16**, 408–418, https://doi.org/10.1002/cem.750

Arteaga F. and Ferrer A. (2005). Framework for regression-based missing data imputation methods in on-line MSPC. *Journal of Chemometrics*, **19**(8), 439–447, https://doi.org/10.1002/cem.946

Arteaga F., Folch-Fortuny A. and Ferrer A. (2020). Missing data. In: Comprehensive Chemometrics, Elsevier, pp. 615–639. https://doi.org/10.1016/b978-0-12-409547-2.14718-3

Awhangbo L., Bendoula R., Roger J. and Béline F. (2020). Fault detection with moving window PCA using NIRS spectra for the monitoring of anaerobic digestion process. *Water Science and Technology*, **81**(2), 367–382, https://doi.org/10.2166/wst.2020.117

Baklouti I., Mansouri M., Ben Hamida A., Nounou H. and Nounou M. (2018). Monitoring of wastewater treatment plants using improved univariate statistical technique. *Process Safety and Environmental Protection*, **116**, 287–300, https://doi.org/10.1016/j.psep.2018.02.006

Baklouti I., Mansouri M., Hamida A. B., Nounou H. N. and Nounou M. (2021). Enhanced operation of wastewater treatment plant using state estimation-based fault detection strategies. *International Journal of Control*, **94**(2), 300–311, https://doi.org/10.1080/00207179.2019.1590735

Balmér P. and Hellström D. (2012). Performance indicators for wastewater treatment plants. *Water Science and Technology*, **65**(7), 1304–1310, https://doi.org/10.2166/wst.2012.014

Basseville M. (1993). Detection of abrupt changes theory and application. Prentice-Hall, Englewood Cliffs, New Jersey.

Basseville M. (2001). On fault detectability and isolability. *European Journal of Control*, **7**(6), 625–637, https://doi.org/10.3166/ejc.7.625-637

Bickel P. J. and Doksum K. A. (1981). An analysis of transformations revisited. *Journal of the American Statistical Association*, **76**(374), 296–311, https://doi.org/10.1080/01621459.1981.10477649

Bigot J. (2020). Statistical data analysis in the Wasserstein space. *ESAIM: Proceedings and Surveys*, **68**, 1–19, https://doi.org/10.1051/proc/202068001

Bikmukhametov T. and Jaschke J. (2020). Combining machine learning and process engineering physics towards enhanced accuracy and explainability of data-driven models. *Computers and Chemical Engineering* **2020**. https://doi.org/10.1016/j.compchemeng.2020.106834

Bishop C. M. (2006). Pattern Recognition and Machine Learning. Springer, United States of America.

Boger Z. (1992). Application of neural networks to water and wastewater treatment plant operation. *ISA Transactions*, **31**(1), 25–33, https://doi.org/10.1016/0019-0578(92)90007-6

Box G. E. P., Jenkins G. M., Reinsel G. C. and Ljung G. M. (2015). Time series analysis: Forecasting and control. 5th edn, John Wiley & Sons, Hoboken, New Jersey.

Bradley A. P. (1997). The use of the area under the ROC curve in the evaluation of machine learning algorithms. *Pattern Recognition*, **30**(7), 1145–1159, https://doi.org/10.1016/S0031-3203(96)00142-2

Buck S. F. (1960). A method of estimation of missing values in multivariate data suitable for use with an electronic computer. *Journal of the Royal Statistical Society Series B*, **22**, 302–306.

Camacho J. and Ferrer A. (2014). Cross-validation in PCA models with the element-wise *k*-fold (EKF) algorithm: practical aspects. *Chemometrics and Intelligent Laboratory Systems*, **131**, 37–50, https://doi.org/10.1016/j.chemolab.2013.12.003

Candès E., Li X., Ma Y. and Wright J. (2011). Robust principal component analysis? *Journal of the ACM*, **58**(3), 1–37, https://doi.org/10.1145/1970392.1970395

Carlsson B. and Zambrano J. (2016). Fault detection and isolation of sensors in aeration control systems. *Water Science and Technology*, **73**(3), 648–653, https://doi.org/10.2166/wst.2015.529

Carvajal G., Roser D. J., Sisson S. A., Keegan A. and Khan S. J. (2015). Modelling pathogen log10 reduction values achieved by activated sludge treatment using naïve and semi naïve Bayes network models. *Water Research*, **85**, 304–315, https://doi.org/10.1016/j.watres.2015.08.035

Corominas L., Villez K., Aguado D., Rieger L., Rosén C. and Vanrolleguem P. (2011). Performance evaluation of fault detection methods for wastewater treatment processes. *Biotechnology and Bioengineering*, **108**(2), 333–344, https://doi.org/10.1002/bit.22953

Corominas L., Garrido-Baserba M., Villez K., Olsson G., Cortés U. and Poch M. (2018). Transforming data into knowledge for improved wastewater treatment operation: a critical review of techniques. *Environmental Modelling & Software*, **106**, 89–103, https://doi.org/10.1016/j.envsoft.2017.11.023

Corona F., Mulas M., Haimi H., Sundell L., Heinonen M. and Vahala R. (2013). Monitoring nitrate concentrations in the denitrifying post-filtration unit of a municipal wastewater treatment plant. *Journal of Process Control*, **23**, 158–170, https://doi.org/10.1016/j.jprocont.2012.09.011

Crowe C. M. (1996). Data reconciliation – progress and challenges. *Journal of Process Control*, **6**, 89–98, https://doi.org/10.1016/0959-1524(96)00012-1

Debruyne M. and Verdonck T. (2010). Robust kernel principal component analysis and classification. *Advances in Data Analysis and Classification*, **4**(2), 151–167, https://doi.org/10.1007/s11634-010-0068-1

Devlin J., Ming-Wei C., Kenton L. and Kristina T. (2019). BERT: Pre-training of Deep Bidirectional Transformers for Language Understanding. NAACL-HLT (1).

Dutta A. and Sarkar S. (2015). Sequencing batch reactor for wastewater treatment: recent advances. *Current Pollution Reports*, **1**, 177–190, https://doi.org/10.1007/s40726-015-0016-y

Eriksson L., Byrne T., Johansson E., Trygg J. and Wikström C. (2013). Multi and Megavariate Data Analysis: Principles and Applications. ISBN 978-91-973730-5-0. Umetrics Academy.

Fan X. W., Du S. X. and Wu T. J. (2004). Rough support vector machine and its application to wastewater treatment processes. *Control and Decision*, **19**, 573–576.

Fawcett T. (2006). An introduction to ROC analysis. *Pattern Recognition Letters*, **27**(8), 861–874, https://doi.org/10.1016/j.patrec.2005.10.010

Fellini S., Vesipa R., Boano F. and Ridolfi L. (2019). Fault detection in level and flow rate sensors for safe and performant remote-control in a water supply system. *Journal of Hydroinformatics*, **22**(1), 132–147, https://doi.org/10.2166/hydro.2019.002

Ferrer A. (2014). Latent structures-based multivariate statistical process control: a paradigm shift. *Quality Engineering*, **26**(1), 72–91. Huber, P. J. and Wiley, J. Robust statistics. Wiley, New York, https://doi.org/10.1080/08982112.2013.846093

Ferrer-Riquelme A. J. (2009). Statistical control of measures and processes. In: Comprehensive Chemometrics, S. Brown, R. Tauler and R. Walczak (eds.), Elsevier, Oxford, Vol. **1**, pp. 97–126.

Folch-Fortuny A., Arteaga F. and Ferrer A. (2015). PCA model building with missing data: new proposals and a comparative study. *Chemometrics and Intelligent Laboratory Systems*, **146**, 77–88, https://doi.org/10.1016/j.chemolab.2015.05.006

Forouzanmehr F., Le Q. H., Solon K., Maisonnave V., Daniel O., Buffiere P., Gillot S. and Volcke E. I. P. (2021). Plant-wide investigation of sulfur flows in a water resource recovery facility (WRRF). *Science of the Total Environment*, **801**, 149530, https://doi.org/10.1016/j.scitotenv.2021.149530

Frank S., Lin G., Jin X., Singla R., Farthing A., Zhang L., and Granderson J. (2019). Metrics and Methods to Assess Building Fault Detection and Diagnosis Tools. https://www.nrel.gov/docs/fy19osti/72801.pdf

Friedman J., Hastie T. and Tibshirani R. (2001). The Elements of Statistical Learning. Springer Series in Statistics, New York, Vol. **1**, No. 10.

Frogner C., Zhang C., Mobahi H., Araya M. and Poggio T. A. (2015). Learning with a Wasserstein loss. In: Advances in Neural Information Processing Systems, pp. 2053–2061.

Gallegos M. T. and Ritter G. (2005). A robust method for cluster analysis. *The Annals of Statistics*, **33**(1), 347, https://doi.org/10.1214/009053604000000940

García H. L. and González I. M. (2004). Self-organizing map and clustering for wastewater treatment monitoring. *Engineering Applications of Artificial Intelligence*, **17**(3), 215–225, https://doi.org/10.1016/j.engappai.2004.03.004

Gernaey K. V., Wade M. and Jørgensen S. B. (2004). Application of recursive principal component analysis to a continuous biological phosphorus removal process. *IFAC Proceedings Volumes*, **37**(3), 363–368, https://doi.org/10.1016/S1474-6670(17)32608-3

Goldberg Y. and Hirst G. (2017). Neural Network Methods in Natural Language Processing. Morgan & Claypool Publishers. ISBN-13: 978-1627052986.

Goodfellow I., Bengio Y. and Courville A. (2016). Deep Learning (Adaptive Computation and Machine Learning Series). ISBN-13: 9780262035613.

Granata F., Papirio S., Esposito G., Gargano R. and de Marinis G. (2017). Machine learning algorithms for the forecasting of wastewater quality indicators. *Water*, **9**(2), 105, https://doi.org/10.3390/w9020105

Grung B. and Manne R. (1998). Missing values in principal component analysis. *Chemometrics and Intelligent Laboratory Systems*, **42**, 125–139, https://doi.org/10.1016/S0169-7439(98)00031-8

Gustafsson F. (2000). Adaptive Filtering and Change Detection. John Wiley & Sons Ltd, West Sussex, England.

Haimi H., Mulas M., Corona F. and Vahala R. (2013). Data-derived soft-sensors for biological wastewater treatment plants: an overview. *Environmental Modelling & Software*, **47**, 88–107, https://doi.org/10.1016/j.envsoft.2013.05.009

Haimi H., Corona F., Mulas M., Sundell L., Heinonen M. and Vahala R. (2015). Shall we use hardware sensor measurements or soft-sensor estimates? Case study in a full-scale WWTP. *Environmental Modelling and Software*, **72**, 215–229, https://doi.org/10.1016/j.envsoft.2015.07.013

Haimi H., Mulas M., Corona F., Marsili-Libelli S., Lindell P., Heinonen M. and Vahala R. (2016). Adaptive data-derived anomaly detection in the activated sludge process of a large-scale wastewater treatment plant. *Engineering Applications of Artificial Intelligence*, **52**, 65–80, https://doi.org/10.1016/j.engappai.2016.02.003

Haitovsky Y. (1968). Missing data in regression analysis. *Journal of the Royal Statistical Society Series B*, **30**, 67–82.

Hanbay D., Turkoglu I. and Demir Y. (2008). Prediction of wastewater treatment plant performance based on wavelet packet decomposition and neural networks. *Expert Systems with Applications*, **34**(2), 1038–1043, https://doi.org/10.1016/j.eswa.2006.10.030

Harrou F., Dairi A., Sun Y. and Senouci M. (2018). Wastewater treatment plant monitoring via a deep learning approach. IEEE International Conference on Industrial Technology (ICIT), 1544–1548.

Hastie T., Tibshirani R. and Friedman J. (2009). The Elements of Statistical Learning: Data Mining, Inference, and Prediction. 2nd edn, Springer Series in Statistics, Springer.

Hauduc H., Neumann M. B., Muschalla D., Gamerith V., Gillot S. and Vanrolleghem P. A. (2015). Efficiency criteria for environmental model quality assessment: a review and its application to wastewater treatment. *Environmental Modelling & Software*, **68**, 196–204, https://doi.org/10.1016/j.envsoft.2015.02.004

Henze M., Harremoes P., Jansen J. L. C. and Arvin E. (2002). Wastewater Treatment, Biological and Chemical Processes. Springer-Verlag Berlin, Heidelberg, New York.

Higham N. J. (2002). Computing the nearest correlating matrix – a problem from finance. *IMA Journal of Numerical Analysis*, **22**(3), 329–343, https://doi.org/10.1093/imanum/22.3.329

Hubert M., Rousseeuw P. and Verboven S. (2005). A fast method for robust principal components with applications to chemometrics. *Engineering Applications of Artificial Intelligence*, **60**(1–2), 101–111.

inCTRL Solutions (2018). Internal report.

inCTRL Solutions (2020). Internal report.

ISO 5725-1 (1994). Accuracy (trueness and precision) of measurement methods and results – Part 1: General principles and definitions.

IWA Task Group on Good Modelling Practices (2012). Guidelines for Using Activated Sludge Models (STR 22). IWA Publishing. eISBN: 9781780401164.

Jackson D. A. (1993). Stopping rules in principal components analysis: A comparison of heuristical and statistical approaches. *Ecology*, **74**(8), 2204–2214, https://doi.org/10.2307/1939574

Jackson J. E. (1980). Principal components and factor analysis: part I – principal components. *Journal of Quality Technology*, **12**(4), 201–213, https://doi.org/10.1080/00224065.1980.11980967

Jung D. (2015). Diagnosability Performance Analysis of Models and Fault Detectors. Linköping University Electronic Press, Linköping.

Kay S. M. (1998). Fundamentals of Statistics Signal Processing: Detection Theory. Prentice-Hall PTR, Upper Saddle River, NJ, USA.

Kazor K., Holloway R., Cath T. and Hering A. (2016). Comparison of linear and nonlinear dimension reduction techniques for automated process monitoring of a decentralized wastewater treatment facility. *Stochastic Environmental Research and Risk Assessment*, **30**(5), 1527–1544, https://doi.org/10.1007/s00477-016-1246-2

Kim J.-O. and Curry J. (1977). Treatment of missing data in multivariate analysis. *Sociological Methods & Research*, **6**, 215–240, https://doi.org/10.1177/004912417700600206

Kim M., Kim Y., Kim H., Piao W. and Kim C. (2016). Evaluation of the *k*-nearest neighbor method for forecasting the influent characteristics of wastewater treatment plant. *Frontiers of Environmental Science & Engineering*, **10**(2), 299–310, https://doi.org/10.1007/s11783-015-0825-7

Klanderman M. C., Newhart K. B., Cath T. Y. and Hering A. S. (2020). Case studies in real-time fault isolation in a decentralized wastewater treatment facility. *Journal of Water Process Engineering*, **38**, 101556, https://doi.org/10.1016/j.jwpe.2020.101556

Kohonen T. (2001). Self-Organizing Maps. Volume 30 of Springer Series in Information Sciences. Springer.

Kourti T. and MacGregor J. F. (1996). Multivariate SPC methods for process and product monitoring. *Journal of Quality Technology*, **28**(4), 409–428, https://doi.org/10.1080/00224065.1996.11979699

Kourti T., Lee J. and MacGregor J. F. (1996). Experiences with industrial applications of projection methods for multivariate statistical process control. *Computers & Chemical Engineering*, **20**, S745–S750, https://doi.org/10.1016/0098-1354(96)00132-9

Ku W., Storer R. and Georgakis C. (1995). Disturbance detection and isolation by dynamic principal component analysis. *Chemometrics and Intelligent Laboratory Systems*, **30**(1), 179–196, https://doi.org/10.1016/0169-7439(95)00076-3

Kurtoglu T., Mengshoel O. J. and Poll S. (2008). International Conference on Prognostics and Health Management.

Le Q. H. (2019). Mass-balance-based Experimental Design and Data Reconciliation for Wastewater Treatment Processes. PhD thesis, Ghent University. *Chapter 4: Data evaluation for full-scale wastewater treatment plant: linear vs bilinear mass balances*. ISBN 9789463572453.

Le Q. H., Verheijen Peter J. T., Loosdrecht Mark C. M. van and Volcke Eveline I. P. (2018). Experimental design for evaluating WWTP data by linear mass balances. *Water Research*, **142**, 415–425, https://doi.org/10.1016/j.watres.2018.05.026

Le Q. H., Verheijen Peter J. T., van Loosdrecht Mark C. M. and Volcke Eveline I. P. (2022). Application of data reconciliation to a dynamically operated wastewater treatment process with off-gas measurements. *Environmental Science: Water Research and Technology*, **8**(10), https://doi.org/10.1039/D2EW00006G

Li W., Yue H., Valle-Cervantes S. and Qin S. (2000). Recursive PCA for adaptive process monitoring. *Journal of Process Control*, **10**(5), 471–486, https://doi.org/10.1016/S0959-1524(00)00022-6

Little R. J. A. and Rubin D. B. (2002). Statistical Analysis with Missing Data. Wiley, New York.

Lloyd J. R., Duvenaud D., Grosse R., Tenenbaum J. B. and Ghahramani Z. (2014). Automatic construction and natural-language description of non parametric regression models. 28th Association for the Advancement of Artificial Intelligence (AAAI), pp. 1242–1250.

Longo S., d'Antoni B. M., Bongards M., Chaparro A., Cronrath A., Fatone F., Lema J. M., Mauricio-Iglesias M., Soares A. and Hospido A. (2016). Monitoring and diagnosis of energy consumption in wastewater treatment plants. A state of the art and proposals for improvement. *Applied Energy*, **179**, 1251–1268, https://doi.org/10.1016/j.apenergy.2016.07.043

Loosdrecht M. C. M. V., Nielsen P. H., Lopez-Vazquez, C. M. and Brdjanovic D. (2016). Experimental Methods in Wastewater Treatment.

Lopes V. V. and Menezes J. C. (2005). Inferential sensor design in the presence of missing data: a case study. *Chemometrics and Intelligent Laboratory Systems*, **78**, 1–10, https://doi.org/10.1016/j.chemolab.2004.11.004

Madron F., Veverka V. and Vanecek V. (1977). Statistical-analysis of material balance of a chemical reactor. *AIChE Journal*, **23**, 482–486, https://doi.org/10.1002/aic.690230412

Maere T., Villez K., Marsili-Libelli S., Naessens W. and Nopens I. (2012). Membrane bioreactor fouling behaviour assessment through principal component analysis and fuzzy clustering. *Water Research*, **46**(18), 6132–6142, https://doi.org/10.1016/j.watres.2012.08.027

Mallat S. G. (1989). A theory for multiresolution signal decomposition: the wavelet representation. *IEEE Transactions on Pattern Analysis and Machine Intelligence*, **11**(7), 674–693, https://doi.org/10.1109/34.192463

Mamandipoor B., Majd M., Sheikhalishahi S., Modena C. and Osmani V. (2020). Monitoring and detecting faults in wastewater treatment plants using deep learning. *Environmental Monitoring and Assessment*, **192**(2), 148, https://doi.org/10.1007/s10661-020-8064-1

Mark H. and Workman J. (2007). Chemometrics in Spectroscopy. Elsevier. ISBN: 978-0-12-374024-3. https://doi.org/10.1016/B978-0-12-374024-3.X5000-4

Markus Goldstein M. and Uchida S. (2016). A comparative evaluation of unsupervised anomaly detection algorithms for multivariate data. *PLoS ONE*, **11**(4), e0152173.

Maronna R., Martin D. and Yohai V. (2019). Robust Statistics: Theory and Methods (with R). Wiley, New York.

Marsili-Libelli S. (2006). Control of SBR switching by fuzzy pattern recognition. *Water Research*, **40**(5), 1095–1107, https://doi.org/10.1016/j.watres.2006.01.011

Metcalf and Eddy (2004). Wastewater Engineering, Treatment and Reuse. McGraw-Hill, New York.

Miletic I., Quinn S., Dudzic M., Vaculik V. and Champagne M. (2004). An industrial perspective on implementing on-line applications of multivariate statistics. *Journal of Process Control*, **14**(8), 821–836, https://doi.org/10.1016/j.jprocont.2004.02.001

Miletic I., Boudreau F., Dudzic M., Kotuza G., Ronholm L., Vaculik V. and Zhang Y. (2008). Experiences in applying data-driven modelling technology to steelmaking processes. *The Canadian Journal of Chemical Engineering*, **86**, 937–946, https://doi.org/10.1002/cjce.20090

Miletic I., Rieger L., Rosenthal A. and Schraa O. (2019a). Get More Out of Your Data – Advanced Methods for Automated Data Integration, Cleaning and Augmentation. WEFTEC 2019 Proceedings, 3266–3271.

Miletic I., Rieger L., Schraa O. and Wade M. (2019b). Measure What You Cannot Measure: Robust Multivariate Soft-Sensors for Influent PO4-P Prediction. WEFTEC 2019 Proceedings, 3272–3277.

Mitra A. (1998). Fundamentals of Quality Control and Improvement. 2nd edn, Prentice Hall, Upper Saddle River, NJ, USA.

Mitra A. (2008). Fundamentals of Quality Control and Improvement. 3rd edn, John Wiley and Sons, Hoboken, NJ, USA.

Montgomery D. G. (2019). Introduction to Statistical Quality Control. 8th edn, Wiley, New York, 768pp.

Montgomery D. C., Jennings C. L. and Kulahci, M. (2015). Introduction to Time Series Analysis and Forecasting. John Wiley & Sons. ISBN 978-1118745113.

Mulder C. de, Flameling T., Weijers S., Amerlinck Y. and Nopens I. (2018). An open software package for data reconciliation and gap filling in preparation of water and resource recovery facility modeling. *Environmental Modelling & Software*, **107**, 186–198, https://doi.org/10.1016/j.envsoft.2018.05.015

Muteki K., MacGregor J. F. and Ueda T. (2005). Estimation of missing data using latent variable methods with auxiliary information. *Chemometrics and Intelligent Laboratory Systems*, **78**, 41–50, https://doi.org/10.1016/j.chemolab.2004.12.004

Narasimhan S. and Jordache C. (2000). Data reconciliation and gross error detection: an intelligent use of process data, Annals of the New York Academy of Sciences. Gulf Publishing Company, Houston, Texas, US. https://doi.org/10.1111/j.1749-6632.1972.tb54778.x

Nelson P. R. C. (2002). Treatment of Missing Measurements in PCA and PLS Models. PhD dissertation, Department of Chemical Engineering, McMaster University, Hamilton, Ontario, Canada.

Newhart K. B., Holloway R. W., Hering A. S. and Cath T. Y. (2019). Data-driven performance analyses of wastewater treatment plants: a review. *Water Research*, **157**, 498–513, https://doi.org/10.1016/j.watres.2019.03.030

Ng A. Y. and Jordan M. I. (2002). On discriminative vs. generative classifiers: A comparison of logistic regression and naive Bayes. Advances in neural information processing systems, pp. 841–848.

Nomikos P. and MacGregor J. F. (1995). Multivariate SPC charts for monitoring batch processes. *Technometrics*, **37**(1), 41–59, https://doi.org/10.1080/00401706.1995.10485888

Nyberg M. and Frisk E. (2020). Model Based Diagnosis of Technical Processes. Linköping University.

Oakland J. S. (2003). Statistical Process Control. 5th edn, Butterworth-Heinemann, Bodmin, Cornwall, UK.

Ohmura K., Thürlimann C. M., Kipf M., Carbajal J. P. and Villez K. (2019). Characterizing long-term wear and tear of ion-selective pH sensors. *Water Science and Technology*, **80**(3), 541–550, https://doi.org/10.2166/wst.2019.301

Orlova D. Y., Zimmerman N., Meehan S., Meehan C., Waters J., Ghosn E. E. B., Filatenkov A., Kolyagin G. A., Gernez Y., Tsuda S., Moore W., Moss R. B., Herzenberg L. A. and Walther G. (2016). Earth mover's distance (EMD): A true metric for comparing biomarker expression levels in cell populations. PLoS ONE, https://doi.org/10.1371/journal.pone.0151859

Ozenne B., Subtil F. and Maucort-Boulch D. (2015). The precision–recall curve overcame the optimism of the receiver operating characteristic curve in rare diseases. *Journal of Clinical Epidemiology*, **68**, 855–859, https://doi.org/10.1016/j.jclinepi.2015.02.010

Ozyurt D. B. and Pike R. W. (2004). Theory and practice of simultaneous data reconciliation and gross error detection for chemical processes. *Computers & Chemical Engineering*, **28**, 381–402, https://doi.org/10.1016/j.compchemeng.2003.07.001

Peters M. E., Neumann M., Iyyer M., Gardner M., Clark C., Lee K. and Zettlemoyer L. (2018). Deep contextualized word representations. Proceedings of NAACL-HLT 2018, pp. 2227–2237.

Pimentel M. A. F., Clifton D. A., Clifton L. and Tarassenko L. (2014). A review of novelty detection. *Signal Processing*, **99**, 215–249, https://doi.org/10.1016/j.sigpro.2013.12.026

Qin S. J. (2012). Survey on data-driven industrial process monitoring and diagnosis. *Annual Reviews in Control*, **36**(2), 220–234, https://doi.org/10.1016/j.arcontrol.2012.09.004

Rajaraman A. and Ullman J. D. (2011). 'Data Mining' (PDF). Mining of Massive Datasets, pp. 1–17.

Ramsay L. O. and Silverman B. W. (2005). Functional Data Analysis. 2nd edn, Springer Series in Statistics, https://www.springer.com/gp/book/9780387400808.

Rasmussen C. E. and Williams C. K. I. (2005). Gaussian Process for Machine Learning (Adaptive Computation and Machine Learning series). The MIT Press, Cambridge, Massachusetts, London, England.

Rieger L., Takacs I., Villez K., Siegrist H., Lessard P., Vanrolleghem P. A. and Comeau Y. (2010). Data reconciliation for wastewater treatment plant simulation studies-planning for high-quality data and typical sources of errors. *Water Environment Research*, **82**, 426–433, https://doi.org/10.2175/106143009X12529484815511

Roberts S. W. (1959). Control chart tests based on geometric moving averages. *Technometrics*, **1**(3), 239–250, https://doi.org/10.1080/00401706.1959.10489860

Rosén C. (2001). A Chemometric Approach to Process Monitoring and Control – with Applications to Wastewater Treatment Operation. PhD thesis, Lund University, Department of Industrial Electrical Engineering and Automation. Lund, Sweden.

Rosen C. and Lennox J. (2001). Multivariate and multiscale monitoring of wastewater treatment operation. *Water Research*, **35**(14), 3402–3410, https://doi.org/10.1016/S0043-1354(01)00069-0

Ruano M. V., Ribes J., Seco A. and Ferrer J. (2012). An advanced control strategy for biological nutrient removal in continuous systems based on pH and ORP sensors. *Chemical Engineering Journal*, **183**, 212–221, https://doi.org/10.1016/j.cej.2011.12.064

Rubin D. B. (1987). Multiple Imputation for Nonresponse in Surveys. Wiley & Sons, New York.

Rubner Y., Tomasi C. and Guibas L. J. (2000). The earth mover's distance as a metric for image retrieval. *International Journal of Computer Vision*, **40**(2), 99–121, https://doi.org/10.1023/A:1026543900054

Russo S., Besmer M. D., Blumensaat F., Bouffard D., Disch A., Hammes F., Hess A., Moritz L., Blake M., Camille M., Eberhard M., Viet T. and Villez K. (2021). The value of human data annotation for machine learning based anomaly detection in environmental systems. *Water Research*, **206**, 117695, https://doi.org/10.1016/j.watres.2021.117695

Russo (2022). in preparation

Salman B. and Salem O. (2012). Modeling failure of wastewater collection lines using various section-level regression models. *Journal of Infrastructure Systems*, **18**(2), 146–154, https://doi.org/10.1061/(ASCE)IS.1943-555X.0000075

Samuelsson O., Björk A., Zambrano J. and Carlsson B. (2017). Gaussian process regression for monitoring and fault detection of wastewater treatment processes. *Water Science and Technology*, **75**(12), 2952–2963, https://doi.org/10.2166/wst.2017.162

Samuelsson O., Björk A., Zambrano J. and Carlsson B. (2018). Fault signatures and bias progression in dissolved oxygen sensors. *Water Science and Technology*, **78**(5–6), 1034–1044, https://doi.org/10.2166/wst.2018.350

Samuelsson O., Zambrano J., Björk A. and Carlsson B. (2019). Automated active fault detection in fouled dissolved oxygen sensors. *Water Research*, **166**, https://doi.org/10.1016/j.watres.2019.115029

Sánchez-Fernández A., Baldán F. J., Sainz-Palmero G. I., Benítez J. M. and Fuente M. J. (2018). Fault detection based on time series modelling and multivariate statistical process control. *Chemometrics and Intelligent Laboratory Systems*, **182**, 57–69, https://doi.org/10.1016/j.chemolab.2018.08.003

Schafer J. L. (1997). Analysis of Incomplete Multivariate Data. 1st edn, Chapman and Hall/CRC, Boca Raton.

Schölkopf B., Smola A. and Müller K.-R. (1998). Nonlinear component analysis as a kernel eigenvalue problem. *Neural Computation*, **10**(5), 1299–1319, https://doi.org/10.1162/089976698300017467

Schneider M. Y., Carbajal J. P., Furrer V., Sterkele B., Maurer M. and Villez K. (2019). Beyond signal quality: the value of unmaintained pH, dissolved oxygen, and oxidation–reduction potential sensors for remote performance monitoring of on-site sequencing batch reactors. *Water Research*, **161**, 639–651, https://doi.org/10.1016/j.watres.2019.06.007

Sousa A. C., Lucio M. M., Bezerra Neto, O. F., Marcone G. P., Pereira A. F., Dantas E. O., Fragoso W. D., Araujo M. C. and Galvão R. K. (2007). A method for determination of COD in a domestic wastewater treatment plant by using near-infrared reflectance spectrometry of seston. *Analytica Chimica Acta*, **588**(2), 231–6, https://doi.org/10.1016/j.aca.2007.02.022

Tanner M. A. and Wong W. H. (1987). The calculation of posterior distribution by data augmentation (with discussion). *Journal of the American Statistical Association*, **82**, 528–540, https://doi.org/10.1080/01621459.1987.10478458

Thomann M., Rieger L., Frommhold S., Siegrist H. and Gujer W. (2002). An efficient monitoring concept with control charts for on-line sensors. *Water Science and Technology*, **46**(4–5), 107–116, https://doi.org/10.2166/wst.2002.0563

Thürlimann C. M., Dürrenmatt D. J. and Villez K. (2015). Evaluation of qualitative trend analysis as a tool for automation. *Computer Aided Chemical Engineering*, **37**, 2531–2536, https://doi.org/10.1016/B978-0-444-63576-1.50116-3

Thürlimann C. M., Dürrenmatt D. J. and Villez K. (2018). Soft-sensing with qualitative trend analysis for wastewater treatment plant control. *Control Engineering Practice*, **70**, 121–133, https://doi.org/10.1016/j.conengprac.2017.09.015

Tian Z. X., Jiang J. P., Guo L. and Wang P. (2012). Anomaly detection of municipal wastewater treatment plant operation using support vector machine. International Conference on Automatic Control and Artificial Intelligence (ACAI 2012), pp. 518–521.

Vachtsevanos G., Lewis F., Roemer M., Hess A. and Wu B. (2006). Fault Diagnosis and Prognosis Performance Metrics. In: Intelligent Fault Diagnosis and Prognosis for Engineering Systems, John Wiley & Sons, Inc., pp. 355–399. ISBN: 978-0-470-11784-2. https://doi.org/10.1002/9780470117842.ch7

van der Heijden R. T. J. M., Heijnen J. J., Hellinga C., Romein B. and Luyben K. C. A. M. (1994). Linear constraint relations in biochemical reaction systems. 1. Classification of the calculability and the balanceability of conversion rates. *Biotechnology and Bioengineering*, **43**, 3–10, https://doi.org/10.1002/bit.260430103

Villez K., Rosen C., Stjin V.-H., Yoo C. K. and Vanrolleghem, P. (2005). On-line dynamic monitoring of the SHARON process for sustainable nitrogen removal from wastewater. *Computer-Aided Process Engineering*, **20**: 1297–1302, https://doi.org/10.1016/S1570-7946(05)80058-6

Villez K., Rosén C., D'hooge E. and Vanrolleghem P. A. (2010). Online phase length optimization for a sequencing batch reactor by means of the Hotelling's T^2 statistic. *Industrial & Engineering Chemistry Research*, **49**(1), 180–188, https://doi.org/10.1021/ie801907n

Villez K., Vanrolleghem P. A. and Corominas L. (2016). Optimal flow sensor placement on wastewater treatment plants. *Water Research*, **101**, 75–83, https://doi.org/10.1016/j.watres.2016.05.068

Vitorino D., Coelho S. T., Santos P., Sheets S., Jurkovac B. and Amado C. (2014). A random forest algorithm applied to condition-based wastewater deterioration modeling and forecasting. *Procedia Engineering*, **89**, 401–410, https://doi.org/10.1016/j.proeng.2014.11.205

Von Sperling M., Verbyla E. M. and Oliveira S. M. A. C. (2020). Assessment of Treatment Plant Performance and Water Quality Data. IWA Publishing.

Wade M. J., Sánchez A. and Katebi M. R. (2005). On real-time control and process monitoring of wastewater treatment plants: real-time process monitoring. *Transactions of the Institute of Measurement and Control*, **27**(3), 173–193, https://doi.org/10.1191/0142331205tm140oa

Walczak B. and Massart D. L. (2001). Dealing with missing data. Part I. *Chemometrics and Intelligent Laboratory Systems*, **58**, 15–27, https://doi.org/10.1016/S0169-7439(01)00131-9

Walther G., Zimmerman N., Moore W., Parks D., Meehan S., Belitskaya I., Pan J. and Herzenberg L. (2009). Automatic clustering of flow cytometry data with density-based merging. Advances in Bioinformatics. https://doi.org/10.1155/2009/686759

Wang X., Kruger U. and Irwin G. (2005). Process monitoring approach using fast moving window PCA. *Industrial & Engineering Chemistry Research*, **44**(15), 5691–5702, https://doi.org/10.1021/ie048873f

Wang X., Fan Y., Huang Y., Ling J., Klimowicz A., Pagano G. and Li B. (2021). Solving sensor reading drifting using denoising data processing algorithm (DDPA) for long-term continuous and accurate monitoring of ammonium in wastewater. *ACS ES&T Water* **1**(3), 530–541, https://doi.org/10.1021/acsestwater.0c00077

Westerhuis J. A., Gurden S. P. and Smilde A. K. (2000). Generalized contribution plots in multivariate statistical process monitoring. *Chemometrics and Intelligent Laboratory Systems*, **51**(1), 95–114, https://doi.org/10.1016/S0169-7439(00)00062-9

Wold S. (1978). Cross validatory estimation of the number of components in factor and principal component analysis. *Technometrics*, **20**(4), 397–405, https://doi.org/10.1080/00401706.1978.10489693

Wold S., Albano C., Dunn W. J., Esbensen K., Hellberg S., Johansson E. and Sjöström M. (1983). Pattern recognition: finding and using regularities in multivariate data. In: Food Research and Data Analysis, H. Martens and H. Russwurm Jr. (eds.), Applied Science Pub, London, Vol. **3**, pp. 183–185.

Wold S., Sjöström M. and Eriksson L. (2001). PLS-regression: a basic tool of chemometrics. *Chemometrics and Intelligent Laboratory Systems*, **58**(2), 109–130, https://doi.org/10.1016/S0169-7439(01)00155-1

Yoo C., Lee J.-M., Lee I.-B. and Vanrolleghem P. (2004). Dynamic monitoring system for full-scale wastewater treatment plants. *Water Science and Technology*, **50**(11), 163–171, https://doi.org/10.2166/wst.2004.0684

Zambrano J., Samuelsson O. and Carlsson B. (2019). Machine learning techniques for monitoring the sludge profile in a secondary settler tank. *Applied Water Science*, **9**(6), 146, https://doi.org/10.1007/s13201-019-1018-5

Zhang D., Hølland E. S., Lindholm G. and Ratnaweera H. (2018). Hydraulic modeling and deep learning based flow forecasting for optimizing inter catchment wastewater transfer. *Journal of Hydrology*, **567**, 792–802, https://doi.org/10.1016/j.jhydrol.2017.11.029

Zhou P., Li Z., Snowling S., Baetz B., Na D. and Boyd G. (2019). A random forest model for inflow prediction at wastewater treatment plants. *Stochastic Environmental Research and Risk Assessment*, **33**, 1781–1792, https://doi.org/10.1007/s00477-019-01732-9

doi: 10.2166/9781789061154_0241

Chapter 7
Conclusions and outlook

D. Aguado[1], J. Alferes[2], A. Menniti[3], Q. Plana[4,5], M. V. Ruano[6], O. Samuelsson[7], K. Villez[8] and E. Zegers[9]

[1]CALAGUA – Unidad Mixta UV-UPV, Institut Universitari d'Investigació d'Enginyeria de l'Aigua i Medi Ambient – IIAMA, Universitat Politècnica de València, València, Spain
[2]VITO – WaterClimateHub, Ostende, Belgium
[3]Clean Water Services, Tigard, OR, USA
[4]modelEAU, Université Laval, Québec, QC, Canada
[5]Eurecat – Technology Center of Catalonia, Barcelona, Spain
[6]CALAGUA – Unidad Mixta UV-UPV, Departament d'Enginyeria Química, Universitat de València, Burjassot, Spain
[7]IVL Swedish Environmental Research Institute, Stockholm, Sweden
[8]Oak Ridge National Laboratory, Oak Ridge, TN, USA
[9]City of Toronto, Toronto, ON, Canada

In this chapter, we provide notes on the state of metadata collection and organization practices and reflect on the achievements of the MetaCO task group. We also include an utility perspective on how to use this report in daily data management practice and a short guide is provided to show how metadata practices can be initiated. The chapter, and the report, is concluded within an outlook, sketching future progress and opportunities on the horizon.

7.1 OBSERVED METADATA CHALLENGES

This report has provided practical and technical advice for collecting and using metadata, that is, *how* metadata can be recorded and made useful. Less focus has been on *why* metadata should be a strategic topic as part of any digital roadmap. Clear answers are needed to both questions to motivate progress throughout the WRRF organization. Here, we pinpoint some key challenges related to these questions, which have been identified during the writing of this report.

7.1.1 The why question: how to motivate metadata

Metadata are seldom valuable on its own. Rather, their value lies in making data useful for a particular purpose. Indeed, it is not enough that metadata describe sensor signals, but the collected metadata should also support a specific *purpose* within the organization. With well-managed sensor signals and clearly defined purpose, metadata can provide value to an organization (Figure 7.1). In contrast, if any

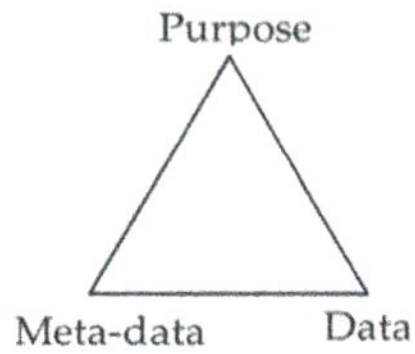

Figure 7.1 Metadata value triangle indicates the needed components for producing useful insights from data.

one of the three components in Figure 7.1 are lacking, this will limit the value of the available data and not provide useful answers in many cases.

Knowing the purpose of the sensor signals is critical for deciding which metadata to record. This should be straightforward for a specific engineering challenge but will be challenging when considering future data use (unknown unknowns). Chapter 3 is a great starting point. In most cases, we recommend that data management plans focus initially on well-defined questions (known unknowns). Commonly, we expect this to require additional attention to metadata describing sensor maintenance actions and sensor data quality.

7.1.1.1 Data sharing

A challenge related to the use of data-intensive models is the increasing need to share data - also outside the fence. Non-water experts will likely be part of the data analysis, where black-box models (such as described in Chapter 6), are used for decision making. Structured metadata are then needed to explain basic sensor data assumptions that are implicitly understood for an experienced water professional. An example would be the expected drift and unreliable data from unmaintained sensors.

The world outside the fence will also open new possibilities providing new data sources. Just as the non-water experts need guidance to interpret water data, the utilities will need a structured way to analyse incoming data in relation to its metadata. Having a systematic metadata strategy in place can facilitate this external data usage.

7.1.1.2 Fostering a metadata organization

Metadata are already used to facilitate data-sharing within the organization, where sharing often means reusing data for a different purpose. Basic metadata such as units and data quality can reduce the risk for making wrong decisions by staff that are unfamiliar with how the data were generated. A common example is the short organizational memory of either sensor or process disturbances. Without a structured and future-proof metadata collection, such critical annotations will fade away with staff rotations and effectively only be valid for a short time (weeks).

7.1.1.3 Metadata costs

It can be tempting to store metadata in logbooks or spreadsheets since it can be set up quickly and at a low cost. However, this type of manual registration is time-consuming and prone to error during registration and retrieval. Automated metadata collection is a recommend. However, this usually requires upfront investment in tailored hardware, software, and training. The return of such investments is rarely accrued in the short term, but rather improves the data value over time. Motivating such costs with a fast return of the investment is of course difficult, especially if the purpose is vague.

When trying to implement data or digital transformation plans, one of the main bottlenecks is to communicate the value of the data to the management level when considering this next to other operational priorities. Being able to translate the added value of metadata and in general of data-based applications into business figures is still a big challenge.

An alternative viewpoint is that metadata, to some extent, are mandatory and that, omitting the effort to collect them in business decisions will cause delay and increased cost to projects. Thus, any metadata costs should be covered in the initial decision when deciding to invest in sensors or digitalization efforts.

An upside with metadata is that data-driven models will likely be cheaper to develop. A supervised model (data with metadata, recall Chapter 6) is generally obtained with less data and at a lower training cost, in contrast to unsupervised models based on only unlabelled data. Indeed, modern application machine learning typically starts by exploring large-dimensional data set to discover signals that allow accurate predictions or forecasts. Metadata can assist the data analysis in automatically selecting the right signals and time periods for specific prediction or forecasting purposes.

7.1.1.4 Reduced uncertainty and improved confidence in data-driven decision making

Finally, a strong motivation for why to use metadata is to reduce the uncertainty in decisions based on data. As an example, consider a plant upgrade where the historic sludge production is to be used to expand the nitrification volume. If the TSS-sensor validation metadata indicates a high accuracy, less safety margins are needed to cover the measurement uncertainty. In an opposite situation, historic sensor validation data are lacking, and therefore a dedicated sensor assessment campaign is executed. The assessment indicates an 30% offset in several sensors, in turn indicating that the historic MLSS likely has been overestimated. A large safety factor is then needed to cover for the lack of trust in the historical data. This can be avoided by systematic metadata collection and management.

7.1.2 Challenges with implementing metadata strategies

7.1.2.1 Communicating the unique selling points with metadata

Not all situations are as clear as the TSS-sensor example. In fact, the most difficult challenge is probably how to provide a unique selling point for metadata since it is usually utility-specific and formulated for a loosely defined purpose (Figure 7.1). In general, it is difficult to price data unless the cost of not measuring is high. Nevertheless, a clear purpose with an expected return of investment will help motivate investments in metadata.

7.1.2.2 Fostering a metadata culture

A just as important challenge is to foster a metadata culture within the organization. This involves both social human interactions, as well as human-to-machine aspects. During discussions in the utility groups, it has become clear that new roles are needed. A good example is data stewards that take responsibility to maintain the data structure and quality, that is, data governance. Understanding and accepting the consequences for, that data-driven applications require maintenance, just as a physical pump, is a first step towards a data-driven organization.

A different social aspect is the data language, which can be confusing with varying explanations depending on the context. This report has tried to clarify the commonly accepted terminology needed for communicating metadata questions (see Chapter 2 and the definition summary in the Glossary).

7.1.2.3 Technical gaps for implementing good metadata practices

Technical challenges still constitute a barrier to good metadata collection and organization practices. Commercial systems for maintaining and integrating WRRF metadata are lacking, which requires tailored system integrations coming with even higher demands on specialists. The most evident technical challenges include:

- **Automating metadata collection:** This can be as simple as noting time for sensor cleaning or keeping track of the actual sensor location.
- Metadata storage. Integrating and exchanging data between different systems is key for **creating a useful metadata database:** As emphasized in Chapter 4, both ingestion and retrieval need to run smoothly, which is not standard today.

- **Lack of standardization:** The wide range of commercial solutions available for data storage and data management (including communication protocols, databases, etc.) can be overwhelming, in turn choosing a suitable setup difficult.

7.2 HOW THIS REPORT CAME TO BE AND ADDRESSES OBSERVED METADATA CHALLENGES

The original ideas for the MetaCO task group arose from a commonly experienced itch when using someone else's data. In many machine learning and other modelling projects, it is common to spend a significant amount of time in collecting metadata. This often proves essential to acquire data that are simultaneously relevant, voluminous, and of sufficient quality quite often, this requires conversation with multiple staff members, covering aspects of information technology (IT), sensor management, operations, and control. Even when executed carefully, the produced metadata may be uncertain still, thus requiring subjective decisions by data analysts and modellers in subsequent steps. This is the challenge that motivated our initiative.

In the proposal stage, we identified four challenges that we hoped to address with this STR, which are: (a) a lack of understanding of what metadata is or can be; (b) a lack of understanding of how metadata is best organized, (c) a lack of shared practices for metadata describing the condition of a sensor, and (d) a lack of a comprehensive overview of data analytical methods that can describe sensor signals and how they can be used to produce metadata. These challenges are addressed in Chapters 3–6 of this STR. We discuss these chapters next and aim to give a subjective assessment whether each of these chapters has achieved its goal and how.

7.2.1 Understanding what metadata is or can be

After Chapter 2, which describes the scope of the STR at a high level, Chapter 3 dives into the question *What is metadata?*. We took a comprehensive strategy by defining sensor signal metadata in its broadest sense of the word. It proved straightforward to collect a comprehensive list of types of metadata that satisfy our definition through internal discussions. We found that sensor metadata is sometimes limited to a description of the hardware and where, when, and how this hardware is used. This type of metadata could be derived from a combination process flow sheet, piping and instrumentation diagrams (P&ID), and bill-of-materials. In contrast to this narrow definition, our definition also includes data quality assessments, contextual information, maintenance records, and annotations as metadata. This allows anyone to make a sensible selection of required metadata based on identified priorities or project needs.

7.2.2 Understanding of how metadata is best organized

In Chapter 4, we answered the question *How to manage metadata?*. The chapter functions as an introduction to both legacy and modern systems for data storage and management. Central to this chapter is the idea that adequate data governance is key to ensuring longevity and accuracy of sensor signal data. Initially, collecting wastewater specific information proved difficult as the design and operation of database systems is often implemented through third-party vendors and service companies. In addition, this type of technology tends to be generic and not specialized to wastewater treatment operations specifically. This generality means that satisfying the need to manage metadata requires careful explanation and discussion during database design and implementation. For this reason, the chapter provides an overview of database technologies, recommendations for implementation and use, as well as some database design examples. This should permit process experts and plant managers to initiate meaningful conversations with database engineers and other IT professionals.

7.2.3 Shared practices for metadata describing the condition of a sensor

Sensors are often considered the most sensitive component for the design, tuning, and use of dynamic process models and advanced process monitoring and control systems. Despite this widely recognized need, we found that sensor quality management is often based on practices that are not documented well. For this reason, our original idea for Chapter 5 was to collect standardized operating procedures (SOPs) for sensor validation from multiple utilities and then condense these into unified procedures, one for each sensor type. Such sensor validation produces are regularly used today to compare a sensor signal to a reference measurement or a medium with a known value for the measured variable (e.g., standard solution or spiked water sample). We found that execution is often considered simple but that the comparative results are rarely recorded for further analysis or inspection of trends. In addition, we discovered that the procedures for sensor validation are rarely available in written form.

This situation appears very different from a laboratory setting, which typically follows written SOPs or analytical standards that are revised on a regular basis. SOPs for sensor validation should be very helpful however to bring sensor data quality assessment to a level that is comparable to laboratory data and keep records of sensor performance. We expect this to remain a pain point in the immediate future as the systematic execution of SOPs for sensor validation may be complicated or tedious.

While a consensus among utilities regarding the use and implementation of such SOPs is still lacking, Chapter 5 includes multiple SOPs for sensor validation. These were based on initial information provided by the utilities regarding their existing procedures and have been modified with the specific aim of rigorously collecting metadata describing sensor signal accuracy, bias, precision, and more. The chapter offers a starting point for sensor validation by providing design principles and templates for a select set of sensors. We encourage anyone to try these out and contribute to improved sensor management practices.

7.2.4 A comprehensive overview of data analytical methods that can describe sensor signals and how they can be used to produce metadata

Data analytical techniques for sensor validation tend to fall into two groups. One is based on physico-chemical knowledge about the monitored process and focuses on the use of redundant relationships between sensors to discover anomalies and data of poor quality. Flow balances and mass balances are regularly used today for data validation. The other group is based on a wide range of data-intensive techniques, based on statistical process control techniques, data mining and pattern recognition, and/or machine learning. These two groups of techniques are rarely presented together as they are rooted in distinct scientific communities. For this reason, Chapter 6 was focused on providing an overview of the most common techniques used for data validation in a single text. As expected, many data experts wanted to contribute, and the chapter turns out lengthy yet comprehensive as a result. To our knowledge, each of the presented techniques has been tested at least once with full-scale process data. Based on preliminary reports, we provided guidance on the selection and implementation of these techniques for data validation purposes. It therefore serves as an excellent starting point for data validation at full scale.

7.3 UTILITY GROUP PERSPECTIVE: GETTING STARTED WITH DATA

Creating and managing metadata falls under the larger umbrella of data management, which is a challenge for many utilities. For this reason, metadata discussions among utilities often turn into larger discussions about sensor management, data management systems and processes, and enterprise data architecture. Discussions among the utility advisory group during the development of this STR highlighted three common themes for challenges and success factors with metadata management:

- Creating a culture that supports and values data quality.
- Creating systems to manage data quality
- Data architecture and integration.

7.3.1 Creating culture: data governance

A data governance framework is an essential part of making sure utility data are fit-for-purpose. Data governance is the process of setting policies about how data are gathered, stored, and processed. Basic elements of data governance generally include:

- Assigning responsibility for each subset of data.
- Communicating expected data quality to data consumers (see, e.g., Chapter 3),
- Documenting data sources, calculations or other data manipulations, and making this information available to data consumers (see, e.g., Chapter 4).

Once data governance policies go into implementation, typical data governance roles might include:

- **Data steward:** Person who represents data assets and will be responsible for investigating data issues and maintaining the data source (generally assigned for all datasets).
- **Data validator:** Person who checks the accuracy, validity, consistency, integrity, timeliness, and completeness of the data (generally only applicable for select datasets that require a higher quality).
- **Report owner:** Person responsible for developing a report, making modifications to it, answering any questions about the report, and keeping the report up to date.

Utilities with well-developed data governance policies that have been translated into ongoing implementation systems to bring those policies into execution typically have a strong data culture where the value of data is communicated and where managers support the effort involved. This cultural component across an organization is a critical success factor to allowing the benefits of the technical ideas in this report to be fully realized.

7.3.2 Systems for data quality management

Many utilities value high-quality data and recognize the importance of metadata. However, even utilities with significant experience successfully deploying sensors struggle with managing those sensors and assuring the quality of the data they produce. While online instruments can be a powerful tool for operations, a successful sensor installation generally requires all the following conditions to be satisfied:

- Confirmation that an online instrument is required (continuous or near-continuous data are important for the application; the effort required to maintain an online instrument is justified by the value of having real-time information).
- Judicious sensor selection (references and/or instrument trials) and representative placement of the sensor in the process.
- Appropriate maintenance processes that are systematized with clear definition of roles and responsibilities among work groups.
- Routine sensor validation (e.g., comparing the sensor reading against a reference, such as a laboratory measurement, see Chapter 5).
- Ability to share sensor data and verification data easily and widely within the organization.
- Data governance policies to guide those efforts.

7.3.3 Data architecture and integrations

Data generated by utilities generally resides in a few different systems. Some of the typical systems most relevant to sensor data management and metadata include:

- **SCADA historians** store data generated by the SCADA system, including sensor data.
- **Works management systems (WMS)/asset management systems** store data about equipment calibration and maintenance performed by maintenance staff.

- **Laboratory information management systems (LIMS)** store results of laboratory analyses.
- **Data management system** is a label that can describe many quite different types of systems, but some utilities have an aggregation system that brings in information from other systems to enable more intelligent operation. This kind of system might include:
  - Data querying capabilities
  - Storage of data manually collected by operators (e.g., handheld sensor readings)
  - Reports
  - Real-time dashboards

Aggregating data from disparate sources so it can be analysed holistically and efficiently is an important step towards improving the use of metadata to identify and diagnose problems and to increase awareness of data quality. The inability to analyse and visualize data from a range of sources is a commonly cited challenge for creating and managing metadata successfully. There are a range of approaches for integrating data sources (see below, Section 7.4). Whatever the architecture, successful leveraging of metadata, and efficient management of sensors in general, relies on integration between these different data sources.

Examples of useful integrations include:

- **LIMS and SCADA:**
  - Relevant to sensor data validation (see Chapter 5), particularly to compare sensor readings with lab analysis results. Digitizing and automating this validation process can facilitate the storage and communication of the sensor validation actions and results to multiple roles in the organization.
- **SCADA and data management system:**
  - Again, relevant to sensor data validation (e.g., comparing sensor readings against a reference value, as explained in Chapter 5, or automatic outlier detection, as explained in Chapter 6). Digitizing and automating this validation process minimizes utility effort and disseminates the information to more roles in the organization.
- **WMS and data management system:**
  - Record of maintenance activities were completed (what and when).
  - Record of calibration (type of calibration, date and time, calibration parameters).
- **Information in SCADA/OT system that is of interest in the works management system:**
  - Instrument run hours
  - Diagnostic information from the sensor on status.

Not only is it important that systems be integrated, but each integration must be managed over time and success hinges on the degree of effort required to do this. Exemplary questions that are relevant to long-term ownership include:

- How easy is it to maintain the integration?
- What happens when physical assets are added or replaced?

Historical data of assets that have been removed may still prove useful in the future.

7.3.4 Digital maturity

All utilities today are generally on a path towards more and/or better use of digital tools (often called a *digital transformation process*). The stage of an individual utility in its journey is sometimes characterized as its *digital maturity*. Digital maturity can be characterized as a utility's ability to effectively harness technology and digital tools. Although a comprehensive discussion of utility digital transformation is beyond the scope of this book, we touch on the concept here because digital maturity underlies and supports implementation of the concepts explored in the sections above, like data governance and integration between IT systems. Ultimately, a higher level of digital maturity would support more effective use of metadata.

In the last few decades, the amount of data collected by utilities has increased dramatically. SCADA/DCS systems were introduced to treatment facilities in the mid-1900s, and the number of data records they collect continues to increase. In addition to SCADA data, utilities also collect laboratory data, water metering and customer usage data, engineering and construction data, and asset performance and maintenance data. With this explosion of data, however, have come challenges at every level. On the most basic level, maintaining instruments at facilities is a challenge that has only grown with the number of sensors installed. As we move up the chain of data analytics and visualization, lack of integration between source systems has limited the possibilities. With all the data being generated, many utilities reported that they lack easy access to their data and lack confidence in its quality.

A recent study from the Water Research Foundation (2022) explores the concept of digital maturity, and emphasizes that it is not just about technology, but underlying business process and people elements. The following are some factors relevant to enterprise digital maturity:

- Staffing levels sufficient to support digital applications.
- Staffing competencies to support digital applications (e.g., business analysts, data analysts, enterprise architects, cybersecurity specialists).
- Defined governance frameworks for enterprise technology applications.
- Existence and effective implementation of data governance and document management policies.
- IT procurement policies.
- Internal training and support resources for digital applications.
- Links between water utility IT and corporate IT groups.
- Links between operational technology (OT) and IT groups.
- Links between job roles (operators, engineers, senior management).

As noted above, all utilities are at one stage or another of a digital transformation. Individual utilities might be at very different places along that path, and each utility's path is not the same. No matter where a utility is in the digital transformation process, getting started creating systems to manage data quality and the associated metadata is a worthwhile effort. With experience, these processes can be automated or otherwise reworked to be made more efficient. Gaining experience will allow a broader basis of experience to guide future decisions in a digital journey.

7.4 DATA TRANSFORMATION PLAN

The above needs expressed from the utility side were used to draft a data transformation plan focused on metadata. Data transformation within the wastewater treatment plants is part of the bigger picture on digital transformation, and actually a core part of the digital journey that utilities, and in more larger sense, the water sector has to accomplish to face the current challenges: ageing infrastructure, decrease of water sources, population growth, climate-change-related events, and stricter governmental regulations.

In practice, discussions about creating and managing metadata are normally part of broader debates about data management, sensors and process management, information systems, data flow, and enterprise data architecture. For water facilities, this digital journey is not easy. Competitive priorities make it difficult to know which direction to take. Several initial questions may be posed:

- How to start the data transformation?
- What investments or upgrades from the current situation are needed?
- What is the right priority for the different steps to be taken?
- What is the budget available for the data/digital transformation?
- How would this affect the general business plan?

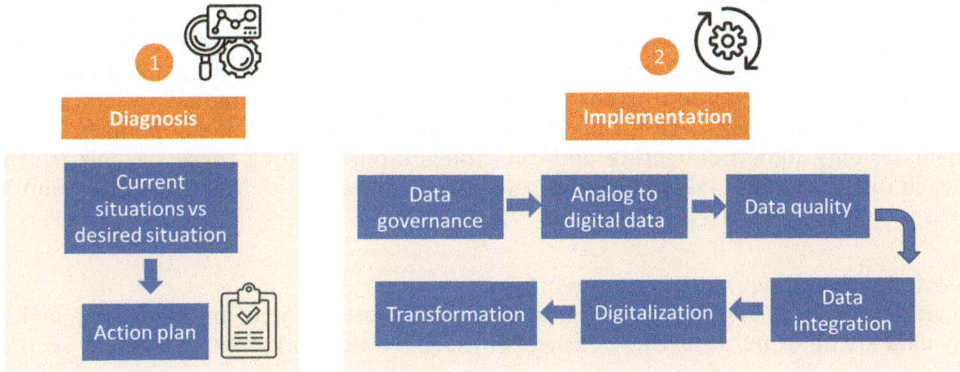

Figure 7.2 Guideline with main steps towards a data and metadata transformation.

- How can data and metadata be brought along without interfering with the main management goals? (Meet standards, reduce costs, minimize environmental impact.)

The latest should be seen as an opportunity and a disruptive action, instead of a threat, but in many cases, this is not the case or at least not enough clear at the operational and management level. Being able to measure the retour of investment (ROI) is not an easy task but would justify investments to reduce costs and improve performance in the long term.

As a guideline, any data or digital transformation at the utility level should follow a two-step procedure as described below and is illustrated in Figure 7.2.

7.4.1 Step 1: diagnosis

The first step is to critically evaluate the current situation and specific needs, what may be dependent on the plant. This will allow to define the *as-is-situation* and define the *to-be-situation*. Aspects to be considered are as follows:

- What is the size, capacity, and age of the existing infrastructure?
- What is the level of instrumentation control and automation?
- What is the level of digital maturity?
- What is the internal expertise? Does additional technical profiles or specific trainings required?
- What are the current regulations? Community, country, region wise.
- What is the final goal and the timeline adapted to the reality of the plant?

Answers to these questions should provide as a result of a customized action plan to meet the desired goals, considering limitations and local conditions.

7.4.2 Step 2: implementation

Once the current situation has been identified and goals are setup, the adoption of the digital and data journey encompasses the following six steps.

7.4.2.1 Transforming manual data and metadata collection into automated processes

This entails the transformation of analog information into a digital form and applies for all the types of data and metadata described in Chapters 2 and 3. A good example consists of the data collected by operators (normally with pen and paper) while performing sensors maintenance activities. If this information is available as digital data, it can be made more accessible, to further make easier analysis and accelerate decision making, what at the end will be reflected in avoiding risks and costs.

As an example – traditional ways of managing sensor data may require a significant and ongoing routine effort by utility staff. The more routine effort required, the less likely a data quality system is to be maintained over time. However, it may be difficult to know what elements can or should be automated to save time without experience implementing a more manual system to start. The experiences can provide a use case for planning changes to data governance policies, data quality management systems, data architecture and data integration. Building these systems in ways that work for each different utility takes time and patience but the result of having greater confidence in the data used to make decisions is worthwhile.

7.4.2.2 Design a data governance framework

This involves setting internal policies about how data are gathered, stored, and processed, making sure that utility data are fit-for-purpose. Basic elements of such a framework are explained in Section 7.3.

7.4.2.3 Setup a system for data quality management

Independently of the level of instrumentation, data collected from sensors will only be useful if the quality of the data are good enough for the intended application. Many utilities value high-quality data and recognize the importance of metadata; however, the reality is that numerous utilities also struggle with assuring the quality of the produced data. Establishing a methodology for assessing and ensuring data quality is crucial. These encompass:

- Standard operational procedures (SOPs) as described in Chapter 5, with appropriate maintenance routines to verify and validate the status of the sensor, with a clear definition of roles and responsibilities among work groups. Proper definition and application of these SOPs may require special training.
- Routines for data quality assessment as described in Chapter 6. Several methods are available, ranging from basic calculations to more complex techniques. Implementation of such methods may also require special training, depending on the selected algorithm approach.
- Collect all the data and metadata produced in (a) and (b) from the sensors validation and data quality assessment, in such a way that can be easily and widely shared within the organization (see Chapters 3 and 4).

7.4.2.4 Data integration and implementation of a data management system

Enabling the integration of data from different sources is a crucial step towards improving the use of data and metadata (see Chapter 4). The following aspects should be accounted for:

- A centralized data management system will allow data to be analysed holistically and efficiently to follow-up operations, diagnose problems on time and increase awareness about the data quality.
- Apart from the sensors data themselves, metadata about sensor validation and signal data quality must be integrated in such a system. As an example, having integrated data from lab measurements (LIMS), maintenance activities (CMMS) and sensor signals (SCADA/PLC), would facilitate the sensor validation (e.g., comparison against lab measurements) and track effects of calibration actions.
- Data integration also sets the basis for data visualization and data analysis, even though it has been identified as one of the main challenges at the utilities level, what jeopardizes creating and managing metadata successfully.
- From the implementation point of view, there are a range of approaches for integrating data sources (Chapter 4) such as data warehouses, data lakes, or managed integrations (i.e., data are kept in distributed systems but can be retrieved and copied to other systems on demand).

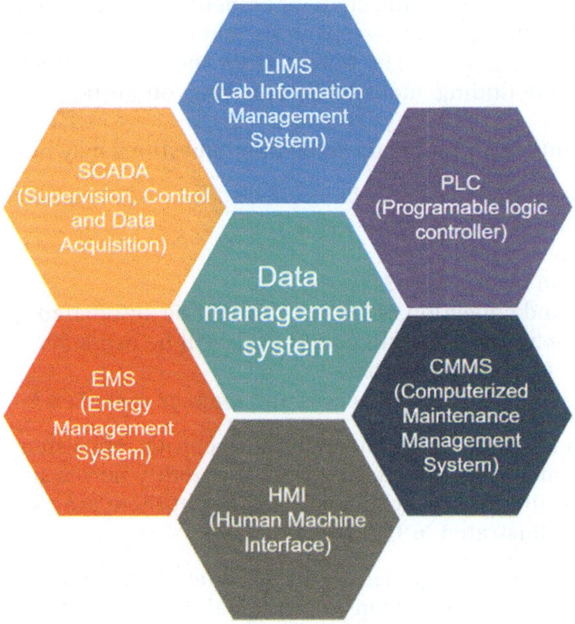

Figure 7.3 Integration of data sources at WWTP into a data management system.

- Whatever the integration architecture chosen, successful leveraging of metadata, and efficient management of sensors in general, relies on integration between the main data sources collected at the utility level as illustrated in Figure 7.3.

7.4.2.5 Intelligent decision-making

Once a data management system is in place, data and their metadata along the organization and its assets can be used within digital technologies that act as data consumers (e.g., cloud computing, internet of things, artificial intelligence). This can lead to new business models and added value for the recorded data. The success of this will depend on the level of instrumentation and digital maturity at each plant, affordable investment, and the ambition to use digital tools in day-to-day operations.

7.4.2.6 Continued transformation

After the previous steps are taken along the day-to-day operations, the digital transformation is a continuous journey in which processes, technologies and business models are continuously evolving towards becoming a *digital water utility*. A good example is the development of the *digital twins* (digital representation of a physical entity). For this to be successful, a data and metadata culture must be fostered where the value of the data is communicated and where managers support the effort involved. This data culture across an organization is a critical success factor to allow the benefits of the technical ideas shared in this report to be fully realized.

7.5 OUTLOOK

Given the explosion of interest in digital technologies and the increasing production of data, the generation and use of metadata becomes even more indispensable for a successful digital evolution of the water industry.

The previous sections have discussed the challenges identified in the adoption of data and metadata practices, and also provided with the main steps to be taken to facilitate that adoption at the utilities level. But this is just the starting point. The potential for metadata is huge and it is clear that in the future metadata will facilitate finding, identifying, selecting, obtaining, and exploring information at different levels.

When considering the future for metadata some open questions may be posed:

- What type of additional metadata will be needed?
- How will the metadata be shared across systems?
- Will metadata become a standard?
- How will the metadata be managed?
- How the generation and collection of metadata could be automated?
- What changes in the paradigm of data governance will be needed?
- What value will be allocated to metadata?
- Who in the organization will need to access and manage metadata?

Upcoming efforts should be focused on creating accurate, consistent, complete, linked, user-centred, and high-quality metadata, facilitating interoperability across units, departments, utilities, and stakeholders. There is still a long way to achieve this, but there are clear areas that will need and generate metadata soon as illustrated in Figure 7.4:

- Data orchestration: Bringing together, in an automated way, data from different sources and storage locations, and combining them in a rational manner will only be possible by means of a metadata-driven orchestration, so that data can be made available for future analysis in an organized and consistent way, avoiding metadata silos.
- Data valuation: Estimating the value of the data and metadata as a business asset will be crucial to achieve a digital transformation and become a data-driven organization. How to get the best ROI from data will be possible by valuing the data based on the role it plays in creating value.
- Data fabric: In the context of data fabric, metadata will support data discovery, data governance, data tracking, and data quality (both monitoring and improving data accuracy and reliability).

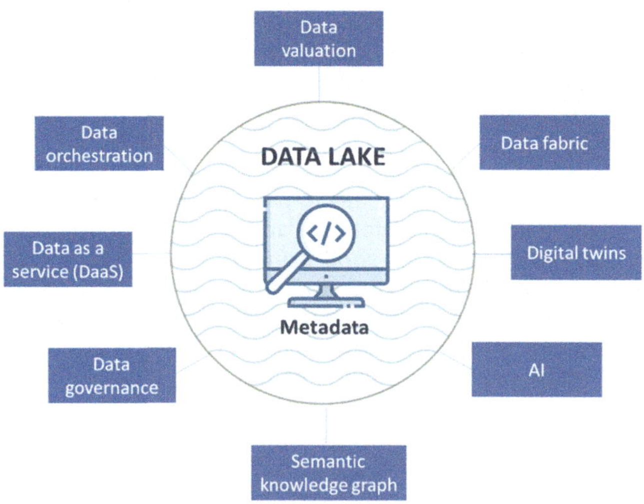

Figure 7.4 Future areas for the water industry where metadata will play a relevant role.

Data fabrics will rely on a metadata-driven architecture to deliver integrated enriched data to the different data consumers.

- Digital twins: The exponential increase of digital twins as digital equivalents of unitary or combined physical entities (assets, sensors, processes, utility) is supported by the data and metadata generated and associated with that particular entity. Metadata as part of digital twins encompasses supporting information (e.g., who created the model, what is the governance around it) and technical information (e.g., data availability and format requirements, what parameters are used, how the model can be used). Such digital versions will be more and more used soon to test operational scenarios, optimize and improve processes performance.
- Data governance: Currently in the middle of a paradigm shift in which the data needs and uses are being defined bottom-up and not enforced top-down. This modern user-led data governance will need a completely new kind of metadata management platform where assets are managed in a more holistic way.
- Semantic knowledge graphs: Knowledge models (collection of interlinked descriptions of entities put into context) provide a smooth way to integrate heterogenous data from different sources by using semantic metadata to describe these data. The resulting knowledge graph provides a framework for data integration, unification, analytics, and sharing.
- AI: It is clear that AI and machine learning (ML) are and will transform the future of the water industry. Good examples are the generation of insights and forecasting of crucial parameters. Similar to digital-twins, every stage on the development and implementation of a ML-based solution generates metadata (datasets used, experiment runs, training parameters, model version, parameters, performance metrics, etc.). Successful deployment of such solutions will require to maintain an always up-to-date, complete, trustable, rich and well managed metadata layer.
- Data as a service: A new business model in which data access, data integration, data processing and/or data analytics are provided on request via a connection to the cloud. It delivers data/applications to the end users over the network rather than running them locally. Metadata as a service would also allow users to search for useful datasets for example (raw data, processed data, etc.).

The water sector is currently at an inflection point concerning metadata management. Metadata will power the entire data environment, allowing for a significant and sustainable transformation in the long term. Good metadata practices will be crucial in this journey turning future data (and current data graveyards and silos) into valuable information to better operate the available assets within WRRFs. This will be encouraged with a structured and integrated data management system, incorporating metadata collection and organization through automatic logging and algorithmic analysis.

REFERENCE

Water Research Foundation (2022). Definition, Framework, and Maturity Assessment for Intelligent Water Systems: A Review of Existing Frameworks and Maturity Models.

doi: 10.2166/9781789061154_0255

Glossary

Absent	*Definition D2.5*: A data record, timestamp, or data point is considered absent if having access to this information is considered valuable at the time of data analysis but no procedure to collect this information was ever initiated.
Accuracy	*Definition D3.18*: Accuracy is the degree of closeness to a test result with the accepted value of reference (ISO, 2006).
Annotation of sensor signal data	*Definition D3.32*: Annotation of sensor signal data is the process of providing additional information to data records or timestamps to inform future use of the sensor signal data.
Automated annotation	*Definition D3.43*: An automated annotation is an annotation that has been assigned through data analytics or algorithmic tools without any human intervention.
Bias	*Definition D3.20*: Bias is the expected difference between a set of data points and the accepted reference value (ISO, 2006).
Calibration	*Definition D5.10*: Calibration is the act of adjusting a sensor to reach the desired accuracy by matching the sensor output to one or more accepted reference values.
Causal annotation	*Definition D3.38*: A causal annotation is an annotation that includes information about the root cause for the observed symptom.
Change of units	*Definition D3.15*: A change of units is the deliberate change of units from the unit of measurement provided by the instrument to a unit considered appropriate for long-term storage and use.

Circumstantial annotation	*Definition D3.40*: A circumstantial or indirect annotation is an annotation based on information residing outside of the data set subject to annotation.
Collective annotation	*Definition D3.36*: A collective annotation is an annotation assigned to multiple data records at once. The annotated data records may have the same or distinct timestamps and may belong to one signal only or multiple signals. Collective annotations are typically used for events that start and end at a different time.
Comment	*Definition D3.34*: A comment is a piece of free-form text used for annotation purposes. A single comment can be assigned to any number of data records, any number of timestamps, or any number of time windows.
Compression	*Definition D4.6*: Data compression is the process of encoding information using fewer bits than the original representation. This reduces the volume required on the physical storage device.
Computer-aided annotation	*Definition D3.42*: A computer-aided annotation is an annotation that is assigned following an inspection by a human expert of data records selected for inspection through data analytics or algorithmic tools.
Data lineage	*Definition D4.10*: Data provenance or data lineage refers to all metadata describing the processes that influenced the data set of interest, thereby providing a historical record of the data and its origin. It includes context on data origin, transformations, and storage. As a result, additional transparency is provided which simplifies root cause analysis in the data analytical process.
Data management	*Definition D4.2*: Data management consists of methods, architectural techniques, and tools for gaining access to and managing delivery of data in a consistent way across different data types in a company. The purpose of data management is to fulfil all data requirements for use cases, applications, and business processes.
Data management technology	*Definition D4.1*: A data management technology is the combined set of hardware and software used for the systematic collection, description, exploration, modification, storage, and retrieval of sensor signals and their metadata.
Data pipeline	*Definition D4.9*: A data pipeline is the combined sequence of software processes and tools used to transfer and transform data between one or more source systems and a target repository. Along its way, the data are structured in such a way that its analysis is facilitated such that new insights can readily be obtained.

Data point	*Definition D2.3*: A data point consists of one or more values that satisfy the following three requirements: (1) All values within the data point are produced by the same instrument. (2) All values are associated with the same time of measurement or, where relevant, the same time of water sample collection. (3) All values describe the same type of variable and are produced by the same measurement principle.
Data provenance	See: *Data lineage*
Data record	*Definition D2.2*: A data record is a pair or triplet of constituting elements, which are (a) a timestamp (required), (b) a data point (required), and (c) a qualifier (optional).
Device history	*Definition D3.14*: The device history consists of all documentation describing important events in the history of the signal-producing device.
Direct annotation	*Definition D3.39*: A direct annotation is an annotation based only on inspection or analysis of the data records being annotated.
Direct measurement	*Definition D5.11*: A direct measurement is obtained when the sensor directly measures the component of interest.
Drift	*Definition D3.27*: Drift is a slow change of bias, variance, or response time as time progresses. It is a symptom that evolves smoothly and continuously over time.
End-to-end testing	See: *transmission testing*
Error	*Definition D3.17*: An error is a deviation between a measured or computed value (of an output variable) and the true, specified, or theoretically correct value.
Expert annotation	*Definition D3.41*: An expert annotation is an annotation that is assigned following the inspection by a human expert of data records without assistance of data analytics or algorithmic tools.
Governance	*Definition D4.4*: Data governance is the exercise of authority and control (planning, monitoring, and enforcement) over the management of data assets.
Gross error	*Definition D3.30*: A gross error is caused by a phenomenon that occurs rarely and during short periods, often affecting a single data point only.
Hysteresis	*Definition D5.7*: Hysteresis is the phenomenon where the sensor measurement value depends on the direction of change for the measured value.
Indirect annotation	See: circumstantial annotation
Intercept	See: *offset*

Label

Definition D3.33: A label is a structured and pre-determined type of annotation with the following properties:

- It is used repeatedly to annotate similar events or phenomena.
- It features at least two elements:
- A unique identifier (alphanumeric string) describing the type of event or phenomenon.
- A description of a type of action, causal event, or symptom.
- In addition, the label may also include the following:
    - Identification of the relevant instruments.
    - Identification of the relevant data records described by the label.
    - Identification of the relevant time point or window.
    - Identification of the expert providing the description.
    - Time the label was created.
    - Purpose of annotation.
    - Software and version used for annotation.

Laboratory data

Definition D2.7: Laboratory data (also: laboratory measurements) consist of values that are (a) recorded through a sequence of steps that include at least one manual water sample preparation step and (b) describe observed values of qualities and quantities in a physical system of interest.

Linearity

Definition D5.6: The linearity of a sensor indicates to what extent a change in the true component (input) is reflected as a linear change in the measurement (output) throughout its measurement range.

Lossless compression

Definition D4.7: Lossless compression is data compression that ensures that the original data can be reconstructed perfectly from the compressed data, that is, storage without loss of information. Compression is achieved by eliminating redundant parts in the data.

Lossy compression

Definition D4.8: Lossy compression is data compression that allows certain information to be lost. As a result, one can only construct an approximation of the original data based on the compressed data. While information will be lost in the process, high levels of compression can often be achieved with small losses in information.

Lossy modification

Definition D3.16: A lossy modification of a signal is any procedure that transforms the original signal into a derivative signal such that the original signal cannot be reconstructed perfectly from the derivative signal, that is, some information is lost in the process.

Maintenance

Definition D5.3: Sensor maintenance includes all actions required to maintain the sensor in a fully functional mode throughout its life cycle.

Manufacturer information

Definition D3.12: Manufacturer information consists of the instrument brand or make, model number, and serial number.

Measurement principle	*Definition D3.9*: The measurement principle of a sensor is a description of the fundamental physicochemical relationship between the variable of interest and the raw signal produced by the sensor.
Measurement range	*Definition D3.6*: The measurement range is the interval defined by the minimum and maximum value that can be produced by a sensor that is configured and operated correctly.
Measurement unit	*Definition D3.5*: The measurement unit is the chosen standard measurement by which the magnitude of the sensor reading is expressed.
Metadata	*Definition D2.8*: Metadata is defined as any piece of information that is used to describe signal data or laboratory data.
Missing	*Definition D2.6*: A data record, timestamp, or data point is considered missing if a procedure to collect it was initiated but failed to produce the intended numerical result.
Numeric resolution	*Definition D3.3*: The numeric resolution is the smallest increment in the monitored variable that can be recorded after digitization (i.e., analog-to-digital conversion).
Offset	*Definition D5.4*: A sensor's offset is a constant deviation from the accepted reference value.
Original data	*Definition D2.11*: Original data are the data as produced by a signal source, for example, sensors.
Outlier	*Definition D3.28*: An outlier is a member of a data set that is inconsistent with the other members of that set.
Point annotation	*Definition D3.35*: A point annotation is an annotation assigned to one data record only.
Precision	*Definition D3.21*: Precision is the degree of closeness of individual data points to each other.
Primary data	*Definition D2.12*: Primary data are the data consisting of the most accurate, comprehensive, and up-to-date version of a given set of data. This data set is used as a single-source-of-truth, that is, a data set that every data user accepts as the best source of information.
Product specifications and guides	*Definition D3.11*: Product specifications and guides consist of documentation detailing the installation, use, and maintenance of an instrument.
Purpose of measurement	*Definition D3.10*: The purpose of measurement describes the way(s) the signal is intended to be used.
Qualifier	*Definition D2.4*: A qualifier is a single number or text produced by an instrument indicating a level of trust in the timestamp and/or data point.
Quality assessment	*Definition D5.2*: Sensor quality assessment is a wide term that describes the combination of several actions, which are performed to verify that a sensor is fully functional and produce data with a desired accuracy.

Random error	*Definition D3.31*: A random error is caused by unpredictable yet persistent variations in the data collection process and affects all measurements by the same sensor in the same or similar way.
Repeatability	*Definition D3.23*: Repeatability is the precision under exactly repeated conditions, that is, obtaining independent test results using the same method in identical tests in the same laboratory, by the same operator, using the same equipment, within a short time interval.
Reproducibility	*Definition D3.24*: Reproducibility is the precision obtained from test results using the same method on identical test items in different laboratories, operators, and equipment (ISO, 2006).
Response time	*Definition D3.25*: The response time is the time it takes for an instrument to reflect a new value when subjected to a step change. It can also be defined as the time required to achieve a percentage of a step change in the true value (typically 90 or 95%).
Retention policy	*Definition D4.5*: A data retention policy clarifies what data should be stored or archived, where that should happen and for how long before it is disposed of.
Secondary data	*Definition D2.13*: Secondary data are the data that have been generated by selection and modification of primary data for a specific context or purpose.
Sensitivity	*Definition D5.5*: A sensor's slope/sensitivity is the amount of change of the measured variable (output) for a standardized change in the variable of interest (input).
Sensor	*Definition D5.1*: A sensor is a device that communicates the value of a process variable.
Sensor location	*Definition D3.4*: The sensor location is the physical position of the sensing element of the sensor.
Shock	*Definition D3.26*: A step-change or shock is a drastic and momentary change of the sensor bias from one timestamp to the next.
Signal	See: *signal data*
Signal data	*Definition D2.1*: Signal data (also: signals) consist of values produced by an online instrument of qualities and quantities in a physical system of interest. Signal data consist of series of data records.
Slope	See: *sensitivity*
Software versioning	*Definition D4.3*: Software versioning is the process of documenting incremental changes in software and assigning unique version names or unique version numbers to refer to a unique, historical state of the software. The goals of software versioning include (a) effective collaboration between programmers, (b) managing compatibility with other software, and (c) guaranteeing reproducibility of results obtained with the software.
Step-change	See: *shock*

Structured data

Definition D2.9: A signal is said to contain structured data when every data point in every data record satisfies the following requirements:

(1) The number of dimensions is constant.
(2) The number of values along all dimensions is constant.
(3) The position of a value along every dimension is meaningful and its meaning does not change over time.
(4) All data point values are restricted to belong to a single number type, such as boolean, integer, or continuous.

Symptom annotation

Definition D3.37: A symptom annotation is an annotation that describes observed symptoms only and does not provide any causal explanation.

Systematic error

Definition D3.29: A systematic error is persistent in time and is likely to produce a measurement error that is constant or changes slowly.

Temporal resolution

Definition D3.2: The temporal resolution is the time interval between subsequent data records produced by a sensor.

Timestamp origin

Definition D3.1: The timestamp origin is the device that produces the timestamp in a data record.

Traceability

Definition D2.14: Data traceability consists of the application of systems that enable monitoring of the entire process of data collection, storage, and processing for a proper and an effective use.

Transmission mode

Definition D3.8: The transmission mode is the way a signal produced by a sensor is carried to the physical storage device (e.g., PLC, SCADA).

Transmission verification

Definition D5.9: Signal transmission verification or end-to-end testing refers to a systematic procedure to verify that the sensor signal remains consistent during transfer from the sensor to the location for data storage such as a database.

Trueness

Definition D3.19: Trueness is the degree of closeness of the mean value obtained from a large set of data points to an accepted reference value (ISO, 2006).

Unique identifier

Definition D3.13: A unique identifier is a name or number unique to a particular instrument or component thereof, allowing a detailed description of the signal source.

Unstructured data

Definition D2.10: Data are said to be unstructured as soon as one of these conditions is satisfied:

(1) The number of dimensions of a data point can vary within the same signal.
(2) The number of values along one or more dimensions of a data point can vary within the same signal.
(3) The position of a value along one or more dimensions of a data point can be changed without consequences for further analysis.
(4) Data point values may belong to multiple number types, such as boolean, integer, or float.

Validation

Definition D5.8: Sensor validation or sensor verification refers to a systematic procedure for (a) collecting and recording measurement values under controlled conditions to quantify quality measures such as accuracy, bias, precision, offset, slope, and/or linearity and (b) to evaluate whether the quality measures satisfy predetermined specifications for the intended application. The simplest form of validation is to compare the sensor reading to a reference measurement.

Variable range

Definition D3.7: The variable range is the interval defined by the minimum and maximum value that is acceptable or typically observed during normal operation, either in general or for a particular context (e.g., location and/or time).

Variance

Definition D3.22: The variance of measurement error is the expected value for the squared deviation between an individual measurement and the expected value (e.g., mean) for the measurement.

Verification

See: *validation*

IWA Publishing's authorised EU representative for General Product Safety Regulations is Diane D'Arras, 15 rue Duret, 75116 Paris, France, e-mail: safety@iwap.co.uk.

Printed and bound by CPI Group (UK) Ltd, Croydon, CR0 4YY
01/04/2026
02082208-0003